Safety, Environmental Impact, and Economic Prospects of Nuclear Fusion

ETTORE MAJORANA
INTERNATIONAL SCIENCE SERIES
Series Editor:
Antonino Zichichi
European Physical Society
Geneva, Switzerland

(PHYSICAL SCIENCES)

Recent volumes in the series:

Volume 38 **MONTE CARLO TRANSPORT OF ELECTRONS
AND PHOTONS**
Edited by Theodore M. Jenkins, Walter R. Nelson, and
Alessandro Rindi

Volume 39 **NEW ASPECTS OF HIGH-ENERGY PROTON–PROTON
COLLISIONS**
Edited by A. Ali

Volume 40 **DATA ANALYSIS IN ASTRONOMY III**
Edited by V. Di Gesù, L. Scarsi, P. Crane,
J. H. Friedman, S. Levialdi, and M. C. Maccarone

Volume 41 **PROGRESS IN MICROEMULSIONS**
Edited by S. Martellucci and A. N. Chester

Volume 42 **DIGITAL SEISMOLOGY AND FINE
MODELING OF THE LITHOSPHERE**
Edited by R. Cassinis, G. Nolet, and G. F. Panza

Volume 43 **NONSMOOTH OPTIMIZATION AND RELATED TOPICS**
Edited by F. H. Clarke, V. F. Dem'yanov,
and F. Giannessi

Volume 44 **HEAVY FLAVOURS AND HIGH-ENERGY COLLISIONS
IN THE 1–100 TeV RANGE**
Edited by A. Ali and L. Cifarelli

Volume 45 **FRACTALS' PHYSICAL ORIGIN AND PROPERTIES**
Edited by Luciano Pietronero

Volume 46 **DISORDERED SOLIDS: Structures and Processes**
Edited by Baldassare Di Bartolo

Volume 47 **ANTIPROTON–NUCLEON AND ANTIPROTON–
NUCLEUS INTERACTIONS**
Edited by F. Bradamante, J.-M. Richard, and R. Klapisch

Volume 48 **SAFETY, ENVIRONMENTAL IMPACT, AND ECONOMIC
PROSPECTS OF NUCLEAR FUSION**
Edited by Bruno Brunelli and Heinz Knoepfel

A Continuation Order Plan is available for this series. A continuation order will bring delivery of each new volume immediately upon publication. Volumes are billed only upon actual shipment. For further information please contact the publisher.

Safety, Environmental Impact, and Economic Prospects of Nuclear Fusion

Edited by

Bruno Brunelli and Heinz Knoepfel

Euratom—ENEA Association
Frascati, Italy

Plenum Press • New York and London

Library of Congress Cataloging-in-Publication Data

Safety, environmental impact, and economic prospects of Nuclear fusion
/ edited by Bruno Brunelli and Heinz Knoepfel.
    p.   cm. -- (Ettore Majorana International science series ; v.
48)
  Includes bibliographical references.
  ISBN-13:978-1-4612-7895-5   e-ISBN-13:978-1-4613-0619-1
  DOI:10.1007/978-1-4613-0619-1

  1. Controlled fusion--Congresses. 2. Controlled fusion--Safety
measures--Congresses. 3. Controlled fusion--Environmental aspects-
-Congresses. 4. Controlled fusion--Economic aspects--Congresses.
5. Fusion reactors--Congresses.   I. Brunelli, B. (Bruno)
II. Knoepfel, Heinz, 1931-   . III. Series.
QC791.7.S34 1990
333.792'4--dc20   90-33317
                                                                    CIP

ECSC—EEC—EAEC (Euratom), Brussels and Luxembourg, 1990

EUR 12580 EN

Neither the Commission of the European Communities (CEC) nor any
person acting on behalf of the Commission is responsible for any
use which might be made of the following information.

Proceedings of the Ninth Course of the International School of
Fusion Reactor Technology, held August 6–12, 1989,
in Erice, Sicily, Italy

© 1990 Plenum Press, New York
Softcover reprint of the hardcover 1st edition 1990

A Division of Plenum Publishing Corporation
233 Spring Street, New York, N.Y. 10013

PREFACE

This book contains the lectures and the concluding discussion of the "Seminar on Safety, Environmental Impact, and Economic Prospects of Nuclear Fusion", which was held at Erice, August 6-12, 1989.

In selecting the contributions to this 9th meeting held by the International School of Fusion Reactor Technology at the E. Majorana Center for Scientific Culture in Erice, we tried to provide a comprehensive coverage of the many interrelated and interdisciplinary aspects of what ultimately turns out to be the global acceptance criteria of our society with respect to controlled nuclear fusion.

Consequently, this edited collection of the papers presented should provide an overview of these issues. We thus hope that this book, with its extensive subject index, will also be of interest and help to nonfusion specialists and, in general, to those who from curiosity or by assignment are required to be informed on these aspects of fusion energy.

As organizers, we were aware of the fact that the goal of the Seminar was highly ambitious and even somewhat peculiar. In fact, on one hand, the concept to which the discussion is referred, i.e., the fusion reactor, is still evolving and depends on further physical progress and technological development. On the other hand, the criteria for evaluating safety, environmental impact, and economic prospects are themselves also in continuous evolution, as they depend on the perception and acceptance of risk. As modern studies and our daily experience show, this in turn is in relation with the historic, cultural, social, and ethical perceptions of our societies and, last but not least, with the harsh competition in the market place. Suffice it here to recall the tortuous path which nuclear fission energy has followed to enter our society.

Although some people think that an ample analysis on the safety and cost of fusion energy is premature given the present status of controlled nuclear fusion, it is a fact that the public, through its parliaments and other channels, is demanding the opinion and judgement of the specialists about the real prospects and merits of this new energy source.

We should take it as a challenge and opportunity to answer these questions in an appropriate, technically sound and intellectually stimulating way, and we hope that this book represents a contribution.

This Seminar would hardly have been possible without the efficient collaboration of Maria Polidoro - assisted in Erice by Mina Misano and in Frascati by Sabrina Antonacci - who throughout the duration managed all the organizational problems and the assembly of the Proceedings with competence and enthusiasm. Many thanks are also due to the permanent secretariat of the Erice Center, staffed by Mrs. Pinola Savalli and Dr. A. Gabriele. Last but not least, the success of the Seminar and, consequently, the quality of these Proceedings are the merit of the contributors, who presented excellent technical reviews and reports and respected our requests on contents and deadlines.

Bruno Brunelli and Heinz Knoepfel
EURATOM-ENEA Association

Frascati, October 1989

CONTENTS

I. SUMMARY OF GENERAL STUDIES ON SAFETY, ENVI-
RONMENTAL IMPACT AND COST OF MAGNETIC FUSION

Fusion Reactor Economic, Safety, and Environmental Prospects   ... 3
 R.W. Conn, J.P. Holdren, D. Steiner, D. Ehst, W.J. Hogan, R.A. Kra-
 kowski, R.L. Miller, F. Najmabadi, K.R. Schultz

European Studies on Safety, Environmental Impact, and Cost of Magnetic Fusion Power   ... 35
 J. Darvas

General Methodology of Safety Analysis/Evaluation for Fusion Energy
Systems (GEMSAFE) and its Applications   ... 41
 Y. Fugii-e

Summary of the U.S. Senior Committee on Environmental, Safety, and
Economic Aspects of Magnetic Fusion Energy (ESECOM)   ... 67
 B.G. Logan, J.P. Holdren, D.H. Berwald, R.J. Budnitz, J.G. Crocker,
 J.G. Delene, R.D. Endicott, M.S. Kazimi, R.A. Krakowski, K.R.
 Schultz

II. GENERAL PROGRAM EVALUATION

The Energy Scene in the Mid-21st Century   ... 81
 L. Gouni

Review of Plasma Physics Constraints   ... 95
 R.S. Pease

The Status of Inertial Confined Fusion Research in the US   ... 117
 R.W. Conn

III. FEASIBILITY PROBLEMS

Main Issues in Fusion Reactor System Engineering   ... 129
 E. Salpietro

Special Materials for Fusion Reactors   ... 147
 C. Ponti

Feasibility Aspects of the D-^{3}He Fuel Cycle in Tokamak Power Reactor
Plants   ... 159
 G. Casini

Feasibility, Safety and Environmental Aspects of D-^{3}He Fusion ... 173
    M. Heindler

IV. COMPONENT RELATED SAFETY AND ENVIRONMENTAL PROBLEMS

First Wall and Blanket Safety ... 183
    M.S. Kazimi

Tritium Environmental Risk in Future Fusion Reactors ... 199
    Y. Belot and P. Zettwoog

V. SAFETY AND ECONOMY OF FUSION PROTO-REACTORS

Safety and Environmental Impact of ITER/NET ... 231
    J. Raeder and W. Gulden

Cost Analysis of Next Step Devices and the Implications for Reactors ... 279
    W.R. Spears

Pulsed Versus Steady-State Reactor Operation in View of Safety and Economy ... 295
    R. Buende

VI. PANEL DISCUSSIONS AND CONCLUSIONS

Introduction ... 307

Short Contributions: "Materials Selection for Fusion" *(G.J. Butterworth)*; "Lessons from Fission - A Personal Perspective" *(E.C. Brolin)*; "Dose Limits for Fusion Reactors" *(P. Rocco)*; "Public Acceptance of Nuclear Fusion" *(M. Snykers)* ... 309

Panel Discussion on Safety and Environmental Impact of Fusion Reactors ... 319
    K. Tomabechi

Panel Discussion on Economic Prospects of Fusion Reactors ... 323
    R.W. Conn

Concluding Panel ... 327
    R.S. Pease

VII. MISCELLANEA

Participants ... 337

International School of Fusion Reactor Technology ... 339

INDEX, including explanation of abbreviations, acronyms, and radiological units ... 341

I. SUMMARY OF GENERAL STUDIES ON SAFETY, ENVIRONMENTAL IMPACT AND COST OF MAGNETIC FUSION

FUSION REACTOR ECONOMIC, SAFETY,

AND ENVIRONMENTAL PROSPECTS

R. W. Conn[1], J. P. Holdren[2], D. Steiner[3],
D. Ehst[4], W. J. Hogan[5], R. A. Krakowski[6],
R. L. Miller[6], F. Najmabadi[1], K. R. Schultz[7]

[1]University of California at Los Angeles, 6291 Boelter Hall,
Los Angeles, CA 90024-1597.
[2]University of California at Berkeley, One Cyclotron Road,
Berkeley, CA 94720
[3]Rensselaer Polytechnic Institute, Tibbits Avenue,
Troy, NY 12181
[4]Argonne National Laboratory, 9700 South Cass Avenue,
Argonne, IL 60439
[5]Lawrence Livermore National Laboratory, P.O. Box 5511,
Livermore, CA 94550
[6]Los Alamos National Laboratory, P.O. Box 1663,
Los Alamos, NM 87545
[7]General Atomics, P.O. Box 85608, San Diego, CA 92138

ABSTRACT

Controlled fusion energy is one of only a few energy sources available to mankind in the future. Progress in fusion reactor technology and design is described for both magnetic and inertial confinement fusion energy. The projected economic prospects show fusion will be capital intensive and the historical trend is towards greater mass utilization efficiency and more competitive costs. Recent studies emphasizing safety and environmental advantages show that fusion's competitive potential can be further enhanced by specific material and design choices. Fusion's safety and environmental prospects appear to substantially exceed those of advanced fission and coal but will not be achieved automatically. A significant and directed technology effort is necessary. Typical parameters have been established for fusion reactors, and a tokamak at moderately high magnetic field (about 7 T on axis) in the first regime of MHD stability ($\beta \leq 3.5$ I/aB) is closest to present experimental achievement. Directions to further improve economic and technological performance include the development of higher magnetic fields to lower the required plasma current and reactor size, improvement in the beta value in the second stable MHD regime to lower requirements of field and plasma current, and improvement in techniques for plasma current drive to efficiently achieve steady-state plasma operation. For inertial confinement, reactor studies are at an earlier stage but two essential requirements are a high-efficiency ($\geq 10\%$) repetitively pulsed pellet driver capable of delivering up to 10 MJ of energy on target, and targets capable of yielding an energy gain (ratio of energy produced to energy on target) of 100.

Safety, Environmental Impact, and Economic Prospects of Nuclear Fusion
Edited by B. Brunelli and H. Knoepfel
Plenum Press, New York, 1990

3

INTRODUCTION

Fusion energy is one of only a few future energy sources available to mankind. The quest to achieve practical fusion energy continues around the world in programs aimed largely at establishing the plasma physics conditions needed for a burning fusion system. Our understanding of the potential of fusion as an energy source is derived primarily from reactor design and systems studies. These in turn are used to assess the safety, environmental, and economic characteristics of fusion power. Fusion is found to have the characteristics for a desirable energy source including significant potential advantages with respect to accident consequences, waste disposal, and air pollution. This outcome is not guaranteed, however, it requires that fusion research and development achieve the characteristics and performance goals identified through design, safety, and environmental studies. We address here the potential of fusion energy in terms of reactor design, economics, safety, licensing, and environmental issues.

"Fusion energy" in this paper will encompass both magnetic fusion energy (MFE) and inertial confinement fusion (ICF) energy. Section 2 provides a summary of fusion reactor studies. The economic potential of fusion energy is examined in Section 3. Safety, environment, and licensing issues are considered in Section 4. Section 5 presents some concluding remarks concerning the potential of fusion energy.

FUSION REACTOR TECHNOLOGY AND DESIGN

The key features of fusion reactor design and technology are described here to set the stage for a discussion of fusion's economic, safety, and environmental characteristics. Reactor design influences, and is influenced by, economic, safety, and environmental requirements. Because of the technological differences between MFE and ICF reactors, each is addressed separately.

Magnetic Fusion Reactor Technology And Design

Magnetic fusion energy (MFE) power plants will consist of a plasma reaction chamber, magnet coils, a blanket for energy recovery, plasma heating and fueling systems, a technique for controlling plasma purity, and a balance of plant for converting the fusion energy to electricity. The schematic elements of such a plant are shown in Figure 1. In the context of deuterium-tritium (DT) fueled reactors, the blanket provides tritium breeding since tritium is radioactive and has a half-life of 12.6 years. Deuterium occurs naturally in large abundance in water and both deuterium and tritium, isotopes of hydrogen, are the key components of the fuel. Even with alternative fuel cycles that are more difficult to achieve in terms of required plasma parameters [i. e., deuterium-deuterium (DD) or deuterium - helium - 3 ($D-^3He$)], there will remain requirements for operation in a radiation environment and tritium handling. As such, both the MFE and ICF fusion reactors will incorporate radiation shielding, radioactive material handling systems, and remote maintenance equipment.

Conceptual design studies of MFE reactors have been carried out for over two decades. Progress through about 1980 is reviewed in References [1-6]. Several important studies [7-10] have been performed in this decade, but no single review of these has been published.

Many concepts for magnetic confinement of plasma have been studied [e. g., tokamak, reversed field pinch (RFP), and helical systems such as the stellarator] [7-14]. The effort devoted to reactor design for a particular concept is generally related to the relative maturity of that concept, the tokamak (having received the largest and most sustained effort) is the most mature of the magnetic confinement designs. Work on the tokamak has included studies of both commercial reactors and more near-term devices. International studies of near-term devices are the INTOR [2] and, more recently, the International Thermonuclear Experimental Reactor (ITER) activity [16]. Although the focus of this paper is on commercial fusion power, the benefit of the

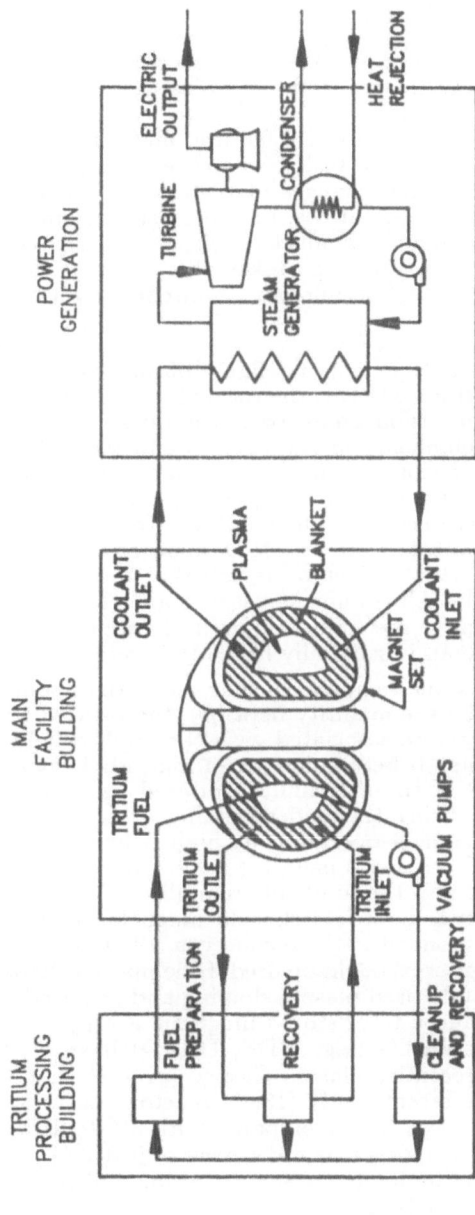

Figure 1. Schematic elements of a magnetic fusion energy (MFE) power plant.

interplay between near-term and long-term studies is crucial. The most thorough and well-documented tokamak reactor study is STARFIRE [7].

Next to the tokamak, the mirror approach has received great attention with the MARS work [8] being the most thorough and well documented tandem mirror reactor study. During this decade the reversed field pinch (RFP) has been examined in several reactor design studies, the most comprehensive of which is the recently completed TITAN study [9]. Conceptual design studies based on a variety of alternative confinement schemes have also been performed [10-14] at a modest level of effort.

Early reactor studies identified engineering problems, technology requirements, and physics implications. Second generation studies developed solutions to issues raised in the early studies. During the past decade, reactor studies have focused upon improving fusion's economic and safety features and achieving design simplicity in order to maximize the potential of fusion as an energy source. The evolution of MFE reactor design is seen in the changes of certain key, generic features: mass power density, magnetic field requirements, pulsed or steady-state operating mode, auxiliary plasma heating techniques, blanket design and energy conversion, plasma purity control, and fuel exhaust. Since the tokamak concept is the most developed, we emphasize here the physics and technology requirements for attractive tokamak reactors.

Mass power density (MPD) is a useful figure-of-merit to assess the economic potential of MFE power reactors. MPD is the ratio of the net electrical power output of a plant to the mass of the fusion power core (the blanket, shield, and magnets). This figure-of-merit places emphasis on the physics and technology of the reactor. Therefore, it is sensitive to fusion-specific cost factors and is independent of costs related to construction and operation of the power plant. Reactor designs show a consistent trend towards increased MPD as studies have matured and economics is emphasized. The studies indicate that an MPD value of about 100 kWe/tonne of fusion-power-core is an important target value and that further increases yield only a moderate improvement [17]. All confinement concepts, which have been examined to date, exhibit the potential to meet this minimum figure-of-merit target, and the RFP systems appear to have an intrinsically high MPD value [9].

Achieving a high MPD value requires efficient use of the applied magnetic field. This efficiency is measured by the quantity beta (β), the ratio of the plasma kinetic pressure to the magnetic pressure associated with the applied field. Beta is limited by plasma physics constraints. If beta is low, then high MPD can only be achieved by operating at high values of the externally produced magnetic field. One early tokamak design study [18], using the understanding of beta limits at that time, adopted superconducting magnets operating at high field (16 T). The result was a very large value of stored magnetic energy (250 billion joules, or 250 GJ). Not only did this field strength strain the limits of credibility at the time, but the stored energy represented a substantial financial risk if the magnets should fail. Both physics and technology programs responded to these concerns. Physicists developed a theory showing that lower fields (higher β) could be used if the plasma shape were elongated. The STARFIRE tokamak study used plasma shaping and achieved a more attractive system with 11 T magnets and a total stored magnetic energy of only 50 GJ. Since then, highly successful experiments (e.g. JET, DIII-D) have verified the benefits of plasma shaping. More recently, plasma theory suggests that further increases in beta may be possible. A recent study [19] of reactor operation in this "second stability regime" with β values of 20% (compared with 3% to 10% earlier) suggests that maximum fields of only 7 T and a stored energy of just 8 GJ would be possible. This exciting prospect is a stimulus to present experimental programs in machines such as TFTR, PBX-M, and DIII-D.

Meanwhile, the worldwide magnet development effort mobilized to demonstrate the design goals identified in reactor studies. The international Large Coil Task [20] successfully constructed and tested an 8 T magnet set arranged in a torus with 1 GJ of stored energy. Other programs [21] are developing small magnets at higher fields (12 T). Reactor studies have identified other key magnet development issues including high strength structural materials, better radiation-resistant insulators, and higher current density superconductors. The High Field Compact Tokamak Reactor study [22] focused attention on a reactor with 13 T magnets having 40 GJ of stored energy.

Both the tokamak and RFP concepts would be limited to pulsed plasma operation if only inductive techniques were available to sustain the plasma current. System studies suggest that steady-state operation will be safer, more reliable, and more economical than pulsed operation. Fortunately, other techniques for sustaining the plasma current have been developed for both the tokamak and the RFP. Although too simplistic, the current-drive technique proposed in the early Mark I tokamak reactor study [23] stimulated theoretical studies of neutral beam (NB) and radio frequency (RF) approaches to current drive. Grounded on a firmer theory, STARFIRE [7] explored the benefits of steady-state operation. Worldwide experimental effort soon verified the prospects for steady-state operation. The main technical issue now concerns the efficiency of the current-drive system. In tokamaks, there is an inherent current drive called the "bootstrap current" which will greatly reduce the circulating power. The bootstrap current has now been experimentally observed and verified in the TFTR and JET tokamaks. The attractiveness of steady-state operation is a goal of the International Thermonuclear Experimental Reactor (ITER) [16]. Other technology development programs [e. g., cooled waveguides, negative-ion-based NB, high-power electron-cyclotron-resonance (ECR) sources] assist in meeting the steady-state goals.

The auxiliary power required for steady-state operation can also provide the external power needed to reach ignition temperatures in fusion plasmas. Early reactor designs considered, for example, neutral beam (NB) and electron-cyclotron resonance RF heating (ECRH) technologies. On the basis of such studies, important directions were identified for technology development. Conventional NB systems are not viewed as attractive for reactors. They are physically large and the penetrations needed make tritium and radiation containment difficult. An alternative NB technology based on negative-ion sources appears to be preferable. For ECRH, gyrotron technology seems impractical if the sources are limited to modest power (100 kW per unit). High power sources at unit sizes above 1 MW [e. g., advanced gyrotrons at 1 MW, cyclotron auto resonance masers (CARMS), or free electron lasers at up to 5 MW per unit] are now under development.

The nuclear performance of a fusion reactor is determined primarily by the blanket materials. Many material combinations have been proposed and early studies focused on achieving an adequate tritium-breeding ratio (> 1.1 tritons produced per triton consumed). Subsequent designs examined thermal-mechanical issues, accident and safety in design, and environmental impact. There are now several desirable choices for breeding materials, structural materials, and coolants in blanket applications. The breeder materials include solid lithium-bearing ceramics or liquid metals; structural materials include low-activation steels, vanadium alloys, and low-activation ceramic composites; coolants include gases, water, and liquid metals. The choice of energy conversion scheme is closely coupled with the choice of blanket materials. The emphasis in reactor studies has been on thermal energy conversion, but several studies of advanced fusion fuel cycles have considered direct energy conversion. Direct conversion can result in dramatic simplifications to the balance-of-plant for fusion reactors [24].

Perhaps the most challenging aspect of MFE reactor design is associated with plasma impurity control and helium and impurity exhaust. In this context, magnetic diversion of helium "ash," impurities, and unburned fuel is the most commonly adopted approach. Unfortunately, the large magnet coils required for diversion can impede reactor maintenance and can also significantly increase the capital cost. Nevertheless, magnetic divertors remain the primary approach in the experimental program, particularly because of recently discovered benefits to plasma energy confinement, namely "H-mode" operation. On a smaller scale, some novel impurity control and exhaust schemes proposed in reactor studies are being investigated experimentally. These include the pump limiter [7] and the self-pumped limiter concept [19].

In summary, MFE reactor studies have identified the technical features, the development needs, and the prospects for MFE systems. Attractive solutions for many problems have been found, and proposed solutions are being examined within the physics and technology programs.

Inertial Confinement Fusion Reactor Technology And Design

Inertial confinement fusion (ICF) [25,26] refers to the approach in which laser or charged particle beams deliver energy to compress and ignite small capsules of deuterium and tritium fuel. The ICF power plants will consist of a driver to implode and ignite the pellet target, a target factory to manufacture and deliver the targets to the center of the reactor core, one or more reaction chambers in which the targets are burned, and the balance of plant in which the fusion energy is converted to electricity. The schematic elements of an ICF reactor plant are shown in Fig. 2. The reaction chamber includes components for energy recovery, tritium breeding, and radiation shielding [27,28]. Differences between ICF and MFE power reactors make some technological issues easier and others more difficult to solve. Design flexibility is gained in ICF systems because of the relaxed vacuum requirements in the reaction chamber and in the separability of the driver, the fusion target manufacture, and the design issues of the reaction chamber. On the other hand, the extreme pulsed nature of ICF's energy production (i. e., three to four 1 GJ explosions per second in a 1000-MWe plant), the very large yields anticipated per shot (up to 1 GJ or 1/4 ton of TNT equivalent) and the manufacture and emplacement of target capsules are design challenges which differ from those of MFE systems. Also, at present, details of the design and operation of ICF fuel targets is classified. A power plant using classified targets is probably unacceptable so that progress on this unique ICF issue is needed. Thus, while ICF shares many technological development issues with MFE, it truly represents a very different approach to commercial fusion energy.

A number of drivers are possible for ICF power reactor applications including heavy ion beams, light ion beams, KrF, and solid-state lasers [29]. While some former driver candidates have been eliminated because of poor target results (e. g., pulsed electron beams and CO_2 lasers), proposals for other new drives (e. g., free electron lasers, compact torus accelerators, and other laser concepts) have taken their place. Two types of targets are in contention: indirect drive target (whose detailed design is classified and in which the drive energy is converted to X-ray energy before interacting with the fuel capsule) and direct drive targets [30,31] in which the fuel capsule is directly illuminated. There has been much work on target fabrication but only cursory attention has yet been paid to the design of a target factory (since the basic target design is still in question), and to target injection, tracking and positioning. In principle, any of the drivers could be used with either target type, illustrating the separability of driver and target issues.

Target performance in general does not depend on the surrounding chamber configuration. (This is not entirely true for pulsed power techniques.) Thus, many reaction chamber concepts have been developed to deal with the effects of target explosions. Also, the large number of driver, target, and chamber combinations, and the evolving emphasis on various design criteria (reliability, practicality, safety, environmental impact, and cost) have led to a large number of reactor studies over nearly two decades. A review of ICF reactor studies through 1984 is given in Reference [29].

Early ICF studies established two main themes: (1) there are several ways to deal with the very large peak-power-density incident on the first wall of the chamber as a result of the target explosion; and (2) self-consistent power plant concepts can be developed for each of the candidate drivers. In an ICF capsule, the energy is produced in a few tens of picoseconds (i. e., $\sim 3 \times 10^{-11}$s). Thus, for a capsule yield of 1 GJ, the instantaneous power is $\sim 3 \times 10^{19}$ W, far above the average fusion power of about 10 GW. This fact requires that the ICF reactor first-wall structure be very different from those in magnetic fusion reactors. The ICF first wall will either be at a very large radius, or it must be designed to tolerate ablation. Most designs have followed the latter principle in order to keep the size of the reaction chamber moderate, consistent with anticipated economic goals. Designs have been proposed in which the permanent

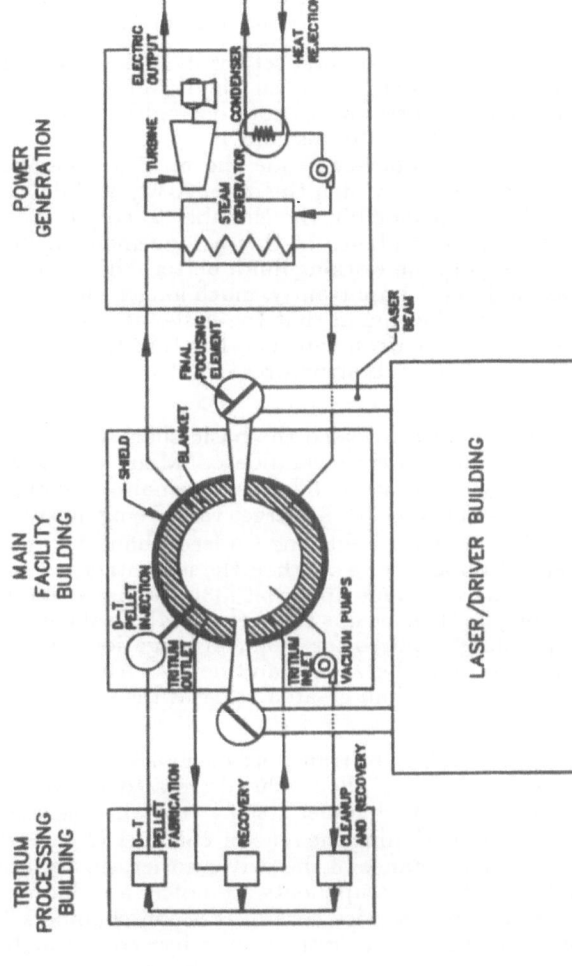

Figure 2 Schematic elements of an inertial confinement fusion (ICF) power plant.

structure is protected by a variety of self-renewing first walls. Materials suggested for use between the target and the chamber structure include gases, liquid sprays, thin liquid-metal layers (i. e., wetted walls), thick liquid-metal layers (i. e., liquid lithium falls), and thick cascading layers of sand-like granules.

In all the liquid and solid first-wall reactors, some material will be vaporized with each pulse (up to a few kg). The material just beyond the vaporized region must be energy absorbing so that large shocks will not be transmitted to the permanent structure. Recondensation of the vaporized material before the next pulse (to reestablish the vacuum needed to inject the next target pellet and to propagate the driver beams to the target) is needed and has been identified as a critical issue for all these designs. The early studies indicated (largely through calculations) that self-renewing layers could be designed which would lengthen the effective time of the short energy pulse so that the peak power on the permanent structure is tolerable. In most studies, however, a paucity of experimental data is noted.

Vacuum requirements in ICF reactors, set by the driver beam propagation, are not stringent (1 to 1000 Pa). Therefore, even hot liquid-metals can be used inside the chamber (i. e., their vapor pressure is not too high). Once this fact was fully appreciated, many studies in the late 70s and early 80s moved the entire energy transport and tritium-breeding blanket inside the reaction chamber, providing far more material than was needed to stretch the short X-ray and debris pulses. In fact, enough cascading material is included in the chamber to stop essentially all the high energy neutrons produced. Advantages of this arrangement include more efficient deposition of thermal energy in the working fluid, better tritium-breeding ratio, lower activation of materials and, most importantly, much longer lifetime of the permanent structure. Previous fusion studies found that the first-wall structure lifetime will be, at most, a few years because of neutron damage. The HYLIFE study [32] showed that with a thick liquid-lithium layer, all structures could be made to last the assumed 30-year lifetime of the plant.

The ICF reactor studies also revealed the basic characteristics of self-consistent reactor concepts for each major driver candidate. Many characteristics proved to be independent of driver type but some differences remain. Vacuum requirements to transport the various types of beams to the target vary, being most stringent for heavy ions (about 10^{-3} Pa) and least for light ions (indeed about 1 kPa of gas is needed) with lasers in between (at about 10 Pa). When the advantages of tolerating a higher background chamber pressure became apparent [33], the heavy ion advocates sought and found ways to propagate their beams through it [34]. In the light ion designs [35], the higher chamber pressure required for beam propagation automatically protects the first wall from the effects of the X-rays and debris. All this energy is deposited in the chamber gas where the fireball created releases the energy to the first wall at an acceptably slow rate.

On the other hand, the ability to transport heavy ion and laser beams over long distances without significant losses allows the drivers to be located in a separate building, away from the reactor chamber itself. This has several advantages: it provides greater flexibility in designing the reactor chamber (i. e., very large chambers are possible) and makes maintenance of the driver easier. In turn, reliability should be improved and high-cost driver components are not in a radiation environment. It is also possible for one driver to service several reaction chambers, making modular construction possible as well as lowering the unit driver cost. Finally, separability of driver and target chamber reduces the target containment chamber volume.

In the mid-to-late 1980s, ICF system studies represented either an evolution of earlier concepts or more radical approaches that went beyond first generation designs. The earlier studies attempted to show that ICF reactors can be practical. The studies assumed the simplest designs and as much existing technology as possible. They showed that once high target gain is achieved, a practical power reactor can be built with existing technology. Nonetheless, design difficulties became apparent. The reactors were relatively expensive in both cost of electricity and capital requirements, they produced induced radioactivity at significant levels, and they raised safety issues because of the large tritium inventory and the large amount of liquid metal used. Present work has focused more on the use of advanced ceramic and composite materials (granules, fabrics, and structures) [36-38], but their viability must be established.

Major unresolved issues for ICF reactors are a cost-effective rep-rated driver that can deliver 2 to 10 MJ to a target, and a target which will yield an energy gain of at least 100. The four major candidates at this time are solid state-lasers, KrF lasers, heavy ion beams, and light ion beams. Of these, only solid-state lasers have actually been used to implode ICF capsules. Recent systems studies [39] and new experimental work [40,41] indicate that conceptual solid-state-laser reactor-driver designs are plausible based on identified technologies. The solid-state laser's natural ability to achieve the high peak-power densities necessary to drive a capsule, its flexible frequency-conversion ability, and its operation at a frequency at which optical materials have a high damage threshold are some of the features that make it attractive as a driver, while rep-rate ability and cost are key outstanding issues. KrF lasers are being actively pursued [42] because of attractive features that include good energy coupling to the target, the use of a gaseous lasing medium, and the potential to be rep-rated. Issues of concern for KrF lasers include efficiency, cost, and optical complexity and size (associated with techniques such as optical multiplexing to shorten and shape the pulse). Heavy ion beams are the favorite candidate of some because of their naturally high wall-plug-to-beam conversion efficiency and their demonstrated, reliable repetitive-pulse operation. Recent systems work [43] showing the ability to generate and accelerate beams of multiply-charged ions reduces the cost of this option, although it still remains high. Light ion beams (lithium ions) generated by pulsed power techniques are being pursued because they are highly efficient (25% to 40%) and are low cost. Light-ion-beam reactor-systems work [35] has focused successfully on identifying plausible pulsed power elements which have the potential of repetitive operation and to transport and focus very high-current light-ion beams. All drivers except solid-state lasers must yet demonstrate that they can successfully drive ICF capsule implosions. A key open issue is whether any laboratory driver is able to attain the threshold energy to achieve high gain.

ECONOMIC ASSESSMENT OF FUSION ENERGY SYSTEMS

The development of an energy policy is subject to large uncertainties [44] in supply-and-demand forecasts for energy end-use, in environmental considerations (e. g., global warming and waste disposal), in variations among national regulations, and in the characteristics of future, competitive energy sources. Accordingly, the ability to project the timing of the market penetration and cost competitiveness of a new energy source for coming decades is difficult. Nevertheless, it is important to make the effort. The search for cheaper, more abundant fuels in recent times has led to systems (such as fission reactors) where the fraction of total cost devoted to buying the fuel is reduced but the cost to implement the technology is increased. Fusion's fuel costs will be negligibly small but the technology could turn out to be excessively costly or complex, or the potential environmental advantages may be offset by difficulties in implementing the technology. Because of this, recent reactor studies have emphasized simplicity and reliability as well as safety and capital-cost improvements.

Several dozen economic assessments of both MFE and ICF central-station electric-power plants have included capital-cost estimates (based on various levels of design detail), cost-of-electricity (COE) projections [based on capital costs, operation and maintenance (O&M) costs, fuel-and-other-consumables costs, decommissioning costs, and assumed plant availability], assessments of potential resource needs, development of energy-payback-time models, and comparisons with competitive energy technologies. Such economic assessments contribute to the identification of attractive avenues for magnetic or inertial confinement fusion research.

The present status of the fission power industry (in the USA), as well as future projections [45], provide lessons for fusion. Problems in the fission power industry can be traced to the increased technological sophistication needed to reduce from 50% (coal) to 30% the fraction of the total energy cost attributable to fuel and, in the U.S., to the large number of independent utilities, each specifying individual power-plant requirements. Now, improved and safer fission power plants are being emphasized in the U.S. [45]. Their features include: (1) passive stability to uncontrolled energy releases; (2) simplification to reduce the number of plant components; (3) ruggedness of design to enhance critical design margins and extend plant longevity (>60 years); (4) ease of operation to reduce the "human factor" as was seen at the 1979 Three-Mile-Island accident; and (5) improved licensing through greater design margins, passive-safety features, factory fabrication/construction, standardization, and reduced probability of a severe accident (both in terms of public safety and investment protection).

Fusion-reactor fuel costs are expected to be less than the fuel costs of both present-day fossil fuel and nuclear-fission plants. The total plant capital costs, however, will likely be greater since the reactor costs associated with the heat-generating fusion power core (reaction chamber, blanket, shield, magnets or target driver, and structure) will probably be larger than the counterpart systems in fossil-fuel or fission-power plants. Plasma-current drivers (e. g., neutral beams or RF) for tokamak or RFP MFE systems or implosion drivers (e. g., heavy/light ion beams or KrF/solid-state lasers) and target factories for ICF systems contribute additional significant costs. Furthermore, most conceptual ICF designs require 10% to 20% of the gross electric power to be used within the plant to supply power to the driver system. This will yield a correspondingly larger balance-of-plant (BOP) and increase the unit's capital cost.

Table 1 is a comparison of estimates of future fission reactor costs and estimates made for magnetic fusion based upon conceptual reactor designs. Cost estimates for ICF reactors have also been made but the designs are at an earlier stage and the cost basis is different. For both MFE and ICF reactors, the results indicate that fusion reactors will be capital intensive and, with significant developments, can provide economically acceptable power [46]. We emphasize, however, that fusion's most attractive attributes as a central station power source are its environmental and safety characterisitics.

More broadly-based studies (including the recent ESECOM study [47,17]) point the way to economically competitive and environmentally acceptable fusion power through a careful choice of enhanced mass power density (MPD), blanket materials, and reactor configurations, together with improvements in plasma physics performance and the development of efficient plasma sustainment systems (heating, current drive, magnets, etc.).

Pending detailed conceptual design of fusion reactors for which systematic "bottom-up" costing analysis can be performed, the MPD (kWe/tonne) is suggested [48] as a useful indicator of progress in reactor design, provided recirculating power costs are kept to acceptable levels (e. g., <10% to 20%). Figure 3 is a plot of the mass power density (MWe/m^3) in the fusion power core (FPC) as a function of the MPD for a number of fusion reactor designs and a typical PWR fission reactor. The MPD values for the UWMAK-I tokamak reactor study (1973), the STARFIRE design (1980), and the ESECOM/GENEROMAK Li/Li/V base case (1988) [47] tokamak designs are, respectively, 20, 50, and 106 kWe/tonne. Although the latter value exceeds the target value of 100 kWe/tonne, it remains below the 500-1000 kWe/tonne values that characterize PWR fission systems. Fusion's more massive (costly) system can be tolerated in economic terms only to the extent that its reduced fuel-cycle costs and improved environmental, licensing, and safety characteristics lead to competitive overall costs.

Advances in plasma physics (e. g., higher beta) and in engineering (e. g., higher power density blankets, high-field coils, efficient ICF drivers, etc.) can lead to high values of MPD and, hence, to competitive costs. Studies such as the recent TITAN Reversed-Field-Pinch (RFP) Study [9] (MPD = 760 kWe/tonne) and the ongoing ARIES Tokamak Study [49] (MPD $\sim$ 84 kWe/tonne) continue this trend to higher MPD (see Fig. 3). Achievement of MPD values in excess of 100 kWe/tonne can be expected to result in lowering the FPC cost below one half the total plant capital costs, reducing the leverage of the most speculative aspect of the fusion plant. This conclusion assumes that the cost of a steam-cycle balance-of-plant is well understood

Table 1. PROJECTED COSTS [a] OF FISSION AND FUTURE MAGNETIC FUSION POWER STATIONS

	FISSION		MAGNETIC FUSION		
	PWR-ME [b]	PWR-BE[c]	ESECOM GENEROMAK [d] Li/Li/V (Tokamak)	TITAN-I[e] Li/Li/V (RFP)	ARIES-I [f] (Tokamak)
Net electrical power (MWe)	~1140	~1140	1200	970	1000
Mass power density, MPD (kWe/tonne)	800 - 1000	800-1000	106	760	85
Direct cost, DC (M$)	1250	905	1800	1470	2271
Unit direct cost, UDC ($/kWe)	1120	825	1500	1604	2271
Cost of electricity, COE (mill/kWeh)					
-- Capital	57	30	37	39	56
-- Operational and Maintenance	13	9	10	7	7
-- Fuel and scheduled replacement item	8	7	10	3.5	7
-- Decommisioning	0.6	0.6	--	0.5	0.5
Total	60	38	57	40	69

a. 1988 U.S. dollars.
b. Refs. 3,4: Pressurized-Water Reactor, Median Experience (Today's fission)
c. Refs. 3,4: Pressurized-Water Reactor, Better Experience (Today's fission, but with regulatory and improved construction practice).
d. Refs. 3,4
e. Ref. 7
f. Ref. 8: High-field, first stability regime.

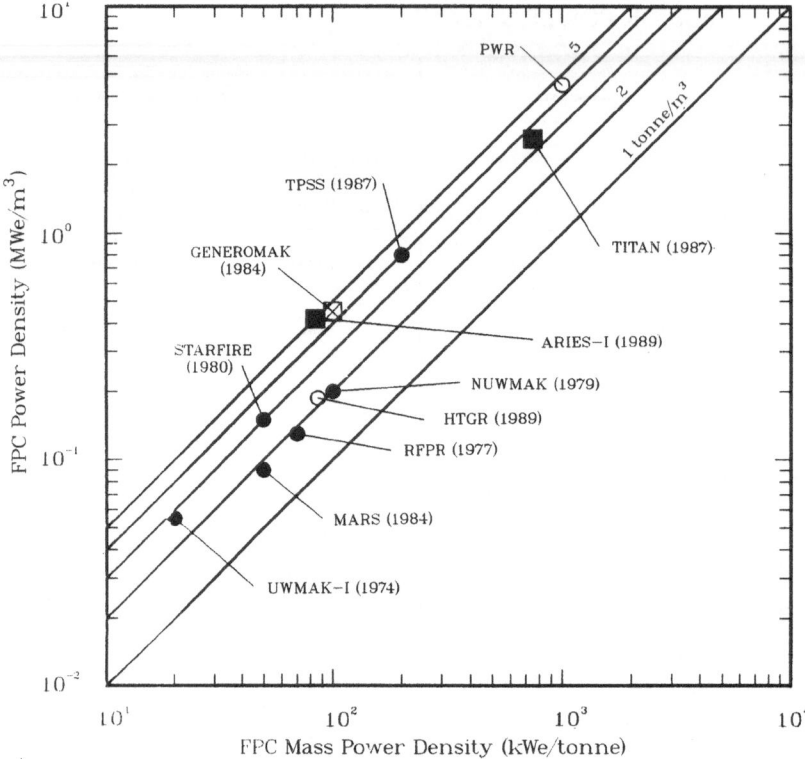

Figure 3. Mass power density for various magnetic reactor designs
showing a trend towards systems with MPD values > 100
kWe/tonne of fusion power core. The HTGR and PWR
are thermal fission reactors. MARS is a tandem mirror
fusion design, TITAN is a reversed filed pinch (RFP) fusion
design, and the remainder are fusion designs based on the
tokamak system.

and comparable to those of fission and fossil plants of similar power output. The MPD may also be a useful figure-of-merit for ICF reaction chambers, although the dominance of the implosion-driver system and target costs have historically led to an emphasis on improving the pellet gain, G, and driver efficiency, η (see reference [32]).

Cost estimates in mills per kWeh for magnetic fusion reactors are given in Fig. 4 from various reactor design studies as a function of mass power density. More recent studies have higher values of MPD. The general trend has been towards higher MPD designs with lower values of COE as both designs and knowledge have advanced. The cost comparison (shown in Fig. 5 of tokamak reactors designed to produce 1200 MW of electricity as a function of the aspect ratio of the device) is indicative of progress that can still be made. The aspect ratio is the ratio of the major and minor radii of the torus. The cost shown is relative and excludes the cost of the balance-of-plant. However, it does include the effect of recirculating power to maintain a steady-state system. The curves in Fig. 5 are for two assumptions on the maximum attainable toroidal magnetic field at a coil, 14 T versus 24 T. Clearly, lower cost systems favor higher field systems at larger aspect ratio.

In summary, fusion power continues the historical trend towards more capital-intensive power-generating systems that use less expensive, more abundant fuel and have additional advantages in areas such as safety, air pollution, waste disposal, and other environmental and licensing features (see Section 4). Cost studies show fusion energy for central-station electric power may come at a premium relative to advanced fission or coal. However, fusion's safety and economic prospects appear to exceed those of either fission or coal by a substantial margin. Recent studies are emphasizing safety and environmental benefits through a broader selection of materials, fuel-cycles, configurations, and energy-conversion schemes. The achievement of a passively-safe reactor can lead to a significant cost credit [47,50] for both MFE and ICF systems.

SAFETY, ENVIRONMENT, AND LICENSING ASSESSMENT OF FUSION ENERGY SYSTEMS

One of the main incentives for investing in the development of fusion energy is the prospect that the safety and environmental (S&E) characteristics of fusion will be less troublesome, both in fact and in public perception, than those of fission and fossil fuels have been. This expectation of S&E advantages for fusion is not a guarantee however (i. e., tritium, neutrons, and neutron-activation products in fusion reactors represent radiological hazards similar in kind if not in magnitude to those of fission reactors). Assuring that fusion technology is developed in ways that exploit the potential for minimizing these hazards deserves high priority. Furthermore, the developers of fusion must address the need to minimize non-nuclear risks (e. g., chemical hazards, exposures to non-ionizing radiation, thermal impacts, etc.).

Background

The first substantial analyses of the safety and environmental characteristics of fusion reactors appeared in the period 1969-70 and focused mainly on tritium hazards in routine operation and in accidents [51], with secondary attention given to problems posed by neutron-activation products [52]. By the mid-1970s, activation-product hazards (occupational radiation doses, radioactive-waste problems, and possibilities for release in reactor accidents) were receiving attention comparable to that given to tritium [53]. Reactor-design studies featuring "low-activation" materials were being conducted [54], and fusion S&E issues had become the focus of international working groups and reviews sponsored by the IAEA [55].

The evolution of fusion S&E studies in the second half of the 1970s reflected the influence of several trends that had been taking place in S&E assessments of other energy options. The first was a trend toward increasing comprehensiveness in the kinds of energy related activities and environmental phenomena included in S&E assessments meant, for fusion, examining not only the most obvious radiological hazards but also the occupational and public hazards of fuel acquisition and transportation, accident risks to workers in component fabrication and power-plant

15

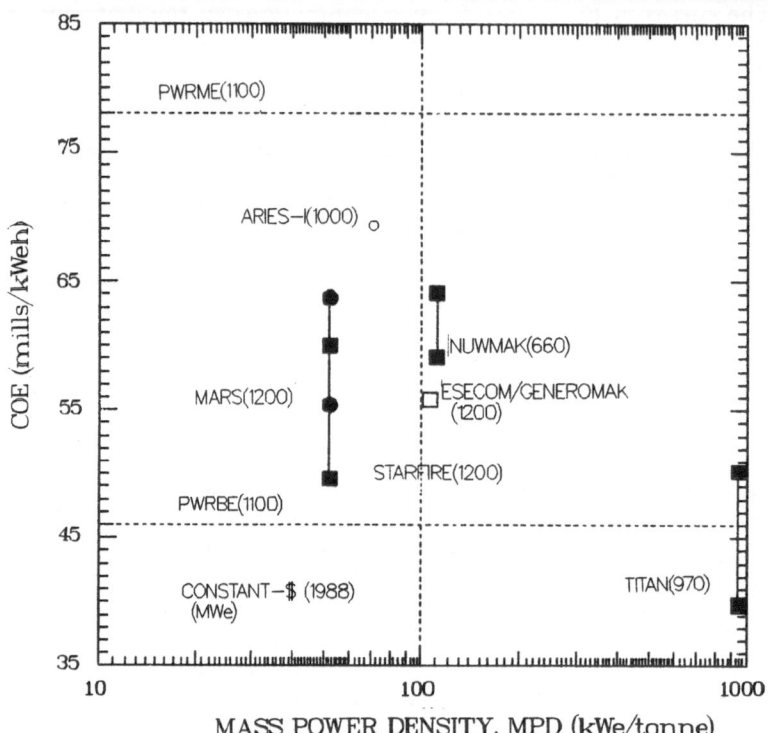

Figure 4. Estimated cost of electricity reported by different magnetic fusion reactor design studies carried out over the years, plotted in 1988 dollars as mills/kWeh versus the mass power density .

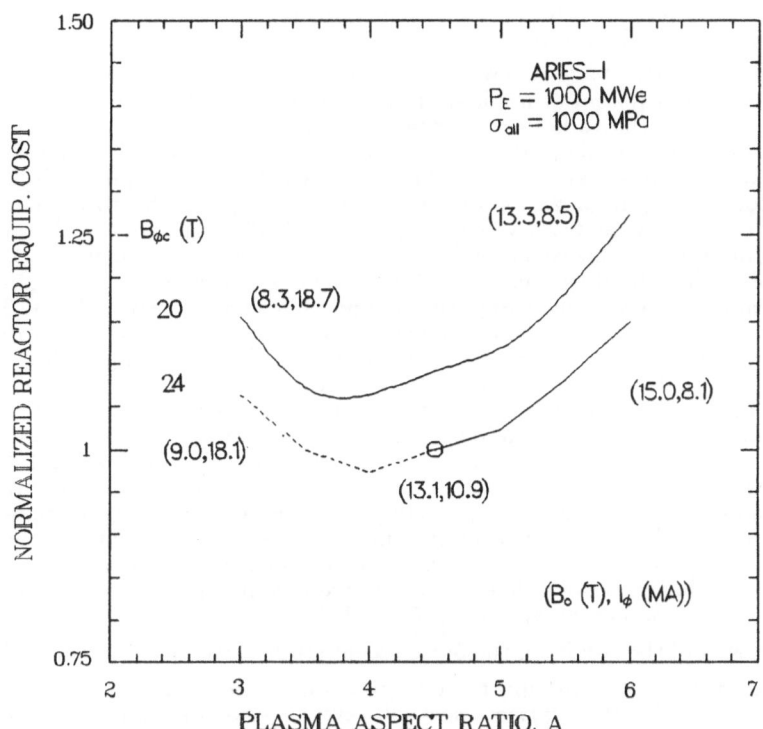

Figure 5. Relative costs of tokamak reactors as a function of aspect ratio for two different values of achievable maximum fields at the coils. The β-limit is that of first stability. The (B_o, I) values are the on-axis magnetic field in Tesla and the plasma current in megamperes, respectively.

construction, chemical hazards in fusion technology, and magnetic-field hazards [56]. A second trend toward comparative assessment of technologies with similar applications, in light of recognition that estimates of S&E risks and impacts are meaningful only in relation to the risks and impacts of alternative ways to obtain the same societal benefits, led to comparative S&E assessments of fusion and fission - perhaps most notably the study of the S&E characteristics of fusion and fission fast-breeder reactors conducted under the auspices of the International Institute for Applied Systems Analysis in 1975-77 [57]. Finally there was a trend, in the study of fission-reactor accident hazards toward analysis of physically plausible pathways by which radioactive material could be released and estimation of radiation doses that could actually result (as opposed to indices based on inventory alone) [58]. This trend led to a similar focus on accident phenomenology in fusion-reactor safety assessments [59,60]. The findings of the fusion S&E assessments by the late 1970s, while obviously preliminary and incomplete, were not wholly reassuring regarding issues such as tritium inventory, accident analysis, and waste disposal.

Expanded efforts have taken place in the 1980s to ensure that fusion's safety and environmental potential is maximized. These efforts included a more accurate characterization of the radiological hazards of fusion technology through the use of more detailed reactor designs [61], new experimental data on accident phenomenology [62], and more sophisticated analytical models (including computer simulation) of accident conditions, mobilization of radioactivity, and pathways to human exposure from reactors as well as from waste repositories [63,64,65]. This work was used to construct indices of relative hazard (or relative assurance of safety) that strike a reasonable balance between clarity and comprehensiveness, that lend themselves to inter-design and inter-source comparisons, and that convey some information about probability of harm (albeit far short of what would be contained in a full probabilistic risk assessment) [64-66]. More extensive studies of the feasibility of low-activation structural materials [67] and advanced (low-neutron) fuel cycles [68] are being carried out and there is a trend towards the systematic exploration of the trade-offs in fusion-reactor blanket design, among traditional performance criteria, minimization of tritium inventories, and maximization of safety margins against mobilization of activation products in accidents [69]. Finally there is a trend towards the integration of the results from the preceding elements, together with engineering-economic modeling of overall system performance and costs [70] to investigate the potential of a range of fusion-reactor concepts to achieve combinations of environmental, safety, and economic characteristics that are attractive compared to those of contemporary and advanced fission-energy systems [47,71,72].

The thrust of the results of the more recent studies can be summarized as follows:

- Fusion reactors of contemporary design will have advantages over fission reactors with respect to the consequences of severe accidents and the magnitude of radioactive-waste burdens, even in the relatively unfavorable case (for fusion S&E characteristics) of stainless-steel structure and liquid-lithium coolant/breeder.

- The magnitude of these advantages becomes more impressive for fusion-reactor designs that employ lower-activation structural materials and/or contain less stored energy than the stainless-steel/liquid-lithium combination. (Examples of lower-activation materials, in order of increasing S&E attractiveness and, but decreasing assurance of technological feasibility, are: elementally tailored ferritic steel, vanadium alloy, and silicon carbide. Coolant and breeders with less stored energy than liquid lithium include FLiBe molten salt, lead-lithium alloy, and helium coolant combined with a solid ceramic breeder material such as Li_2O).

- For some fusion-reactor designs, it appears that avoidance of off-site deaths from acute radiation syndrome in severe accidents can be assured without reliance on active safety systems or containment buildings. Increased experience in the last several years with the handling and control of tritium supports the belief that fusion will be able to meet the same tight standards on routine emissions with which fission reactors now comply.

- Fusion-energy systems will present smaller problems than fission in respect to unwanted links to nuclear weaponry: fusion systems (other than fusion-fission hybrids) would contain no fissile material, and the introduction of means to produce it will be relatively easy to detect.

- Fusion will create certain non-nuclear risks and impacts similar to those of other energy sources (e. g., waste heat, hazards of materials transportation and facility construction, chemical hazards to workers) and at least one non-nuclear hazard not shared by other sources [intense magnetic fields in MFE reactors (powerful pulses of laser light or particle beams in ICF)], but none of these issues currently appears to be serious enough to weigh importantly in societal decisions about fusion's attractiveness compared to other energy options.

- The present detailed nuclear-power-plant licensing regulations are strongly tied to fission reactor designs and will not be applicable to fusion. The underlying regulations in the U.S. for radiation protection (10CFR20, 10CFR50, and 10CFR100), however, are quite general and are expected to apply. It appears that fusion reactor designs can readily satisfy these regulations.

- Fusion has the potential to guarantee the safety of the public by limitations on radioactive material and stored energy inventories. This is expected to result in capital cost savings due to fewer licensing requirements for nuclear-stamp components. The potential for achieving a demonstrably safer reactors may also reduce the time and costs associated with the licensing process.

More information about the results on which these conclusions rest is provided in references [47] and [73]. The following section draws on the work in reference [73].

Hazards In Fusion Energy Systems

The discussion in the following subsections cover: tritium, activation products, radioactive waste management, and accident pathways.

Tritium

Tritium (H^3, in the fusion literature often denoted T) is radioactive with a half-life of 12.3 years, emits a beta particle of rather low energy (maximum 18 keV), and yields stable helium-3 (He^3). Tritium will occur in necessarily sizable quantities as an ingredient of the fuel in fusion reactors operating on the deuterium-tritium (D-T) reaction, and in smaller quantities as the product of D-D reactions in "advanced fuel" systems that contain deuterium. It is also formed as the result of neutron bombardment of light elements that may be present in fusion-reactor components.

The tritium in a D-T fusion reactor will reside in two quite different inventories. The "active" inventory consists of T in the plasma, tritium- breeding blanket, tritium-extraction system, impurity-control and vacuum- pumping system, fuel injectors, and any pipes and intermediate reservoirs linking these components. The magnitude of the active inventory depends on the characteristics of all these subsystems, including especially the mean residence time of T in the breeding blanket and the fraction of injected T that burns on each pass through the plasma. The "inactive" tritium inventory is that part kept in storage for use in the event of breakdown of the tritium extraction system or for eventual shipment to start up other reactors. Its magnitude is a matter of choice, but since it can be kept separate from the fusion power core and very well confined and protected, it usually is not given much attention as a source of routine emissions or accidental releases.

Estimates of the active tritium inventory in various designs for D-T reactors range from around 200 g to several kgs [15,28,47,69,74]. A D-T-fueled fusion reactor operating continuously at 4000 thermal megawatts (1200 to 1500 electrical megawatts) burns about 500 g of tritium per day. At this burn rate, a fractional burnup of 0.1 corresponds to an injection rate of 5 kg/d of T and a fractional burnup of 0.01 corresponds to 50 kg/d. Estimates of the amount of tritium in the breeding blanket have tended to fall over the years as designs of blankets and tritium-removal systems became more sophisticated. However estimates of the inventory in MFE systems that recycle tritium from the plasma have been growing recently because of physics results that imply a low fractional burnup in tokamak reactors. Estimated fractional burnups around 0.3 for ICF reactors lead to lower tritium throughput and, therefore, inventories. Fusion reactors based on advanced fuels (D-D or D-He^3) should be able to achieve T inventories below 100 g [47,65,75].

An objective is to keep the tritium inventory in the range of 100 to 2,000 g. This would correspond, at tritium's specific activity of 9700 curies (Ci) per gram, to between 1 and 20 megacuries (MCi). Issues are radiation exposure to workers inside the plant and to members of the public off site, in normal operation as well as in accidents.

International guidelines, as well as national regulations in most countries, limit occupational radiation exposures to 50 mSv (5 rem) per year to an individual worker [76,77]. The corresponding Recommended Concentration Guidelines (RCG) for tritiated water in air for continuous occupational exposure is 5 MCi/m^3 [76]. (HTO is about 20,000 times more hazardous per curie than HT gas and is usually assumed to be the form present unless there is good reason to believe otherwise. The HT is converted to HTO in air with a variable characteristic time ranging from hours to days). If follows that 3 Ci of HTO (representing on the order of 10^{-6} to 10^{-7} of the active T inventory) would be enough to contaminate all the air in a 5,000,000 m^3 reactor building to the RCG. There is already considerable capability for and experience with tritium cleanup indicating that these problems are manageable [78].

National regulations in most countries limit the radiation exposure by airborne effluents to a hypothetical member of the public who spends full time at the boundary of a nuclear-energy facility to about 50 microsieverts (μSv) per year [77]. For a boundary at a distance of 1 km from the reactor, and considering annual average meteorological conditions at typical sites, complying with such regulations implies restricting routine releases of airborne tritium to the range of 100 to 200 Ci of HTO per day, which represents on the order of 10^{-5} to 10^{-4} of the plant inventory [79]. (Releases of other forms of radioactive material, of course, would reduce the releases of tritium that could be permitted.) Recent experience with handling large quantities of tritium in fusion-technology test facilities (e. g., the TSTA [78] indicate that the required degree of tritium control is indeed feasible.

Tritium, either as HT gas or as HTO liquid or vapor, is highly mobile even under normal operating conditions, and under accident conditions it would surely be one of the more likely radioisotopes to escape in quantity. How large a release of tritium would be tolerable in an accident depends on the "threshold" dose to the most-exposed member of the public that one must not exceed. One such dose threshold used by regulatory authorities in many countries is a 50-year dose commitment of 100 mSv (10 rem) [75], which corresponds approximately to an increase of 1% in the exposed individual's pre-existing probability of dying of cancer. This dose could be received by an individual remaining for several days at the site boundary, 1 km from an accidental release of 100 to 200 g of tritium as HTO under highly unfavorable weather conditions (accounting for inhalation and skin absorption, resuspension of ground-deposited tritium, and a quality factor of 2 Sv/Gy in computing the dose per curie of tritium intake [77,79]). Such a release is in the range of 10% to 100% of the active tritium inventory in recent fusion-reactor designs.

How much tritium could be released in an accident without exceeding the 2 Sv critical-dose threshold below which no fatalities from acute radiation syndrome would be expected [26]? (The critical does is that received in the first 7 days from the beginning of exposure plus half of that received in the 8th through the 30th days; it correlates well with acute radiation effects [58].) Taking into account that the critical dose from an accidental exposure to tritium is about half of the 50-year dose commitment, the tritium release (as HTO) corresponding to a critical dose of 2 Sv to an individual at the site boundary, 1 km away from the reactor is 4 to 8 kg, neglecting the possibility of evacuation. If the individual leaves within the first few hours after the accident, the release needed to give a critical dose of 2 Sv is 2 to 3 times as large, say 10 to 20 kg.

Thus, a key conclusion is that even a 100% release of the active tritium inventory in a fusion reactor could not produce an off-site dose large enough to lead to early fatalities unless that inventory exceeds 3-4 kg (no evacuation from the site boundary) and more probably 10 to 20 kg (most exposed individual leaves the site boundary after a few hours).

Activation Products

Neutron-activation products are formed when fusion neutrons (at 14 MeV from every D-T reaction and at 2.5 MeV from about half the D-D reactions) strike the main constituents and impurities in the reactor structure, coolant, tritium breeder (if any), neutron-multiplier materials (if any), and reactor-building atmosphere. The variety of neutron-activation reactions and resulting isotopes is very large when one takes into consideration the diversity of materials that may be found in fusion reactors, as well as the multistep reactions in which activation products or their decay products are themselves bombarded by further neutrons. On the other hand, the array of activation products that actually arise is controllable by design through the choice of structural and other materials to be used in the reactor (subject to other criteria these materials must satisfy).

The types and quantities of neutron-activation products that will be present in a fusion reactor can be predicted with considerable confidence using computer programs that have been improved and checked against experiments over a period of many years [80]. The neutron activation calculations reveal that typical D-T MFE reactors of contemporary design, built of metals such as austenitic or ferritic steels or refractory-metal alloys, and sized to deliver 1200 MWe, would contain in the range of 0.6 to 4 GCi of activation products when shut down after a long period of operation, and 3 to 10 times less a day after shutdown [47,61,69]. A low activation D-T MFE reactor with SiC structure would have comparable activity to the metal structure systems at shutdown but 10^2-10^3 times less than these more conventional systems a day after shutdown [47]. The Cascade ICF reactor design, using a variety of SiC components, has a calculated inventory of 0.24 GCi at shutdown and 0.008 GCi a day later [28]. A fusion reactor design using the harder-to-ignite D-He^3 reaction has been predicted to have an activation product inventory of about 0.04 GCi at shutdown (the neutrons in this case come from D-D side reactions) [47].

Between 15% and 60% of the activation-product inventory in typical fusion-reactor designs will be embedded in the solid first wall separating the plasma chamber from the rest of the tritium-breeding and energy-absorbing blanket, with most of the remainder in the other solid- blanket components and the radiation shield. Some ICF D-T reactor designs achieve a substantial reduction in total activation-product inventory by interposing a flowing layer of low-activation liquid- or granular-coolant/breeder material between the reaction zone and the innermost solid structure [28,81].

The total activation-product radioactivity inventories for a typical 1200 MWe D-T fusion reactor at shutdown are 1 to 10 times smaller than the 5 x 10^9 Ci of fission products in a fission reactor with the same electrical output. Use of the D-D fuel cycle brings little or no advantage in the quantity of activation [57,28], but the inventory in a D-He^3-fueled fusion reactor could be 3 to 100 times smaller than that of its fission counterpart [47,75].

Hazard Measures Relating to Accidental Releases

Table 2 presents a number of measures of the accident hazard associated with the inventories of radioactivity in fusion and fission reactors. The measures are arranged in ascending order of informatives, starting with numbers of curies. (The more informative the indicator, unfortunately, the more difficult it is to calculate.) The estimates shown are drawn largely from references [47], [65], and [82].

The integrated risk is by far the most informative hazard measure but also the most difficult to obtain. Very detailed design information and extensive operating experience are needed in order to catalog all the possible accident modes and to predict their probabilities of occurrence; and separate consequence calculations must be carried out for all the accident modes [58].

The maximum plausible doses (MPD) from "worst-case" accidents are of considerable interest (to the public and to licensing authorities) and are easier to calculate

Table 2. SOME QUANTITATIVE MEASURES OF ACCIDENT HAZARD[60,74,87]

Measure	Units	Fission/Fusion Ratios[a]
Radioactivity Inventory	Ci	1.2 to 8, various D-T 100, V-alloy D-He^3
Biological Hazard Potential	m^3	5 to 20, steel D-T 15 to 80, V-alloy D-T
Complete Release Dose Potential	Sv	2 to 4, stainless steel D-T 3 to 6, ferritic steel D-T 40 to 20, V-alloy D-T 20 to 100, SiC D-T 200, V-alloy D-He^3
Partial Release Dose Potentials, e.g., for 100% of T (as HTO) and noble gases	Sv	10 to 100, various D-T[b] 300, V-alloy D-He^3
Maximum plausible release fractions[c] for all isotopes (giving maximum plausible dose, MPD)		4, ferritic steel D-T 6 to 100, V-alloy D-T 50 to 100 SiC D-T 150, V-alloy D-He^3
Integrated risk (probability x population dose summed over all possible accidents)	man-Sv per reactor/yr	not yet available

a. Based on LMFBR characteristics in the numerator and characteristics of indicated fusion-reactor types in the denominator.

b. Assumes active T inventories from 200 to 2000 g and 0.2 Sv critical dose at 1 km per kg of T released as HTO.

c. Based on division of elements into 5 mobility categories with maximum plausible release fractions = 1.0, 0.3, 0.1, 0.03, and 0.01, respectively, for both fusion and fission; some fusion designs (and some future fission designs) may warrant smaller MPRFs for given elements.[60]

than the integrated risk, but doing so convincingly still requires detailed analysis of specific designs combining sophisticated models of accident phenomenology and experimental data on mobilization of radioactivity under extreme conditions [60,62,73,86]. Estimates of MPDs can be obtained by dividing the radioactive inventories into categories according to the relative mobility of the associated elements under accident conditions and associating each category with a maximum plausible release fraction (MPRF) based on whatever mobilization data and analysis are available for the relevant materials and configuration [47]. For the most preliminary comparisons of different reactor types, one may choose to use a common, conservatively defined "envelope" of MPRFs for all of the designs considered (as in Table 2).

In addition to such approaches to estimating maximum permissable doses, it is useful in preliminary safety assessments to be able to characterize in a general way the relative ease or difficulty of accident prevention in different reactor designs. Some important conclusions are:

a) Fusion reactions and the fusion plasma are rather easy to quench, and the amount of fusion fuel in the reaction chamber at one time is small, so designing fusion reactors to preclude nuclear reactivity excursions large enough to do serious damage should not be very difficult.

b) The chemical energy of liquid lithium is a troublesome source of stored energy in D-T fusion reactors that use this material as the coolant and tritium breeder. This contribution to accident risk can be greatly reduced by using lithium-lead alloy in place of liquid Li and can be avoided altogether by using solid or molten-salt Li compounds as the tritium breeder (or by using advanced fuel cycles that do not need to breed tritium). Liquid Li should only be used in combination with structural materials highly resistant to the release of activation products at temperatures attainable in lithium fires. Lithium-lead unfortunately creates some activation problems.

c) Apart from liquid Li, the most potentially troublesome source of stored energy in fusion reactors is the decay heat from structural activation products. Decay heat from the radioactivity in fission reactors is the major source of concern about core melt in loss-of-coolant accidents (LOCAs) and the reason for requiring engineered emergency-core-cooling systems in fission reactors of current types. For most fusion-reactor designs and materials, however, the afterheat power levels and power densities are enough lower than those in fission to substantially ease the emergency cooling problem. In fact, with suitable attention to this point in materials selection and design, D-T fusion reactors can be designed so that passive mechanisms of heat removal alone suffice to prevent structural damage and significant radioactivity releases in LOCAs; and D-He^3 fusion reactors should be able to achieve this desirable characteristic with ease.

d) The energy stored in the confinement system (magnets for MFE reactors, lasers or particle beams in ICF reactors) and its power supplies will be sufficient to damage the reactor if released suddenly. (Such releases could generate projectiles capable of breaking coolant pipes or damaging other safety-related barriers.) These systems can probably be designed, however, with passive mechanisms to dissipate their energy gradually in accidents [47,28].

A useful way to summarize the most important safety-related characteristics of a given fusion-reactor or fission-reactor design is the level-of-safety-assurance (LSA) concept developed by Piet [66] and applied in the U.S. ESECOM study [47]. The LSA evaluations are based on the extent and nature of passive versus active design features in a reactor in order to assure public safety, perhaps most importantly to assure that early off site fatalities from release of radioactivity are precluded. An LSA value of 1 means that safety is assured by the sizes of the radioactivity inventories and stored energy sources alone without reference to any specifics of reactor configuration or accident scenario; an LSA of 4 means that assurance of safety requires demonstrating that active-safety systems will perform satisfactorily; and the intermediate LSA values mean assurance of safety depends on demonstrating that passive-design features can maintain the large-scale configuration of the reactor (LSA = 2) or that they can maintain both large-scale and small-scale configuration (LSA = 3). Although LSA = 4 systems may be adequately safe, lower LSA values mean that safety is easier to prove and, therefore, that siting and licensing should be easier.

The ESECOM study concluded that many D-T fusion reactor designs could achieve LSA = 2 or 3 and that use of very-low-activation materials (such as silicon carbide) or advanced fuels (such as D-He^3) could bring LSA = 1 within reach [47].

Measures Relating to Radioactive Waste Management

Radioactive wastes from fusion reactors will consist mainly of activated structural material, part of it leaving the reactor when fusion-core components are replaced at intervals during the reactor's lifetime and part of it resulting from the dismantling of the reactor at the end of its service. This material will total between 400 and 3000 m^3 over a 1200 MWe reactor's nominal life cycle of 30 years, compared to perhaps 400 m^3 of "high-level" (>10,000 Ci/m^3) fission-product and structural waste from an LMFBR of the same output [21,47,28]. Associated with the reprocessing plant and fuel fabrication plant for the LMFBR, however, will be another 100 to 400 m^3/y of "medium-level" (0.1 to 10,000 Ci/m^3) and transuranic contaminated wastes [57,88] that will require long-term management. Both the fusion and LMFBR fuel cycles will generate solid and liquid low-level wastes (<10,000 Ci/m^3) amounting to perhaps 1000 m^3/y [57,71], which are more easily managed.

In any case, other measures convey much more information about the relative difficulty of the waste management task than volumes do. Two such measures are [47,65]:

(i) The integrated biological hazard potentials (IBHP), obtained by dividing curies of each isotope by its RCG [83] for public water supplies, multiplying by its mean life, and summing over all isotopes in the wastes (units are m^3/reactor year).

(ii) The annualized intruder hazard potentials (IHP) [86], obtained from standardized scenarios of the consequences of inadvertent intrusion into shallow burial sites after periods of 100, 500, and 1000 years (rem-m^3/reactor year).

Some ratios of these indices for fission and fusion reactors are presented in Table 3. Use of the shallow burial scenarios provide a systematic way to account for the relative mobilities and toxicities of the constituents of wastes of different types. That wastes from some fusion reactor designs would qualify for shallow burial under current U.S. regulations [47] is an impressive indicator.

Hazard Measures Relating to Routine Emission and Exposures

Most of the activation products in a fusion reactor remain embedded in solid structure during routine operation. Only the far smaller quantities of activated material in the coolant and the building atmosphere have any chance of escaping to the environment under these normal conditions, and various studies have indicated that the doses expected from these sources can easily be held within the range of 10% to 50% of the total exposure guidelines (50 μSv/y each from airborne and aqueous effluents) for the most exposed individual at a 1-km site boundary [71,73,74]. This would leave room for the anticipated levels of tritium emissions without exceeding the guidelines.

Activation products will pose larger problems for the control of worker exposures inside the power plant. Approximate calculations of the contact dose rates to be expected at the surfaces of various fusion reactor components [47] give values exceeding the usual guidelines for unlimited hands-on maintenance (0.025 mSv/h contact dose rate) by seven to ten orders of magnitude at the first wall at 1 hour after shutdown and by four to nine orders of magnitude at 1 year after shutdown. Even at the shield, the factors are four to eight orders of magnitude at 1 hour and three to seven orders of magnitude at 1 year. Clearly, all maintenance in the immediate vicinity of the fusion power core will have to be done remotely. Proper choice of coolant,

Table 3. SOME MEASURES OF RADIOACTIVE WASTE HAZARD[55,74]

Measure	Units	Fission/Fusion Ratios[a]
Integrated BHP_w	m^3-yr	5,000, D-T stainless steel 10,000, D-T ferritic steel 100,000, D-T V-alloy
Annualized Intruder Hazard Potential	rem-m^3 per yr	250 to 1,000, D-T ferritic steel[b] 50,000, D-T tailored[c] ferritic steel 10,000 to 100,000, D-T V-alloy 300,000 D-T silicon carbide 2,000,000, V-alloy D-He^3

a. Based on LMFBR characteristics in the numerator and characteristics of indicated fusion-reactor types in the denominator.

b. The lower figure comes from a design with a lithium-lead blanket.

c. HT-9 ferritic steel modified to replace molybdenum with tungsten.

however, may allow hands-on maintenance of some parts of the coolant system; and use of low-activation structural material can also expand the scope for limited contact maintenance.

Non-nuclear Hazards

Previous studies of the fuel cycles of coal, fission, fusion, and renewable energy sources have indicated that, aside from emissions and accident risks from power plant operations, the highest risks to health and safety tend to be those of more-or-less routine accidents in fuel and materials acquisition, processing, manufacturing, and transport. These hazards for fusion systems appear to be in the same range as those of the renewable electricity sources, hydropower and wind, and well below those of coal [89]. Fusion reactors and fuel-cycle operations would not release significant quantities of greenhouse gases or acid-rain precursors, and fusion's land-use requirements would be smaller than those of most fossil and renewable energy sources [89]. Thermal pollution from D-T fusion reactors would not be significantly different in proportion to electrical output than that from fossil or fission plants. Advanced-fuel fusion reactors with a high proportion of the reaction energy carried by charged particles have the potential for significantly increased conversion efficiencies and thus of markedly reduced thermal pollution [68].

CONCLUDING REMARKS

Fusion energy has the potential to be a safe and environmentally attractive power source at economically acceptable levels. The safety and environmental advantages are substantial enough that they are a major part of the rationale for fusion development: fusion would have no counterpart to the problems of mining, air pollution, acid rain, and climate change associated with coal use; it offers the prospect of increased assurance of safety against major accidents; it has diminished linkages with military applications; and it presents a smaller waste-management task than fission. Achieving the full potential of fusion will not happen automatically but will require success in plasma physics and the technology of magnets and ICF drivers, the development of low-activation materials for the construction of reactors, and the development of low-tritium inventory designs. The size of the challenge and the present inadequacy of the resources make the task formidable. It also underlines the importance of increased international coordination and collaboration to optimize the utilization of diverse national resources in the common goal to achieve practical fusion energy.

Acknowledgement

The authors wish to thank Dr. Shahram Sharafat for his great help in preparing this manuscript. The research is partially supported by the U.S. Department of Energy, Office of Fusion Energy.

REFERENCES

1. Proc., Workshop on Fusion Reactor Design Problems, (Culham, 1974) Nucl. Fusion, Special Supplement International Atomic Energy Agency, Vienna, (1974).

2. Proc., 2nd Workshop on Fusion Reactor Design Concepts, (Madison, 1977) (International Atomic Energy Agency, Vienna, 1978); see also R. W. Conn, T. G. Frank, R. Hancox, G. L. Kulcinski, K. H. Schmitter, and W. M. Stacey, Jr., Nucl. Fusion 18, 1985 (1978).

3. Proc., 3rd Technical Committee Meeting and Workshop on Fusion Reactor Design and Technology, (Tokyo, 1981) (International Atomic Energy Agency, Vienna, 1982); see also Y. Iso, W. M. Stacey, Jr., G. L. Kulcinski, R. A. Krakowski, G. A. Carlson, C. Yamanaka, G. Casini, and N. Igata, Nucl. Fusion, 22, 671 (1982).

4. W. M. Stacey, Jr., "Fusion Reactor Development – A Review," Adv. Nucl. Sci. Technol., 15, (1981).

5. C. C. Baker, G. A. Carlson, and R. A. Krakowski, "Trends and Developments in Magnetic Confinement Fusion Reactor Concepts," Nucl. Technol./Fusion, 1, 5 (1981).

6. R. W. Conn, "Magnetic Fusion Reactors," in Fusion, E. Teller, Ed. (Academic Press, New York, 1981).

7. C. C. Baker, et al., "STARFIRE - A Commercial Tokamak Fusion Power Plant Study," Argonne National Laboratory report ANL/FFP-80-1 (September1980).

8. "MARS, Mirror Advanced Reactor Study," Lawrence Livermore National Laboratory report UCRL-53480 (1984).

9. The TITAN Research Group, "The TITAN Reversed-Field-Pinch Reactor Study – The Final Report," University of California at Los Angeles, GA Technologies Inc., Los Alamos National Laboratory, and Rensselaer Polytechnic Institute, UCLA report, PPG-1200 (1989).

10. R. L. Miller, C. G. Bathke, and R. A. Krakowski, et al., "The Modular Stellarator Reactor: A Fusion Power Plant," Los Alamos National Laboratory report LA-9737-MS (1983).

11. B. Badger, et al., "UWTOR-M: A Conceptual Modular Stellarator Power Plant," University of Wisconsin report UWFDM-550 (1982).

12. O. Motojima, A. Iiyoshi, and K. Uo, "The Design of Heliotron Steady Reactor," Proc., 9th Int. Conf. Plasma Physics and Controlled Nuclear Fusion Research, Baltimore, Maryland, IAEA-CN-41/L-3 (International Atomic Energy Agency Baltimore, MD, September 1982).

13. R. F. Bourque, "OHTE Reactor Concepts," Proc., 9th Symp. Engineering Problems of Fusion Research, Vol.II, (Institute of Electrical and Electronics Engineers, Chicago, IL, October 1981 p. 1851).

14. M. Katsurai and M. Yamada, "Studies of Conceptual Spheromak Fusion Reactors," Nucl. Fusion, 22, 1407 (1982).

15. R. A. Krakowski, "Identification of Future Engineering Development Needs of Alternative Concepts for Magnetic Fusion Energy," Los Alamos National Laboratory report LA-UR-82-1973 (1982).

16. ITER Definition Phase Report, International Atomic Energy Agency (1988).

17. R. A. Krakowski and J. G. Delene, "Connections Between Physics and Economics for Tokamak Power Plants," J. Fusion Energy, 7, 49 (1988).

18. R. G. Mills, et al., "A Fusion Power Plant," Princeton Plasma Physics Laboratory report MATT-1050 (1974).

19. C. C. Baker, et al., "Tokamak Power System Studies," Argonne National Laboratory report ANL/FPP/85-2 (1985).

20. Fusion Engineering and Design, 7, Special Issue on The IEA Large Coil Task (1988).

21. T. Ando, paper E.4.7, Specialist's Meeting on Tokamak Concept Innovation, International Atomic Energy Agency report IAEA-TECDOC-373 (1986).

22. D. R. Cohn, et al., "High Field Compact Tokamak Reactor (HFCTR) Conceptual Design," Massachusetts Institute of Technology report MITRR-79-2 (1979).

23. J. T. D. Mitchell and R. Hancox, Proc., 7th Intersociety Energy Conversion Engineering Conference, (San Diego, CA, 1972) 1275.

24. G. L. Kulcinski, et al., "Apollo – An Advanced Fusion Fueled Reactor for the 21st Century," University of Wisconsin, Madison report UWFDM-780 (1988).

25. J. J. Duderstadt and G. A. Moses, Inertial Confinement Fusion (J. Wiley and Sons, New York, 1982).

26. T. H. Johnson, Proc., Inertial Confinement Fusion: Review and Perspective, IEEE 72 (1984) 548.

27. M. J. Monsler, et al.,"An Overview of Inertial Fusion Reactor Design," Nucl. Technol./Fusion VI, 302-358 (1981).

28. W. J. Hogan, G. L. Kulcinski, "Advances in ICF Power Reactor Design," Fusion Technol., Vol. 8, 17-726, (1985).

29. W. J. Hogan, "ICF Drivers: A Comparison of Some New Entries and Old Standbys," Fusion Technol., 10, (3, 2a), 649-655 (1986).

30. R. L. McCrory, et al., "Laser Driven Implosion of Thermonuclear Fuel to 20 to 40 g-cm^3," Nature, 335, 225 (1988).

31. J. D. Lindl, "Progress on Achieving the ICF Conditions Needed for High Gain," Fusion Technol., 15, (2, 2a), 227-238 (1989).

32. J. A. Blink, W. J. Hogan, J. Hovingh, W. R. Meier, and J. H. Pitts, "The High-Yield Lithium-Injection Fusion-Energy (HYLIFE) Reactor," Lawrence Livermore National Laboratory report UCRL-53559 (1985).

33. B. Badger, et al., "SOLASE-A Laser Fusion Reactor Study," University of Wisconsin report UWFDM-220 (1977).

34. D. J. Dudziack and W. B. Herrmannsfeldt, "Heavy-Ion Fusion Systems Assessment Study," Proc., AIP Conf. No. 152 (N.Y., 1986) pp. 111-121.

35. D. L. Cook, "Pulsed Power Driver Technologies for Inertial Confinement Fusion Power Reactor," Proc., Third International Confinement Fusion Systems and Applications Colloquium, University of Wisconsin report UWFDM-749, (1988) pp. 106-127.

36. J. H. Pitts, "Cascade: A Centrifugal-Action Solid-Breeder Reaction Chamber," Nucl. Technol./Fusion, 4, (2, 3), 967-972, (1983); also, I. Maya, K.R. Schultz, "Inertial Confinement Fusion Reaction Chamber and Power Conversion System Study," GA Technologies report GA-A17843 (1985).

37. G. A. Moses, et al., "LIBRA-A Light Ion Beam Fusion Reactor Conceptual

Design," Fusion Power Associates report FPA-88-3 (1988).

38. H. Pitts, "Turbostar: An ICF Reactor Using both Direct and Thermal Power Conversion," Fusion Technol., Vol. 10, 695-703 (1986).

39. W. F. Krupke, "Solid State Laser Driver for an ICF Reactor," Fusion Technol.,15, (2), 377-382 (1985).

40. G. F. Albrecht and S. B. Sutton, "Gas Cooling of Laser Disks," Energy and Technology Review, Lawrence Livermore National Laboratory, (May 1988) pp. 25-34.

41. W. Streifer, et al., "Advances in Diode Laser Pumps," IEEE J. Quantum Electronics, 24, 883, (1987) and R. G. Waters, et al., "High Power Conversion Efficiency Quantum Well Diode Laser," Appl. Physics Lett., 51, 1318 (1987).

42. D. B. Harris, et al., "Conceptual Design of a Large Electron-Beam-Pumped KrF Laser for ICF Commercial Applications," Fusion Technol., 10 (1986).

43. E. P. Lee and J. Hovingh, "Heavy Ion Induction Linac Drivers for Inertial Confinement Fusion," Fusion Technol., 15, (2) 369-376 (1989).

44. A. M. Weinberg, "Energy Policy in an Age of Uncertainty," in: Issues in Science and Technology, 5 (2) 81 (Winter 1988-1989).

45. J. J. Taylor, "Improved and Safer Nuclear Power," Science, 244, 318 (1989).

46. J. Sheffield, R. A. Dory, S. M. Cohn, et al., "Cost Assessment of a Generic Fusion Reactor," Oak Ridge National Laboratory report ORNL/TM-931 (1986).

47. J. P. Holdren, D. H. Berwald, R. J. Budnitz, J. G. Crocker, J. G. Delene, R. D. Endicott, M. S. Kazimi, R. A. Krakowski, B. G. Logan, and K. R. Schultz, "Exploring the Competitive Potential of Magnetic Fusion Energy: The Interaction of Economics with Safety and Environmental Characteristics," Fusion Technol. 13, 7 (1988); also, by the same authors, "Report of the Senior Committee on Environmental, Safety, and Economic Aspects of Magnetic Fusion Energy," Lawrence Livermore National Laboratory report UCRL-53766 (to be published 1989).

48. R. W. Conn (Chairman), "Panel X Report to the USDOE Magnetic Fusion Advisory Committee," University of California at Los Angeles (May 8, 1985).

49. F. Najmabadi and the ARIES team, "The ARIES Tokamak Reactor Study," in 13th Symposium on Fusion Engineering, Knoxville, TN, October 2-6, 1989.

50. I. Maya, et al., "Inertial Confinement Fusion Reaction Chamber and Power Conversion System Study," GA Technologies report GA-A17842 (1985).

51. F. Morley and J. W. Kennedy, "Fusion Reactors and Environmental Safety," Nuclear Fusion Reactor Conference Proceedings (British Nuclear Energy Society, London, 1969), pp. 54-65; also, A. P. Fraas and H. Postma, "Preliminary Appraisal of the Hazards Problems of a D-T Fusion Reactor Power Plant," Oak Ridge National Laboratory report, ORNL-TM-2822, (1969).

52. J. D. Lee, "Some Observations on the Radiological Hazards of Fusion," Fusion Technol.: Presentations from the 5th Intersociety Energy Conversion Engineering Conference (Las Vegas, September 1970), pp. 58-60; D. Steiner and A. P. Fraas, "Preliminary Observations on the Radiological Implications of Fusion Power," Nucl. Safety, 13, 353-362 (1972).

53. See, e. g., G. L. Kulcinski, "Fusion Power: An Assessment of Its Potential Impact in the USA," Energy Policy (June 1974), pp. 104-125; D. J. Dudziak and R. A. Krakowski, "Radioactivity Induced in a Theta-Pinch Fusion Reactor," Nucl. Technol., 25, 32 (1975); R. W. Conn, T. Sung, and M. A. Abdou, "Comparative Study of Radioactivity and Afterheat in Several Fusion Reactor Designs," Nucl. Technol., 26, 391 (1975); J. S. Watson and F. W. Wiffen, Proc., International Conf. on Radiation Effects and Tritium Technology for Fusion Reactors, U.S. Energy Research and Development Administration report CONF-750989 (4 vols.) (March 1976).

54. J. R. Powell, F. T. Miles, A. Aronson, and W. E. Winsche, "Studies of Fusion Reactor Blankets with Minimum Radioactivity Inventory and With Tritium Breeding in Solid Lithium Compounds," Brookhaven National Laboratory report BNL-18236 (1973); also L. H. Rovner and G. H. Hopkins, "Ceramic Materials for Fusion," Nucl. Technol. 29, 274 (1976).

55. See, et al., J. T. D. Mitchell, "Fusion and the Environment," Proc., IAEA Workshop of Fusion Reactor Design Problems (Culham, UK, 29 Jan – 15 Feb 1974), IAEA-STI/PUB/23, (International Atomic Energy Agency Vienna, 1974); also, F. N. Flakus, "Fusion Power and the Environment," Atomic Energy Review 13 (IAEA, Vienna, 1975) pp. 587-614.

56. The most impressive example is the 19-volume assessment, "An Environmental Analysis of Fusion Power to Determine Related R&D Needs," prepared in 1976 under the direction of J. R. Young at the Battelle Pacific Northwest Laboratories for the Energy Research and Development Administration; in addition to the overview and fusion technology survey (BNWL-2010, -2011, -2013, -2014), the reports cover fuel procurement (BNWL-2012), siting issues (BNWL-2015), materials availability (BNWL-2016), thermodynamic efficiencies (BNWL-2017), tritium and radiocarbon releases and effects (BNWL-2018, -2020, -2022), non-tritium radioactive wastes (BNWL-2019, -2023), magnetic-field issues (BNWL-2021), safety (BNWL-2024), transportation requirements (BNWL-2025), social impacts (BNWL-2026), comparisons with nonfusion energy systems (BNWL-2027), and environmental cost/benefit analysis for fusion (BNWL-2028).

57. W. Haefele, J. P. Holdren, G. Kessler, and G. L. Kulcinski, with contributions by A. M. Belostotsky, R. R. Grigoriants, D. K. Kurbatov, G. E. Shatolov, M. A. Styrikovitch, and N. N. Vasiliev, Fusion and Fast Breeder Reactors International Institute for Applied Systems Analysis, RR-77-8 (Laxenburg, Austria, July 1977) 506 pp.

58. "Reactor Safety Study," U.S. Nuclear Regulatory Commission report WASH-1400 (NUREG-75/014), (Washington, D.C., 1975): 200 pp. main report plus 11 appendices; also A. Bayer and J. Ehrhardt, "Risk Oriented Analysis of the German Prototype Fast Breeder Reactor SNR-300: Off-Site Accident Consequence Model and Results of the Study," Nucl. Technol. 65, 232 (1984); also "Reassessment of the Technical Basis for Estimating Source Terms," U.S. Nuclear Regulatory Commission report NUREG-0956, (1985).

59. W. E. Kastenberg and D. Okrent, Principal Investigators, "Some Safety Considerations for Conceptual Fusion Power Reactors," Electric Power Research Institute report EPRI ER-546 (1978) 332 pp; also D. Okrent, W. E. Kastenberg, T. E. Botts, C. K. Chan, W. L. Ferrell, T. H. K. Frederking, M. J. Sehnert, and A. Z. Ullman, "On the Safety of Tokamak-type, Central Station Fusion Power Reactors," Nuclear Eng. and Des., 39, 215-238 (1976).

60. See, et al., M. S. Kazimi, et al., "Aspects of Environmental and Safety Analysis of Fusion Reactors," MIT Department of Nuclear Engineering report MITNE-212 (1977); also, D. A. Dube and M. S. Kazimi, "Analysis of Design Strategies for Mitigating the Consequences of Lithium Fire Within Containment of Controlled Thermonuclear Reactors," MIT Department of Nuclear Engineering report MITNE-219 (1978).

61. Especially important in this respect was the Starfire design: Argonne National Laboratory, McDonnell-Douglas Astronautics Co., General Atomic Co., and The Ralph M. Parsons C., "STARFIRE: A Commercial Tokamak Fusion Power Study," Argonne National Laboratory report ANL/FPP-80-1 (1980).

62. See, et al., D. W. Jeppson, "Scoping Studies – Behavior and Control of Lithium and Lithium Aerosols," Hanford Engineering Development Laboratory report HEDL-TME-80-79 (1981).

63. See, et al., M. S. Tillack and M. S. Kazimi, "Modeling of Lithium Fires," Nucl. Technol./Fusion, $\underline{2}$, 233 (1982).

64. S. J. Piet, "Potential Consequence of Tokamak Fusion Reactor Accidents: The Materials Impact," Ph.D. thesis, Massachusetts Institute of Technology (1982); also, S. J. Piet, M. S. Kazimi, and L. M. Lidsky, "The Materials Impact on Fusion Reactor Safety," Nucl. Technol./Fusion, $\underline{5}$, 382 (1984)

65. S. A. Fetter, "Radiological Hazards of Fusion Reactors: Models and Comparisons," Ph.D. thesis, University of California, Berkeley (1985); also, S. Fetter, "A Calculation Methodology for Comparing the Accident, Occupational and Waste-Disposal Hazards of Fusion Reactors," Fusion Technol., $\underline{8}$, 1359 (1985).

66. S. J. Piet, "Approaches to Achieving Inherently Safe Fusion Power Plants," Fusion Technol. $\underline{10}$(1), 7-30 (1986).

67. R. W. Conn, E. E. Bloom, J. W. Davis, R. E. Gold, R. Little, K. R. Schultz, D. L. Smith, and F. W. Wiffen, Report of the DOE panel on "Low Activation Materials for Fusion Applications," University of California Los Angeles report UCLA-PPG-728 (1983); also, G. R. Hopkins, E. T. Cheng, R. L. Creedon, I. Maya, K. R. Schultz, P. Trestler, and C. P. C. Wong, "Low Activation Fusion Design Studies," Nucl. Technol./Fusion, $\underline{4}$, (2), 1251 (1985).

68. G. H. Miley, "Potential and Status of Alternate Fuel Fusion," Proc., 4th Topical Meeting on the Technology of Controlled Nuclear Fusion (King of Prussia, PA, October 1989) pp. 905.

69. D. L. Smith, C. C. Baker, D. K. Sze, G. D. Morgan, M. A. Abdou, S. J. Piet, K. R. Schultz, R. W. Moir, and J. D. Gordon, "Overview of the Blanket Comparison and Selection Study," Fusion Technol., $\underline{8}$, (1,1), 10 (1985).

70. See, et al., J. Sheffield, R. A. Dory, S. M. Cohn, J. G. Delene, L. F. Parsly, D. E. T. F. Ashby, and W. T. Reiersen, "Cost Assessment of a Generic Magnetic Fusion Reactor," Fusion Technol. $\underline{9}$, 199 (1986).

71. "Environmental Impact and Economic Prospects of Nuclear Fusion," Commission of the European Communities, Brussels report EURFU BRU/XII-828/86 (1986).

72. Kenzo Yamamoto (chairman), "Investigation of Feasibility of Nuclear Fusion," report by the Japanese Atomic Energy Society to the Atomic Energy Commission of Japan, Tokyo (1986) (in Japanese).

73. In addition to other references cited on particular points, the reader will find a wealth of detained information on many of these issues in two comprehensive reviews from the mid-1980s: J. B. Cannon, Ed., "Background Information and Technical Basis for Assessment of Environmental Implications of Magnetic Fusion Energy," Office of Energy Research, U.S. Department of Energy report DOE/ER-0170 (1983); and Fusion Safety Status Report IAEA-TECDOC-388 (International Atomic Energy Agency, Vienna, 1986). Data presented here and not otherwise attributed to a specific source are from these volumes.

74. J. Raeder and W. Gulden, "NET Safety Analyses and the European Safety and Environmental Programme," em Proc., 15th Symposium on Fusion Technology (Utrecht, The Netherlands, September 1988); see also [47,69].

75. S. J. Brereton and M. S. Kazimi, "A Comparative Study of the Safety and Economics of Fusion Fuel Cycles," Fusion Eng. and Des., $\underline{6}$, 207 (1988).

76. Office of the Federal Register, Code of Federal Regulations: Title 10. Regulations: Title 40. National Emission Standards for Hazardous Air Pollants, Part 61, U.S. Government Printing Office (1985).

77. J. Raeder, S. Piet, Y. Seki, and L. N. Topilski, "Safety and Environment Work Performed for ITER," Proc., Workshop on Fusion Reactor Safety (Jackson, WY, April 1989) (International Atomic Energy Agency, Vienna, in press).

78. J. L. Anderson, J. R. Bartlit, R. V. Carlson, D. O. Coffin, F. A. Damiano, R. H. Sherman, and R. S. Williams, "Experience of TSTA Milestone Runs with 100 Grams Level Tritium," Fusion Technol., $\underline{14}$, 438-443 (1988); also, B. Hircq, "Tritium Activities for Fusion Technology in Bruyeres-le-Chatel Center, CEA, France," Fusion Technol., $\underline{14}$, 424-428 (1988); also, R. V. Carlson, "Five Years of Tritium Handling Experience at the Tritium Systems Test Assembly," Proc., Workshop on Fusion Reactor Safety (Jackson, WY, April 1989), International Atomic Energy Agency, Vienna, in press).

79. G. Casini, C. Ponti, and P. Rocco, "Environmental Aspects of Fusion Reactors," Joint Research Centre Ispra Technical Note I.04.B1.85.156 (1985); also, L. Devell and O. Edlund, "Environmental Radiation Doses from Tritium Releases," Fusion Reactor Safety, IAEA-TECDOC-440, (International Atomic Energy Agency, Vienna, 1986) pp. 317-326.

80. For detailed descriptions of a set of computer programs for performing such calculations, see et al., E. F. Plechaty and J. R. Kimlinger, "TARTNP: A Coupled Neutron-Photon Monte Carlo Transport Code," Lawrence Livermore National Laboratory report UCRL-50400, Vol. 14 (1976); also, J. A. Blink, R. E. Dye, and J. R. Kimlinger, "ORLIB: A computer Code that Produces One-Energy-Group, Time- and Spacially-Averaged Neutron Cross Sections," Lawrence Livermore National Laboratory report UCRL-53262 (1981); also J. A. Blink, "PORIG: A Computer Code for Calculating Radionuclide Generation and Depletion in Fusion and Fission Reactors," Lawrence Livermore National Laboratory report UCRL-53633 (1985); also A. G. Croff, "ORIGEN2: Isotope Generation and Depletion Code – Matrix Exponential Method," Oak Ridge National Laboratory report ORNL/TM-7175 (1980).

81. J. H. Pitts, "A High-Efficiency ICF Power Reactor," in Laser Interaction and Related Plasma Phenomena, Vol. 7 (Plenum, 1986); also J. H. Pitts and I. Maya, "The Cascade Inertial-Confinement-Fusion Power Plant," Lawrence Livermore

National Laboratory report UCRL-92558 (1985).

82. J. P. Holdren and S. Fetter, "Contribution of Activation Products to Fusion Accident Risk: Part II. Effects of Alternative Materials and Designs," Nucl. Technol./Fusion, 4, 599 (1983).

83. RCGs for all important fission products have long been available from official bodies (et al., U.S. Code of Federal Regulations, Title 10, Chap. 1, Pt. 20, U.S. Government Printing Office, Washington, D.C. 1975), but performing useful fusion-fission comparisons using BHPs became possible only after Fetter developed a consistently calculated set of RCGs for fusion-relevant isotopes not encountered in fission applications (see Note [59]).

84. For an up-to-date treatment, see S. J. Brereton, S. J. Piet, and L. J. Porter, "Offsite Dose Calculations for Hypothetical Fusion Facility Accidents," Proc., Workshop on Fusion Reactor Safety (Jackson, WY, April 1989) International Atomic Energy Agency, Vienna, in press); also, [58, 59, 76].

85. S. J. Piet, "Implications of Probabilistic Risk Assessment for Fusion Decision Making," Fusion Technol. 10, 31-48 (1986); also, H. Djerassi and J. Rouillard, "Fusion Reactor Blanket and First Wall Risk Analysis," Fusion Reactor Safety, IAEA-TECDOC-440 (International Atomic Energy Agency, Vienna, 1986) pp. 123-132; also, Y. Fujii-e, Y. Kosawa, M. Nishikawa, T. Yano, I. Yanagisawaa, S. Kotake, and T. Sawada, "A Function Based Safety Analysis on Fusion Systems," ibid., pp. 155-165; also, S. Sarto, G. Cambi, and G. Zappellini, "A Methodological Safety Analysis for NET Accident Scenarios," ibid., pp. 185-202.

86. D. F. Holland, "Computer Codes for Safety Analysis," Fusion Reactor Safety, IAEA-TECDOC-440 (International Atomic Energy Agency, Vienna 1986) pp.177-184; also, J. Massidda, "Thermal Design Considerations for Fusion Reactor Passive Safety," Ph.D. Thesis, Massachusetts Institute of Technology (1987); also, D. F. Holland, L. C. Cadwallader, S. J. Brereton, S. J. Herring, G. R. Longhurst, R. E. Lyon, B. J. Merrill, and S. J. Piet, "Fusion Safety Program Annual Report," Idaho National Engineering Laboratory report EGG-2561 (1989).

87. S. J. Piet, "Fusion Activation Product Behavior and Oxidation- Driven Volatility," Proc., Workshop on Fusion Reactor Safety (Jackson, WY, April 1989), International Atomic Energy Agency, Vienna, (in press).

88. H. O. Haug, "Anfall, Beseitigung, und Relative Toxizitaet Langlebiger Spaltprodukte und Actiniden in den Radioktiven Abfaellen der Kernbrennstoffzyklen," Kernforschungszentrum Karlsruhe report KfK 2022 (1975).

89. J. P. Holdren, K. B. Anderson, P. M. Deibler, P. H. Gleick, I. M. Mintzer, and G. P. Morris, "Health and Safety Impacts of Renewable, Geothermal, and Fusion Energy," in Health Risks of Energy Technologies, C. Travis and E. Etnier, Eds. (Westview Press, Boulder, CO, 1983); also, M. S. Kazimi, "Risk Considerations for Fusion Energy," Nucl. Technol./Fusion, 4, 527 (1983).

90. J. G. Delene, H. I. Bowers, and B. H. Shapiro, "Economic Potential for Future Light Water Reactors," Trans. Amer. Nucl. Soc. and European Nucl. Soc., 57, 205, (1988).

EUROPEAN STUDIES ON SAFETY, ENVIRONMENTAL IMPACT, AND COST OF

MAGNETIC FUSION POWER

Janos Darvas

Commission of the European Communities
DG XII/Fusion Programme
Brussels

INTRODUCTION

Is fusion an "inexhaustible source of environmentally clean energy"?
The question in this form is oversimplified, rhetoric, and the answer
cannot be simply yes or no.

One of the first European attempts to qualify this question and find
possible answers was a report in 1980 by C.M. Braams /1/. His "Social
Aspects of Nuclear Fusion" considered safety and environmental, economic,
and proliferation aspects. Safety and environmental impact of fusion
reactors was again the subject of studies published in 1985 by R. Hancox
and W. Redpath /2/ and in 1986 by G. Casini, C. Ponti, and P. Rocco /3/.
As a sign of growing preoccupation with regard to the environmental
issue, the European Parliament requested 1985 from the Commission of the
European Communities to report on the "Environmental Impact and Economic
Prospects of Fusion" /4/. This first Commission report and the ensuing
discussion will be summarised in the next section.

Certainly, the discussion is only at the beginning and the progress
in fusion research and the preparation for the decisions on the Next Step
warrant a new, more complete and more detailed assessment of the
"environmental and economic potential of fusion" (EEF), a demand
reiterated 1988 by the European Parliament and by the European Council of
Ministers for the next revision of the Fusion Programme. The Commission
is now undertaking this new study with the help of a group of senior
scientists and economists, called the EEF-Study Group, scheduled to
produce a report by October 1989. Clearly, it is not yet possible to give
a full account of the work, but the main ideas and findings developed so
far by the Group are included in the second section of this paper. The
author wants to emphasize that, since the EEF study is not yet completed,
the conclusions in the last section are his own, based in part on EEF,
and in part on the previous studies.

THE FIRST COMMISSION REPORT AND ENSUING DISCUSSION

The first Commission report of 1986 starts by enumerating criteria
for the social acceptance of fusion power, and then examines in some
detail whether or not presently envisaged fusion reactors would fulfill
the criteria of economic acceptability, low radiological burden to the
environment, and safety potential to exclude large, socially disruptive
accidents.

Safety, Environmental Impact, and Economic Prospects of Nuclear Fusion
Edited by B. Brunelli and H. Knoepfel
Plenum Press, New York, 1990

The reactor envisaged in this study was a Tokamak with net electric power of 1200 MW(el), the "First Commercial-sized Tokamak Reactor", or FCTR. The radioactive inventories were estimated at 3 kg tritium (30 MCi), of which at most 200 g would be mobilised in case of a major accident; and about 9000 MCi of activation products at shutdown, mostly in the first wall (43%) and blanket structures (47%), which are assumed to be of AISI 316 stainless steel. The inventory of induced activity at shutdown would be reduced between 10 and 100 times and also the decay rates would be much faster if a low activation steel or V-alloy would be used instead of AISI 316.

It was found in the study that the technology would be at hand to keep radiation doses from routine emissions of radioactivity well below the natural radioactive dose burden. In case of a major accident it was found that the upper limit of tritium release would be 200 g, which in the most unfavourable wheather conditions would lead to doses to the maximum exposed individual of about 60 - 80 m Sv at 1 km distance from the emission, concluding that no immediate harm was caused and no emergency measures such as evacuation were necessary.

Because of the very restricted mobility, only a few cubic centimeters of solid activation products would be released in a major accident and the resulting dose was negligible compared to the dose from tritium.

From the radioactive characteristics of the fusion waste, it was concluded that their storage and disposal was easier, particularly if low activation materials were used, than that of fission wastes, albeit the volume of low and medium level waste is slightly larger in the fusion case.

The overall conclusion was that fusion reactors will provide a safe, environmentally acceptable energy source.

For the assessment of the economic prospects of fusion, the electricity generation cost of FCTR was calculated using the costing methods developed for NET (costing of a first-of-a-kind device). This cost turned out to be 2-3 times that of today's thermal fission and coal power stations. With current expectations on future improvements, it was concluded that there were good prospects that fusion electricity cost would fall within the range of other large-scale energy technologies of the next century.

Other beneficial aspects quoted by the report are the security of fuel availability along with a low price of fuel and no need for external reprocessing; and, last but not least, the potential for long-term improvements by using advanced fuel cycles, such as D-D and D-^{3}He.

The ensuing discussion and criticism of these findings is summarised in a report edited on behalf of the Scientific and Technological Options Assessment (STOA) of the European Parliament /5/.
The main sections to be recalled here are the critical remarks of the "Sweet-Report" by C. Sweet, economist at the Centre of Energy Studies, UK; and the "Metten-Report", by the rapporteur of the Parliament's Committee on Energy, Research and Technology, MEP A. Metten.

The Sweet-Report maintains that the uncertainties are not sufficiently acknowledged and that there is "undue complacency" about environmental aspects in the Commission's report. Sweet says that fuel resource requirements are not properly researched; he criticizes that no cost-benefit analysis in the context of alternative energy technologies, no assessment in terms of Community R&D priorities, and no sensivity analysis have been made; he feels misled by the fusion vocabulary, such as the terms "break-even" and "scientific feasibility", believes that the Community's strategy approach JET-NET-DEMO is inappropriate and warns that the "learning curve" from DEMO to the commercial rector may be < 1, so that fusion may not be competitive.

The Metten-report draws, for the benefit of the European Parliament, very concise conclusions with regard to the fundamental questions on fusion. Is fusion an inexhaustible energy source ? yes, says the report. Is it clean ? Not entirely. Is it safe ? Maybe; but "inherent safety" is an exaggeration and more R&D is needed to minimise the risks. Is it cheap? Impossible to say, but fusion cannot be dismissed purely on economic grounds.

At the end of this discussion the European Parliament, in considering the Commission's proposal of a continued strong fusion programme, supported the proposal, but requested that, before the next revision of the programme, the Commission shall arrange for a new "appraisal of the potential environmental, safety-related and economic attractiveness of fusion".

To carry out this appraisal, the Commission set up the EEF Study Group in October 1988, asking for a report before October 1989. The work of the Group not being finished, the attempt to present the methodology and main lines of reasoning in what will become the EEF-report is bound to contain errors and give rise to misinterpretations. The author of the present takes the blame for these imperfections, in exchange for the considerable benefit of the criticism of the experts assembled in this seminar.

OUTLINE OF THE EEF-REPORT

Fusion may become a mature energy technology around the middle of the next century, i.e., the point of time when fusion will possibly make a significant impact on the energy market, achieving a share of 5-10%, is assumed to be around the year 2050. The EEF study gives an estimate of the energy market and of the possible constraints on new energy systems by the year 2050.

The Market for Fusion in 2050

Economic predictions for such a distant future are very difficult - in fact they are only possible if one assumes a "tranquil development", unperturbed by "shocks"; and even in case of "tranquil development", the uncertainties will be very large.

Assuming that no major wars, shortages, bankruptcies, natural cataclisms or unexpected revolutionary technical developments will occur in the next 60 years, we expect the economic activity in the European Community to increase between 2 and 20 fold; the energy consumption will increase slower than the GDP, but the share of electricity in the energy consumption may still increase relative to the present situation, so that in the end, electricity consumption may increase between 2 and 10 fold until 2050. We need to remember that the present electricity consumption in the EC is about 1600 TWh (1987).

As fusion is envisaged to generate base-load electricity, we have to find the base-load capacity installed in 2050. This may be about 250-1000 GW, compared to the present 100 GW. Considering the increase in demand and adding estimates for new and replacement base-load capacity, we find the total construction demand in 2050 to be 10-80 GW/year. This is the predicted "energy market" for fusion and other competitive energy systems. The range appears to be wide, reflecting all uncertainties of the analysis, but it should be noted that the probability of the extreme values, particularly at the upper end of this range, is very low. If shocks occur, the most probable consequence will be a sharp decline of economic activity, perhaps followed by a slow recovery. Since shocks are a matter of speculation, and not of probabilistic evaluation, we must disregard them in our analysis.

The Institutional Constraints on Energy Systems in 2050

Essentially, two main constraints on future power systems can be anticipated: the availability/reliability of supplies, and the environmental effects.

Supply is already a serious issue in to-day's decisions. We may state that it is much more relevant for coal, oil, and natural gas plants than for nuclear plants. Utilities accept overcosts of up to 10% for indigenous energy sources, but the emphasis may change in the future.

The environmental effects are expected to become the overriding issue in the next century. Governments are expected to impose either taxes or regulatory constraints in order to achieve much higher environmental standards than at present. One possible example of what the effect on the energy market could be is the case of the Greenhouse Effect. According to present estimates, more than 50% of the emitted "greenhouse gases" is CO_2 (the others are methane, fluorocarbons, sulphur dioxide and nitrous oxides) and about 1/3 of the present global CO_2 emission is attributed to electricity generation by the combustion of fossil fuels. Calculations show that, unless the emission of greenhouse gases is curbed, the average temperature on Earth will increase by 1.5-4.5°C by the middle of the 21st century, with cataclismic effects on human health, habitat, and productivity.

The scientific evidence for these predictions is not yet complete, but if confirmed, the Greenhouse Effect will have a profound impact on future energy technologies. In the meantime, we can anticipate the following as the likely elements of future European energy policy:
- reliability of supplies
- respect of the natural environment
- competitiveness in energy costs
- diversification and structural flexibility.

On the Reliability of Supplies

The primary fuels of DT fusion reactors are deuterium and lithium. Deuterium, obtained from water (100-160 ppm), is plentiful, universally present, and cheap.

Lithium is also present in sea water (about 0.17 ppm), but the cost of Li extraction from sea water is not known, since present production is from brines and pegmatites with much higher concentrations of lithium. The richest known reserves of lithium are pegmatite deposits in Australia, Zimbabwe and in the USA, and brines in Chile; and the exploitation of these is more than sufficient to cover today's small demand for this metal. As a result of this market situation, there is no incentive to search for new deposits and in particular there is virtually no production of Li in the EC, except for a very small amount in Portugal.

Worldwide, known and economically exploitable land-based reserves of

Li amount to 1.1-1.2 million tons; the not yet evaluated potential reserves are certainly much higher, perhaps 10 times higher; and if the market for Li allows extraction from sea water to become competitive, there is a huge reserve of Li in the oceans of about 2×10^{11} tons. Since the effective energy equivalent of 1 t Li in a DT fusion reactor is about 3-4 TWh, DT fusion may not be short of fuel supply for centuries. However, when it comes to domestic supplies, we have to recognise that the potential Li-reserves of the EC are not adequately mapped and certainly not developed.

Reference Reactor

For the EEF study, parameters of a reference Tokamak reactor of 1200 MW(e) were developed. The reference reactor is assumed to work continously, with 20% of the thermal output recirculated in current drive and heating power. The peak magnetic field is close to 15 Tesla, and Beta is assumed to reach 7.6%. The first wall and blanket structure is ferritic steel, the breeder material is lithium-lead and the coolant is water.

The reference reactor will be used as far as possible for the assessment of the safety and environmental aspects. Possible improvements of some of the parameters were also investigated, in particular of the Troyon coefficient, of the efficiency of the current drive, and of the energy confinement time in the plasma.

On the Environmental Quality of Fusion

This is the least advanced chapter of the EEF report, since it requires extensive analysis based on the reference reactor.

Therefore, not more than the tentative content can be given here. The chapter will include radioactive inventories, consequences of accidents, waste disposal costs and risks, operational risks from routine release, benefits of using low activation materials, environmental concerns related to fuel provision, and safeguards. It is the intention to compare fusion vs. fission vs. fossil energy technologies throughout the chapter.

On the competitivity of fusion

What will the cost of energy by the middle of the next century be ? To answer this question, EEF first finds a "reference cost", and then examines whether or not the cost of fusion is commensurate with this reference cost.

Since the price of coal has been remarkably constant in the past, we may assume that the reference cost of energy in 2050 is the one corresponding to the range between a constant price of coal equal to the average price over the last 30 years, and a 50% increased price of imported coal. The cost of energy corresponding to this range is between 100 (arbitrary units) and 130. We shall further consider that the main competitors of fusion in 2050 will be fossil fuel plants, and fission plants: pressurized water reactors (PWR) and fast breeder reactors (FBR). We assume that the electricity cost of these competitors is equal to the reference cost.

We know very little about the future cost of fusion electricity, but we know for sure that fusion will be capital intensive, the cost of fuel will be very small (perhaps 2%) and the cost of operation and maintenance may be similar to fission, say about 25%. This would leave the capital cost of fusion at 73%, as compared to 21% for coal, 43% for PWR and 56% for FBR, as a fraction of the "reference cost".

Present estimates of fusion kWh-costs are all based on "conservative designs", i.e, 20th century technology. Certainly there will be a "learning effect" from now to the year 2050, e.g. high-temperature superconductor technology instead of $NbSn_3$, or some innovative, more efficient plasma heating or current drive systems, or materials with longer lifetimes, which may reduce the capital cost of a fusion reactor in 2050. Looking for analogies in the development of fast breeder fission reactors, we may tentatively assume that the "learning curve" gives a cost reduction by 2.2-2.5. That would make fusion marginally competitive, if the price of coal which determines the reference cost remained constant; and fully competitive if the reference cost was up by 30%.

CONCLUSION

Fusion has the potential to become an important, or perhaps the major electricity generating system by the middle of the next century. Though not completely clean, it has very favourable environmental characteristics. Therefore, even in case it would be only marginally competitive, it cannot be dismissed on purely economic grounds. It should be regarded as an important option for a future which perhaps will be dominated by environmental issues, and for which the industrialized nations, and in particular the European Community, must prepare the ground to-day.

REFERENCE

/1/ C.M. Braams, The Social Aspects of Nuclear Fusion, Report Nr. EUR FU BRU/XII/1212/80-EN (Original: ENERGIESPECTRUM N° 9 September 1980, pp. 214-224 (NE));
/2/ R. Hancox and W. Redpath, Fusion Reactors Safety and Environmental Impact, Nucl.Energy, 1985, 24, pp. 263-272;
/3/ G. Casini, C. Ponti and P. Rocco, Environmental Aspects of Fusion Reactors 1985, Report Nr. EUR 10728 EN (1986);
/4/ Commission of the European Communities DG XII, Environmental Impact and Economic Prospects of Nuclear Fusion, Report Nr. EURFU BRU/XII-828/86, (1986);
/5/ European Parliament Scientific and Technical Options Assessment (STOA), Criteria for the Assessment of European Fusion Research, EP-STOA-F1, (1988).

GENERAL METHODOLOGY OF SAFETY ANALYSIS / EVALUATION FOR FUSION

ENERGY SYSTEMS (GEMSAFE) AND ITS APPLICATIONS

Yoichi Fujii-e

Research Laboratory for Nuclear Reactors
Tokyo Institute of Technology
Tokyo, Japan

INTRODUCTION

Deutrium-tritium fusion reactions must have a potential to realize an attractive energy system in near future. From the result of the conceptual designs and the relevant engineering studies, the safety characteristics of a fusion energy system have been made fairly clear. Namely, in a fusion energy system, we have to handle successfully many kinds of energy sources and a large amount of radioactive materials, so that we should recognize it as having large potential hazards. In the process of introduction and deployment of the fusion system to the human society, therefore, it is desirable to demonstrate the societal acceptability from the early stage of the system development.

In order to assess the societal acceptability, it is important to take into account the following three aspects: technical feasibility, economy and safety. In the development of fusion systems, such three aspects are not satisfied independently, but they must be enhanced in relevance with one another. The fusion energy is under development, therefore the research on safety should be consistent with the system development.

On an investigation as to fusion system safety under the present condition, neither the safety protection methods and the data base are well established, nor the design of a commercial fusion reactor is fixed. Moreover, many subjects are left in R and D field to demonstrate the fusion system functions. However, since the functions essentially required for a fusion system have been made clear by the research and development focused mainly on the core plasma and by some conceptual design studies of the fusion systems, the investigation on safety issues must be possible by using a system model based on the essential functions.

In this context, we will describe the general methodology of safety analysis and evaluation for fusion energy systems, in abbreviation: GEMSAFE, which has been developed for recent 5 years in the light of the above-mentioned importance of the safety assessment in the process of the fusion energy development. In GEMSAFE, we have constructed the framework of the methodology to make sure the logical base in the safety consideration for a whole fusion system. In GEMSAFE new ideas and approaches as follows were introduced: (1) Fusion system model for safety assessment, in which the safety characteristics are taken into account and a large number of subsystems and/or components having complicated interdependency can be integrated into a few system elements as essential functions, (2) Safety ensuring principle for fusion systems, by which the safety requirements from the environment can be satisfied in a reasonable way to cope with the inherent safety features of a fusion system, and (3) Function-Based Safety Analysis, by which the typical events in a system element and the design basis

Safety, Environmental Impact, and Economic Prospects of Nuclear Fusion
Edited by B. Brunelli and H. Knoepfel
Plenum Press, New York, 1990

events for a whole fusion system can be selected, and the logical base of the siting event selection can also be prepared. The selection of these events is principally comparative to that for a typical fission system. In the remained part of this context we will describe the application of our methodology to a D-T reacting plasma experimental device, named as the R-tokamak and the fusion experimental reacter, in abbreviation: FER, in order to confirm the adequacy of the GEMSAFE and to refine it. At the same time, we could point out the safety design requirements for them and the associated safety study items as a result of safety analysis. This is the other important scope in the development of our methodology.

DEVELOPMENT OF GENERAL METHODOLOGY OF SAFETY ANALYSIS AND EVALUATION FOR FUSION ENERGY SYSTEMS

The safety assessment for fusion systems has been performed in several references 1~9. In the design and construction of TFTR, the potential hazard, focused mainly on tritium, was evaluated from both viewpoints of the environmental safety and the system one[1]. The safety assessment for the fusion reactor named STARFIRE was performed qualitatively for the exposure dose due to tritium and activation materials under the normal condition, and also for the potential hazard and the tritium release under the accidental conditions, based on a comparison with a fission reactor[2]. The applicability of the probabilistic safety assessment (PSA) was investigated for next generation tokamak devices by several researches[3, 4, 5]. For the fusion experimental reactor (FER) designed by JAERI, the maximum credible accidents were selected[6].

All of these assessments are considered to be applicable to the design-fixed system. In order to make the methodology of the safety analysis and evaluation meaningful for the fusion system development, the following items must be important:
-without any omission of items to be considered in the fusion safety
-with high generality and logicality, and
-with capability to improve the fusion system design.
We have developed the general methodology of safety analysis and evaluation for fusion energy systems, considering the above-mentioned items[7].

Framework of the Methodology

When we consider the above-mentioned situation concerning the fusion safety, especially the role of safety analysis and evaluation, we can conclude that it is important to select and then systematically consolidate the main themes to be considered in the development of the safety analysis and evaluation methodology. Here, it is useful to construct a "framework" of the safety analysis and evaluation methodology in order to have a reasonable prospect for the logical development of the methodology as a whole. The framework of the methodology was schemed to include key safety issues, as illustrated in Fig. 1. As can be seen in the figure, the fusion system can be grasped both from the external and the internal viewpoints. The former is a standpoint that one assesses the fusion system from the surrounding environment and the latter assesses the fusion system behavior in itself. The safety ensuring principle in the figure is responsible for coordination between both viewpoints.

The matter required from the societal environment to fusion systems is summarized into the safety requirements. In principle, the safety requirements for a fusion system are focused on prevention and mitigation against the radiological burden as well as in the case of fission systems. Therefore, the safety requirements for fission systems which are mainly concerned with radiological burden, such as ALARA and permissible dose in case of an accidental release, can be also used for fusion systems based on their intrinsic ideas. In order to satisfy the safety requirements with practical ways for a fusion system, several principles and practices in the fission system safety can be applied to a fusion system bearing their intrinsic meanings in mind. They are defense-in-depth principle, design basis events (DBEs) and probabilistic risk analysis.

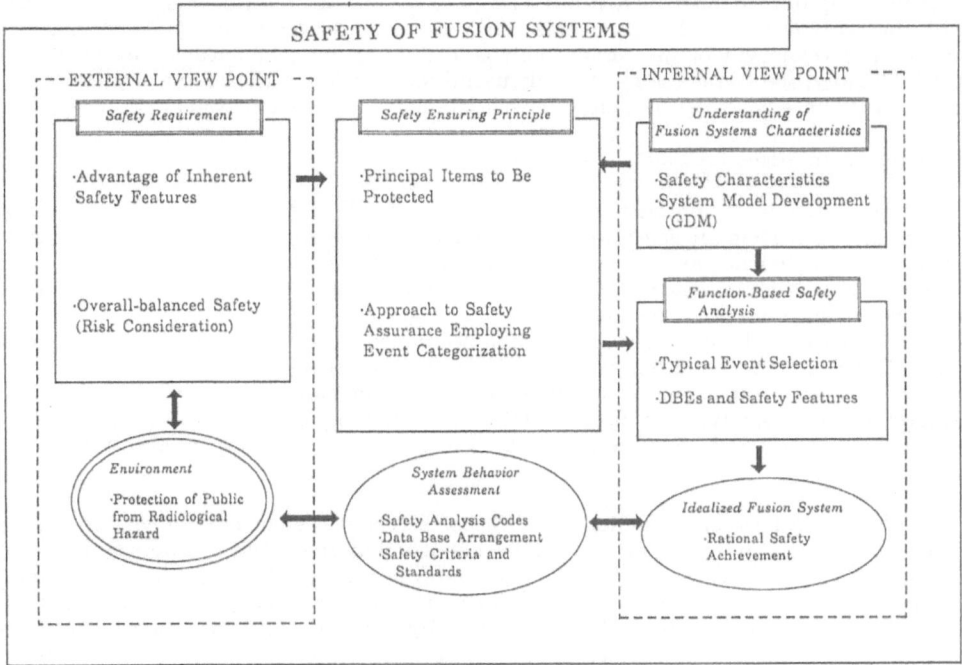

Fig.1 Framework of GEMSAFE

The safety ensuring principle is a basic idea of the fusion safety such that the influence due to introduction of the system to the societal environment can be rationally controlled within an allowable level. The safety ensuring principle should be described concretely to conform with the defense-in-depth principle, the rationality involved in the risk concept and the preferential utilization of inherent and/or passive safety features. With this motif, a better understanding of fusion system characteristics promotes the extraction of the core of the safety ensuring principle, that is, "items to be protected" and "event categorization". A distinctive feature of our methodology development is to set up beforehand the safety ensuring principle of a fusion system.

The first investigation from the internal viewpoint is to understand fusion systems, especially their safety characteristics. This investigation is necessary to prevent rationally the effects of potential hazards from actualizing by taking account of the intrinsic safety characteristics. The most important matter in the safety characteristics of fusion systems is to identify the characteristics of radioactive materials and energy sources as the potential hazards.

In the methodology, we developed a new safety analysis method focused on the functions of a fusion system. The method is called the Function-Based Safety Analysis (FBSA). In the FBSA, the generalized fusion system, which is a kind of fusion system model having no safety feature explicitly is analyzed in conformity with the safety ensuring principle, especially with the items to be protected and the idea of event categorization. By using the method of the FBSA, the possibility of actualization of the potential hazards can be surveyed comprehensively and consequently, a small number of the typical events can be picked up. In the FBSA, the DBEs can be selected by developing the event sequence which starts by regarding the typical event as an initiating event and branches due to a loss of the integrity of the items to be protected. Since the abnormal events in a fusion system can be already summarized into a small number of the typical events, the similarity and the envelope of the event sequence can be also assessed so easily that a set of the design basis events (DBEs) for a fusion system can be finally obtained.

By using our methodology, it becomes possible to search a fusion system endowed with the rational safety. Moreover, the safety research on the fusion system behavior, including development of the safety analysis code, will be conducted according to the DBEs. Finally, the methodology is surely useful for decision in the course of preparation of the safety data base and the safety criteria/standards for fusion systems in the future.

Fusion System Model for Safety Assessment

One of the most interesting characteristics of GEMSAFE is the modeling of a fusion energy system directed to safety analysis and evaluation. We propose a new method of the fusion energy system modeling with paying attention mainly to the functions necessary to realize a fusion system, the energy and mass storage and flow and the integration of distributed nature of radioactive material as well as energy source. Such a model can help us to rationally introduce the safety protection systems into a fusion system based on the safety characteristics of fusion systems. In the fusion system, it is important to install the safety features to be balanced as a whole in order to achieve rational safety. The preferential use of the inherent and/or passive safety features is considered to have potentials both to rationalize the system constitution and to enhance safety.

The General Descriptive Model (GDM) of a fusion system can be constructed as shown in Fig. 2. In Fig. 2, it should be noticed that the GDM is composed of five elements: (1) fusion-reaction related or vacuum-area related, (2) energy-conversion and fuel-production related or blanket-area related, (3) fuel-processing related or fuel-area related, (4) waste-processing related or waste-area related, and (5) containment-area related. These elements are named as the "system elements". Each system element has a set of functions, an area and the associated boundaries. In the GDM, the RI sources and energy ones related to the functions and areas can be mapped, based on the understanding of safety characteristics. The details of the RI sources and the energy one s related to each system element of a fusion system are also shown in Table 1. The GDM can be regarded as a generalized and integrated expression of fusion systems by employing a small number of the system elements.

Safety Ensuring Principle for a Fusion Systems

Safety ensuring principle is a principle necessary to satisfy rationally the safety requirements from the environment. The followings are introduced as to construct the backbones of the safety ensuring principle;
- The items to be protected, "IPs" in abbreviation, in fusion systems are derived from the analysis on the RI leak structure.
- The approach to safety assurance based on the event categorization is proposed to give a way to investigate well-balanced safety features as a whole system.

The formation of the source term and the leak path are two factors which are influential in causing the radiological hazard. Therefore, the items to be protected must be those loss of whose integrity can lead to the formation of the source term and the leak path. Figure 3 shows the RI leak structure in fusion systems. The RI leak structure represents the infrastructure of the environmental RI release in relation with the formation of the source term and the leak path.

The release of most of the RI sources in fusion systems could be prevented if the controllability on the mobility, i.e., the RI controllability, is assured. The RI controllability is the principle item related to the formation of the source term.

Boundaries serve as physical barriers against RI release. The failure of boundaries is a cause of generating a major leak path in the system. Therefore, the boundary integrity is also the principal item. A containment is introduced as a practical choice required for satisfying the defense-in-depth principle. The containment can functionally back up the former two principal items, besides it physically separates the public from a fusion system.

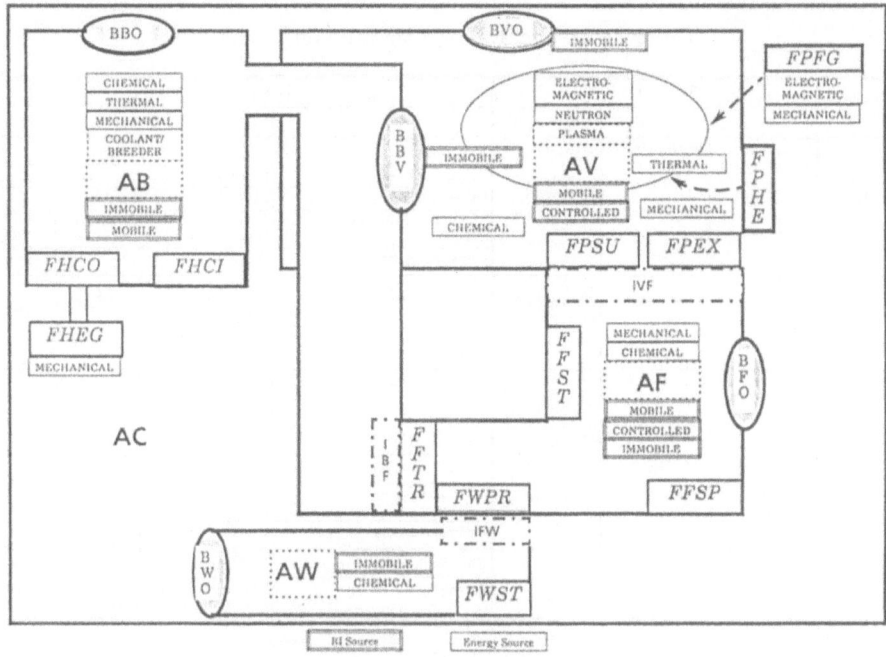

FUNCTION (12 kinds)
- FPFG Function of Magnetic Field Generation
- FPHE Function of Plasma Heating
- FPEX Function of Particle Exhausting
- FPSU Function of Fuel Supply
- FHCI Function of Material Circulation
- FHCO Function of Heat Conversion
- FHEG Function of Electricity Generation
- FPSP Function of Fuel Separation and Purification
- FFTR Function of Fuel Transfer
- FFST Function of Fuel Storage
- FWPR Function of Waste Processing
- FWST Function of Waste Storage

AREA (5 kinds)
- AV Vacuum Area
- AB Blanket Area
- AF Fuel Area
- AW Waste Area
- AC Containment Area

BOUNDARY (6 kinds)
- BVO Vacuum Area Boundary
- BBO Blanket Area Boundary
- BFO Fuel Area Boundary
- BWO Waste Area Boundary
- BBV Boundary between Vacuum and Blanket Areas
- BOE Boundary between Confainment Area and Environment

INTERFACE BOUNDARY (3 kinds)
- IVF Interface Boundary between Vacuum and Fuel Areas
- IBF Interface Boundary between Blanket and Fuel Areas
- IFW Interface Boundary between Fuel and Waste Areas

Fig.2. General Descriptive Model of Fusion System

Table 1. RI and Energy Sources on GDM

System Elements	RI Sources			Energy Sources	
	Mobile	Controlled	Immobile	Form	Example
Vacuum-Area Related	Tritium in Vacuum Area	Tritium in Cryo-Panel	Tritium and Induced-activity in First Wall, Blanket and Shield	Neutron Energy	Plasma
				Electromagnetic Energy	Plasma, Coil Current(FPFG)
				Thermal Energy	Plasma, Neutral Beam(FPHE)
				Mechanical Energy	Cryogenics(FPEX), Non-Tritium gas (FPSU), Rotational Equipment(FPSU)
				Chemical Energy	Hydrogen
Fuel-Area Related	Tritium in Fuel Area	Tritium in Cryo-Panel	Tritium in Getters and Pipings	Mechanical Energy	Cryogenics(FFST,FFSP), Non-Tritium gas (FFST), Rotational Equipment (FFSP)
				Chemical Energy	Chemical Materials(FFTR,FFSP), Hydrogen
Waste-Area Related	Tritium in Waste Area	<NONE>	Tritium in Getters and Radioactivity in Solid Materials	Chemical Energy	Chemical Materials(FWPR), Hydrogen
Blanket-Area Related	Tritium in Blanket Area (If Solid Breeder is selected)	<NONE>	Tritium in Breeding Materials	Thermal Energy	Breeder/Coolant
				Mechanical Energy	Breeder/Coolant, Rotatinal Equipment (FHCI)
				Chemical Energy	Breeder/Coolant, Hydrogen
Containment-Area Related	<NONE>	<NONE>	<NONE>	Electromagnetic Energy	Coil Current(FPFG)
				Mechanical Energy	Cryogenics(FPFG), Field Coil (FPFG), Rotational Equipment(FHEG)

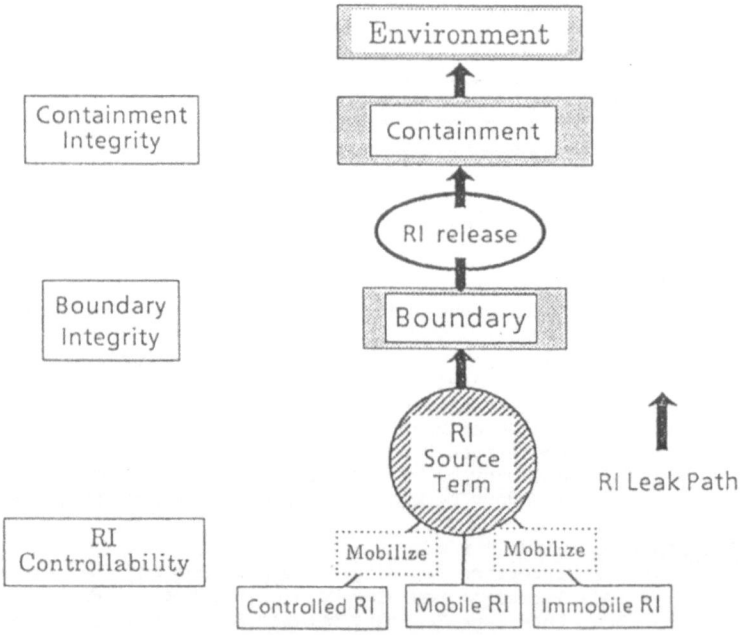

Fig.3. Three Principal Items and RI Leak Structure

Table 2. Definition of Event Categorization

Category	Definition	Frequency (1/year)	Released RI (Ci / event)
Category-1	• includes events which have the possibility of causing Class-1 boundary failure	$\sim 10^{-2}$	comparable to normal condition release
Category-2	• includes events which release Class-1 RI and / or have the possibility of releasing Class-2 RI • includes events which have the possibility of causing Class-2 boundary failure	$10^{-2} \sim 10^{-4}$	$\sim 10^3$
Category-3	• includes events which release Class-2 RI and / or have the possibility of releasing Class-3 RI.	$10^{-4} \sim 10^{-6}$	$\sim 10^5$
Category-4	• includes events which release Class-2 RI • includes events with large scale fire, missile and / or failure of Containment System. *Events of this category are treated as events beyond DBE.*	$< 10^{-6}$	$\sim 10^7$

The consequence limit for each category is provided using the amount of the released radioactivity(Ci) to the environment. The values should be considered as reference values. The mitigating factor of 10^{-2} is assumed for the containment system.

From the RI leak structure, "RI controllability", "boundary integrity", and "containment integrity" have been identified as the three principal items to be protected for ensuring fusion system safety. The major causes that could jeopardize the three principal items are the abnormal energy release and the associated abnormal state in the system. They are influential in the formation of the source term and the leak path through jeopardizing the principal items. However, the energy sources are not included explicitly into the principal items, because there would be a room to prevent the RI release by assuring the integrity of boundary and RI controllability when an abnormal release occurs.

In investigating well-balanced safety features as a whole system, it is well acceptable as a rational goal to realize the safety design where the frequencies of abnormal occurrences are lowered as their consequences increase. In order to achieve the goal, it is crucial to provide a measure to prescribe the necessity of safety features for protecting the principal items. A probabilistic way of thinking can provide useful means for the purpose.

The event categorization proposed here is shown in Table 2. The abnormal events are classified into four categories, from Category-1 to Category-4 events, on the basis of their consequences. The frequency range of occurrences for each category is reduced by 10^{-2} in order, assuming the unavailability of an active safety feature to be 10^{-2}/demand. The relation between the frequencies and the consequences is defined based on an equi-risk curve as a perspicuous expression. The number of the category, four, is determined under the assumption that the events having a frequency lower than 10^{-6}/year (Category-4 event) are considered as incredible events and treated as events beyond the scope of design basis. Here, the consequence is expressed by the amount of the released radioactivity (Ci) to the environment instead of the environmental dose, because the evaluation of the environmental dose is not so much meaning without the site selection.

The classification of the RI sources is shown in Table 3. In the proposed classification, the Class-1, Class-2 and Class-3 mean the limit of the mobile-, the controlled- and the immobile-form inventory, respectively. The classification of boundaries is shown in Table 4. The boundaries are classified into two classes in relation with the anticipated consequences resulting from boundary failures and with the feasibility of countermeasures against the RI releases after the boundary failures.

Table 3. Classification of RI Sources

	Inventory (Ci)	Mobility
class-1	$\sim 10^5$	mobile-form
class-2	$\sim 10^7$	controlled-form
class-3	$\sim 10^9$	immobile-form

Function-Based Safety Analysis

Function-Based Safety Analysis (FBSA) is a new method to conduct a safety analysis by paying attention mainly to functions in a fusion system. The FBSA enables us to carry out a safety analysis on a general fusion system along with the safety ensuring principle. The FBSA gives the way to realize the safety assurance based on the event categorization which holds both the rationality in the probabilistic approach and the realistic outlook in the safety evaluation employing the DBEs.

The FBSA has the following features:

(1) The abnormal events which could jeopardize the IPs are summarized into a small number of typical events in each system element of the GDM without investigation of the detailed system behavior under abnormal conditions.

(2) The DBEs and the necessary safety features can be investigated on the event sequences starting from the typical events. Then the event sequence analysis would not be very complicated, so long as the sequences are expressed by addressing to the integrity/loss-of-integrity of IPs and the development of a sequence can be bounded within the events which are conceived as credible in the sense of their occurrence frequencies. The integrated representation of the GDM and the typical events can give a good perspective in the selection of the DBEs.

(3) The FBSA is applicable to any fusion system from the early stage of the system development, even when the detailed design is not fixed, by the use of a general and integrated system model, i.e., the GDM.

The DBEs for the general fusion system can be summarized as listed in Table 5. Nine Category-1 events, seven Category-2 events and four Catefory-3 events are selected. In Table 5, the judgement criteria in evaluation and related safety features to be considered are also shown for each event. The judgement criteria indicate the IPs to keep the event within the same event category.

Table 4. Classification of Energy Sources

	Definition
class-1	Boundaries whose failure can cause Category-2 events directly
class-2	Boundaries whose failure can cause Category-3 events directly

Table 5. Design Basis Events for General Fusion System

Category	Events	Judgement Criteria	Safety Features
Category-1	Plasma Disruption	Integrity of BBV and BVO, Protection against Mobilization of Controlled-RI	Physical Separation, Physical Barrier(limiter)
	Plasma Power Excursion	Integrity of BBV	Energy Damp due to Power Control, Inherent Safety due to Beta-limit
	Heat Load Transient due to Plasma Heating Function	Integrity of BBV and BVO	Energy Damp due to Interlock, Physical Barrier(beam damp)
	Pressure Transient in Vacuum Area	Integrity of BVO, Protection against Mobilization of Controlled-RI	Pressure Control, Reduction of Released Energy
	Pressure Transient in Fuel Area	Integrity of BFO, Protection against Mobilization of Controlled-RI	Pressure Control, Reduction of Released Energy
	Thermal-hydraulic Transient	Integrity of BBV	Cooling, Pressure Control
	Magnetic Field Transient	Integrity of Boundaries and Functions	Physical Separation, Energy Damp
	Pressure Transient in Containment Area	(Integrity of Containment System)	Pressure Control
	Temperature/Pressure Increase in Waste Area	Integrity of BWO	Cooling, Pressure Control
Category-2	Air-Inleakage (AV)	Protection against Mobilization of Controlled-RI (AV), Protection against Fire	Isolation, Fire Protection
	Pressure Transient with Mobilization of Controlled-RI(AV)	Integrity of Vacuum Boundary	Pressure Control, Recovery of Tritium
	Breeder/Coolant-Inleakage (AV)	Integrity of Vacuum Boundary, Protection against Mobilization of Controlled-RI (AV)	Isolation, Reduction of Released Material
	Air-Inleakage(AF)	Protection/Mitigation for Mobilization of Controlled-RI and Immobile-RI (AF), Protection against Fire	Isolation, Fire Protection
	Pressure Transient with Mobilization of Controlled-RI(AF)	Integrity of Fuel Boundary	Pressure Control, Recovery of Tritium
	Coolant-release into Containment Area(when coolant is not a breeder): Loss of Breeder/Coolant(AB)	Integrity of Blanket-Vacuum Boundary (Integrity of Containment System), Protection against Fire	Pressure Control, Fire Protection
	Breeder/Coolant-Inleakage(AF)	Integrity of Fuel Boundary, Protection/Mitigation for Mobilization of Controlled-RI and Immobile-RI (AF), Protection against Fire	Isolation, Fire Protection
Category-3	Breeder/Coolant-Inleakage(AV) with Vacuum Boundary Failure and Mobilization of Controlled-RI	Protection/Mitigation for Mobilization of Immobile-RI (AB), Protection against Fire	Reduction of Released Material(Isolation), Cooling, Fire Protection
	Air-Inleakage(AF)(or Breeder/Coolant Inleakage) with Mobilization of Controlled-RI and/or Isolation Failure of Immobile-RI(AF)	Protection/Mitigation for Mobilization of Immobile-RI (AF), Protection against Fire	Cooling, Fire Protection
	Breeder-release into Containment Area (Liquid Breeder is used as Coolant)	Protection/Mitigation for Mobilization of Immobile-RI (AB), Protection against Fire	Reduction of Released Material(Isolation), Cooling, Fire Protection
	Air-Inleakage(AW)	Protection/Mitigation for Mobilization of Immobile-RI (AW)	Isolation, Cooling

For the Pressure Transient with Mobilization of Controlled-RI (AV), integrity of vacuum boundary is the judgement criterion to keep the event within Category-2. It is required for safety features to keep the probability of loss of the vacuum boundary integrity less than 10^{-2}/event.

In FBSA, we describe events by combinations of energetic abnormal phenomena and intact-defective of associated IPs; and then classified events referring to the consequences that IPs would not be protected. In safety evaluations, it is required to confirm which the following two items:
- An event's consequence is within the prescribed range of the category the event belong to; and
- Safety protection is assured so that it can prevent the event from proceeding into next category.

In the DBE selection, we can show the qualitative expression of the DBEs, considering similarity in phenomena. However, in order to carry out quantitative evaluations of the DBEs, it is necessary to define initial and boundary conditions by taking into account spectra of initiating events. Here, as an example of events selection in consideration with the spectra of initiating events, we introduce an approach to selecting events in both cases where the effectiveness of safety features depends on severity of phenomenon and not.

Suppose that the consequence brought by an initiating event depends on two IPs: IP1 and IP2, where the consequence drops into the Category-2 upon the loss of the IP1 and, moreover, into the Category-3 upon the loss of the IP2. Additionally the safety features to assure integrities of the IP1 and IP2 are supposed to be SF1 and SF2, respectively. The consequence brought by an initiating event would depend on the magnitude of an energetic phenomenon which could jeopardize IPs. To simplify the discussion, we introduce "severity" as a measure for the magnitude of an energetic phenomenon to represent its influence on IPs. Conceptually, the severity will be a function of the amount of released energy, the release time constant and the location of energy release. We assume that the threshold for the integrity of IPs and ineffectiveness of associated safety features can be expressed using the severity. Denote the thresholds for assuring integrities of the IP1 and IP2 and by S_{th}^{IP1} and S_{th}^{IP2}, while the thresholds where the SF1 and SF2 would become ineffective by S_{th}^{SF1} and S_{th}^{SF2}

The severity(S) of phenomena like plasma disruption and loss of coolants is not constant but has certain variation in range. For instance, the influence on IPs by a small leakage of coolants will quite differ from that by a large break LOCA. We introduce a cumulative probability function "$F(S)$" which indicates the occurrence probability of an event having severity larger than S. As shown in Fig. 4, $F(S)$ is a monotonously decreasing function. Although the thresholds for IP integrity and SF effectiveness depend on initiating events and designs, the minimum requirements according to the event categorization is:

$$\max(S_{th}^{SF1}, S_{th}^{IP1}) \geq S1 \text{ and } \max(S_{th}^{SF2}, S_{th}^{IP2}) \geq S2$$
where, $F(S1) = 10^{-2}/y$ and $F(S2) = 10^{-4}/y$.

Now, we show a way of event selection to confirm these conditions in safety evaluation in both cases where the effectiveness of SF would depend on the severity and not. The Figure 5 shows a case where the effectiveness of SF would depend on the severity. In this model, SF1 is effective against severity less than S1 and SF2 against less than S2. The unavailability of SF1 or SF2 is assumed to be 10^{-2}/demand. As shown in Fig. 5, we can represent the events in Category 2 and Category 3 evolved from an initiating event by combinations of severities and success/failure of safety features. In the event shown here, both the consequences and the occurrence probabilities conform with the requirements by the event categorization. Then we can select the severest event as the evaluation event in each category:

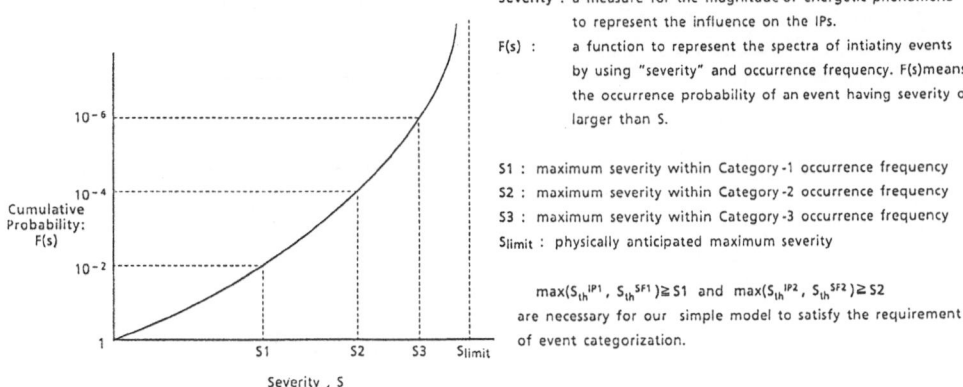

Severity : a measure for the magnitude of energetic phenomena to represent the influence on the IPs.

F(s) : a function to represent the spectra of intiatiny events by using "severity" and occurrence frequency. F(s)means the occurrence probability of an event having severity of larger than S.

S1 : maximum severity within Category-1 occurrence frequency
S2 : maximum severity within Category-2 occurrence frequency
S3 : maximum severity within Category-3 occurrence frequency
S_{limit} : physically anticipated maximum severity

$\max(S_{th}^{IP1}, S_{th}^{SF1}) \geqq S1$ and $\max(S_{th}^{IP2}, S_{th}^{SF2}) \geqq S2$ are necessary for our simple model to satisfy the requirement of event categorization.

Fig.4. Spectra of Intiating Events Expressed of Severity and Occurrence Probability

Category 1 → S = S1,
Category 2 → S = S2 with loss of IP1 and
Category 3 → S = S3 with loss of IP1 and IP2 .

Figure 6 shows a case where the effectiveness of SF is independent of the severity. In this case, the magnitude of consequence depends essentially on success/failure of safety features and not on the severity. An event with failure of SF1 will be selected as a Category-2 event; an event with failures of both SF1 and SF2 as a Category-3 event. Considering event occurrence probability, it is enough to carry out evaluations for events under severity of S=S1 for both Category-2 and 3 events. The DBEs obtained here can provide the requirements on the safety features and could be the candidates for the DBEs of the fusion systems whose design has been fixed in detail.

Selection of Siting Events

In the event categorization, the Category-4 events are treated as beyond design basis events(BDBEs) because their occurrence probabilities are low enough. Although,

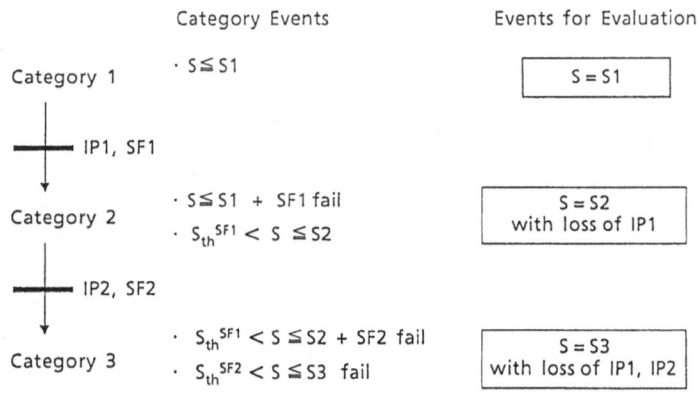

Fig.5. In a Case where Effectiveness of SF depend on S

$$(S_{th}^{IP1} < S_{th}^{SF1} < S2, S_{th}^{IP2} < S_{th}^{SF2} < S3)$$

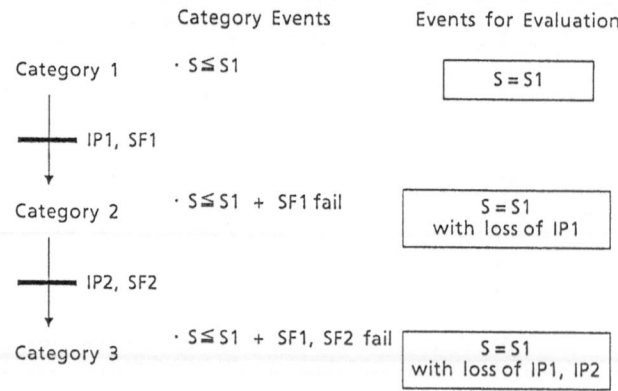

Fig.6. In a Case Where Effectiveness of SF is independent of S

according to engineering judgement, the Catefory-4 events could be assumed not to occur, it is desirable to assure following two items through siting evaluation in order to enhance safety margin against BDBEs and verify the proper isolation between the system and the public:

(1) As an evolved event from a DBE, a BDBE would not result in an event with extraordinary large consequence. In other words, it means that no event would deviate from the objective risk curve in the range just below the criterion of the probability of the Categoty-4 events: 10^{-6}/year. To investigate events in so-called gray zone: 10^{-6}/year to 10^{-8}/year.

(2) Even for an event which would bring about the maximum source term, proper isolation between the system and the public could be assured.

Based on these ideas, we propose two-grade selection of siting events. Figure 7 shows the siting events in relation with a risk curve. In Table 6, we summarize the objectives and selection method of the siting events.

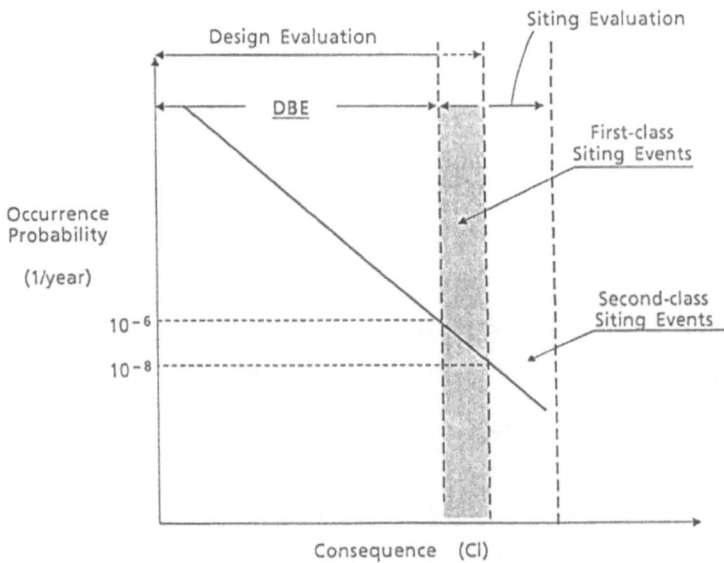

Fig.7. Assignment of Siting Events on Equi-Risk Curve

Table 6. Proposal for Siting Events

The first-class siting events

Objectives: −To assure mitigation measure against the events in the gray zone beyond DBE.

−To assure the isolation between system and public.

Selection: −For DBEs, first assume a loss of an IP that could make the maximum consequence or a failure of an active safety system, and then select events within them that would increase consequences.

The second-class siting events

Objectives: −To assure the isolation between system and public under the event that could bring about the maximum source term.

Selection: −Select an event that could bring about maximum source term under an assumption that we would not expect operation of any active safety system within a fusion system. Here, of course, we can rely on passive safety features that could work under such an event.

The role of the first-class siting events is to assure that those events belonging to the Category 4 by their occurrence probabilities would not lead to remarkably large consequences. The first-class siting events can be cited to investigate effects of events progressed into the range of "one-step ahead" of the DBE zone. The idea is quite different from the case of LWRs, where severer accidents, as siting events, namely major accident and hypothetical one, are defined as DBE-associated events with excessive estimation of source terms. Being different from a fission reactor which is considered as a concentrated system as to its potential hazards, a fusion system has a characteristic as a distributed system. This characteristic indicates that a loss of an IP could cause another new phenomenon in spatially. Therefore, it will not be sufficient to select DBE-associated events with excessive source terms as siting events for evaluating what the first-class siting event results in. The disadvantage of this idea, however, is that it would potentially make evaluations complicated in vain. Although the problem would be rather straightforward if an event evolution beyond DBE zone could simply increase source term, the problem would be complicated if such an evolution could cause another energetic aspect due to the distributed characteristic.

The second-class siting event can be cited to evaluate the siting appropriateness based on the maximum source term of a system. The major issue here is how to define the maximum source term. It might be a straightforward way to assume as releasing the whole inventory involved in a system in order to limit the maximum of potential hazard inherent to the system. Such a way, however, is neither realistic nor reasonable, because it will lead to neglecting inherent safety features inf a system. Although it might be hard to demonstarte the adequacy for identifying the maximum source term as a result of an event with low occurrence probability, we proposed the way described above because we consider that many of ISFs and passive ESFs in the system can contribute to decrease of the maximum source term. In this case we can effectively realize an idea of "partial containment"* by making full incorporation of isolation to the distributed nature. Whether we have to include the activated structural materials into source term, depends on their possibility to become vulnerable to release under energetic phenomena. If we can assume that most of the activated structural materials would not be released, the siting evaluation with the maximum source term will be dealt with with out serious difficulty.

APPLICATION OF THE GEMSAFE

In this chapter, the application of GEMSAFE is described, which was performed in order to confirm the validity and usefulness of the methodology: GEMSAFE. The R-takamak[9], a conceptual design of D-T reacting plasma experimental device, and FER, a Fusion Experimental Reactor, were chosen as two references for the application. Although the R-tokamak design does not include several subsystems important to safety, such as tritium breeding blanket and super conducting magnet, the design includes enough safety-related issues to confirm most of the results of then methodology.

The meaning of the application to FER is to analyze a system that is closer to a future fusion reactor, and the application has following two objectives: (1) to confirm the adequacy of the GEMSAFE and to refine it, and (2) to identify critical safety issues to be investigated for the integtation of safety design requirements.

Application to R-tokamak

Having performed the safety analysis by GEMSAFE, the following recommendations for the R-tokamak design were obtained:
[1] As to magnitude of stored energy in the energy source is generally low in the R-tokamak, most of accidental energy releases were tolerable within the component design. However, since the failure of boundary may occur at the structural weak parts, e.g., diagnostic windows, vacuum seal, bellows and valves, they have to be taken into account as an accident initiator.
[2] The importance of the isolation as active safety features is emphasized. Especially, the isolation of cryo-pumps is crucial to the safety design.
[3] Casing of cryo-pumps in the torus vacuum pumping system and gas reservoir of the gas feed system have to be designed in a high quality, since they are the Class-2 boundary.
[4] Refrigerant materials, such as liquid helium and nitrogen, are the key energy sources which cause the pressure increase in the containment area. The isolating functions to mitigate the pressure increase and the pressure relief functions to prevent the repture of cryo-panels have to be prepared in the design.

Application to FER

Design Description of FER The Fusion Exprimental Reactor (FER)[10, 11, 12] was designed at the Japan Atomic Energy Research Institute (JAERI) to be the next machine of JT-60. The major experimental goal of the FER is to achieve self-ignited D-T

Table 7. Major Paramenters of the FER

Operation mode	quasi-steady state
Current driver	OH coils/LHW
Current drive RF power (MW)	10
Burn time (sec)	~2000
Major/minor radius (m)	5.2/1.12
Plasma current (MA)	5.9
Neutron wall loading (MW/m^2)	0.68
Fusion power (MW)	297
Neutron fluence (MW·y/m^2)	0.3
Breeding blanket	test module
Heating (MW)	50 (RF)

Table 8. Major RI Sources in the FER

Tritium

Subsystem/Component	Inventory
Vacuum System	
Cryopump	2.4×10^5 Ci (24 units) (max. 2.0×10^4 Ci /unit)
Fuel Gas Purification System	
Surge tank	4.0×10^4 Ci
Freezer	3.1×10^4 Ci
Palladium diffuzer	9.5×10^1 Ci
Catalytic oxidizer	8.7×10^1 Ci
Electrolysis cell	1.1×10^3 Ci
Isotope Separation System	
Cryogenic distillation column	4.4×10^5 Ci
Fuel Storage System	
Metal bed (ZrNi bed)	1.5×10^7 Ci (12 unit) (1.25×10^6 Ci/unit)
Fuel Injection System	
Liquid chamber	1.0×10^6 Ci
Blanket Tritium Recovery System	
Cold trap	1.4×10^3 Ci
First Wall/Divertor	2.6×10^6 Ci
Blanket Test Module	1.2×10^2 Ci
Primary Cooling System	4.0×10^4 Ci

Induced-activity

Subsystem/Component	Inventory
Torus Vacuum Vessel System	
Induced-activity of structural material	$> 10^7$ Ci
Sputtered material	2.6×10^7 Ci

Plasma and quasi-steady state operation with long burn pulse. The major design parameters are shown in Table 7. In the FER, full tritium breeding blanket and power generating facilities are not equipped with, which is the major difference between the FER and future fusion reactor.

Table 9. Major Energy Sources in the FER

Magnetic Energy	Plasma	0.55 MJ
	TF coils	22.5 GJ
	PF coils	6.5 GJ
Thermal Energy	Plasma	130 MJ
	Decay Heat	~0.4 W/cc(at First Wall)
Mechanical Energy	Coolant	1.5 MPa, 90°C
	Cryogenics	Liq. He ~300 kl Liq. N_2 ~450 kl
Chemical Energy	Hydrogen	~20 MJ

Major RI sources in the FER are summarized in Table 8. The inventory and the mobility are also shown. The total inventory of tritium is estimated to be 2×10^6 Ci. Most of the inventory exists in the fuel storage system as immobile RI. As for the induced radioactivety, the sputtered dust in the vacuum system is of concern because of its uncertainly in mobility. In the FER, the inventory of the sputtered dust is estimated to be 2.6×10^7 Ci and most of it is treated as immobile RI in this paper. Major energy sources in the FER are shown in Table 9 .

The General Descriptive Model (GDM) for the FER is shown in Fig. 8. The GDM consists of six system elements; related to vacuum-area (AV) , fuel-area (AF) , blanket-area (AB) , coolant-area (AT) , waste-area (AW) and containment-area (AC) . In the FER, the solid breeder blanket design is used for the test module, whereas the liquid breeder is considered in the general discussion. The exsitence of the coolant-area (AT) related system element is the difference from the GDM used for DBEs selection of general fusion systems.

Event Categorization for FER. In GEMSAFE, proposed an event categorization which prescribes the relation between frequencies and consequences of off-normal events (see Table 2). The value of limited frequency and consequence of each category is determined for general fusion systems based on equi-risk curve as a perspicacious expression. The consequence is expressed using the amount of the released radioactivety (Ci) instead of the environmental dose, tentatively.

For the application to the FER, since the FER has some uniqueness arising from the nature of experimental device, we discussed the necessity of changing the risk curve, which is the basis of the event categorization, and the frequency limit for the design basis event ("credible event").For the first point, we judged that the safety goal should not be determined from the nature of unique experimental device, but from the societal

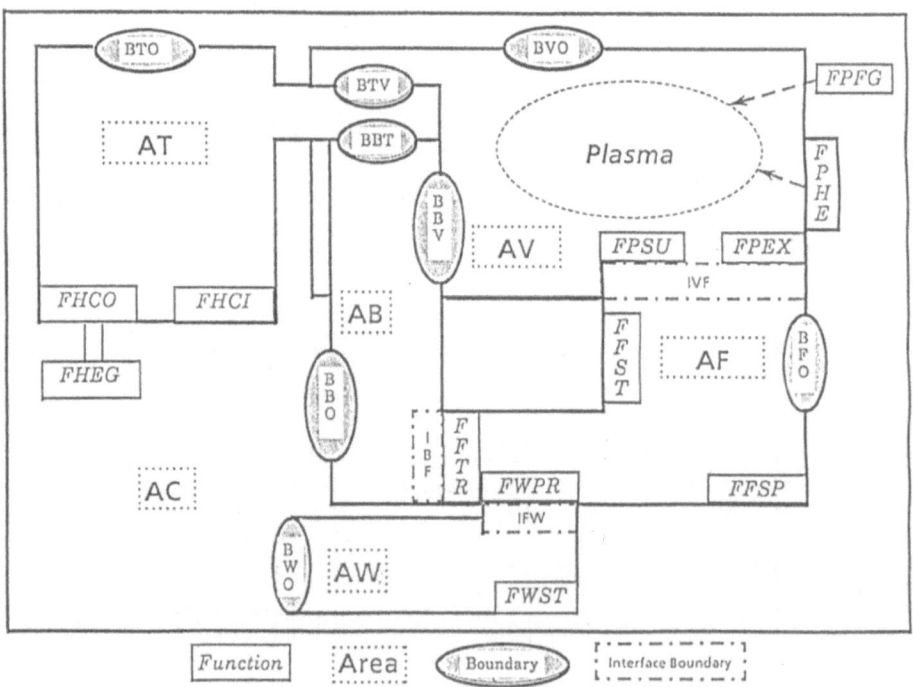

Fig.8. General Descriptive Model(GDM) for the FER

acceptability of future fusion systems. For the second point, it may be possible to treat events beyond Category-3 as "incredible" in the FER considering its low availability, short lifetime and one-plant construction. However, for this application, we decided to use the event categorization without any modification.

In the event categoraization, it is cosidered essential to take the nature and the characteristics of IPs; items to be protected into account. To determine the class of boundaries for example, the feasibility of equipping a safety features to protect and mitigate the RI release upon the boundary failure should be essential in addition the inventory of radioactive material in the system element. In other words, it is necessary to determine the class of a boundary recursively in order to satisfy the requirements of event categorization considering the feasibility of safety features and feasible reliability of the boundaty.

Fig.9 shows the classification of major RI sources in the FER. As for the cryo-pumps and the metal bed, the total inventories are with in the range of Class-2 and Class-3 RI, respectively. However, since the tritium is held separately in a number of units and each unit is expected to operate independently, we judged to classify them using the inventory per unit and the trutium in cryo-pumps and metal beds are treated as Class-1 and Class-2 RI, respectively. The Class-3 RIs in the FER are induced structural material and activated dust in the vacuum vessel.

The boundary and RI class of major components of the FER is shown in Table 10 which is summarized based on the result of the DBEs selection. The boundaries of the cryogenic distillation columns in the isotope separation system and the liquid chamber in the pellet fueling system are considered to be the Class-2 boundary since there seems to be no effective safety features to protect the Class-2 RI release in case of these boundary failures. The failure probability of less than 10^{-4}/year is imposed on the design of these boundaries.

As for the vacuum vessel (vacuum boundary) and the coolant boundary of first wall/divertor, the classification depends on the system behavior after the boundary

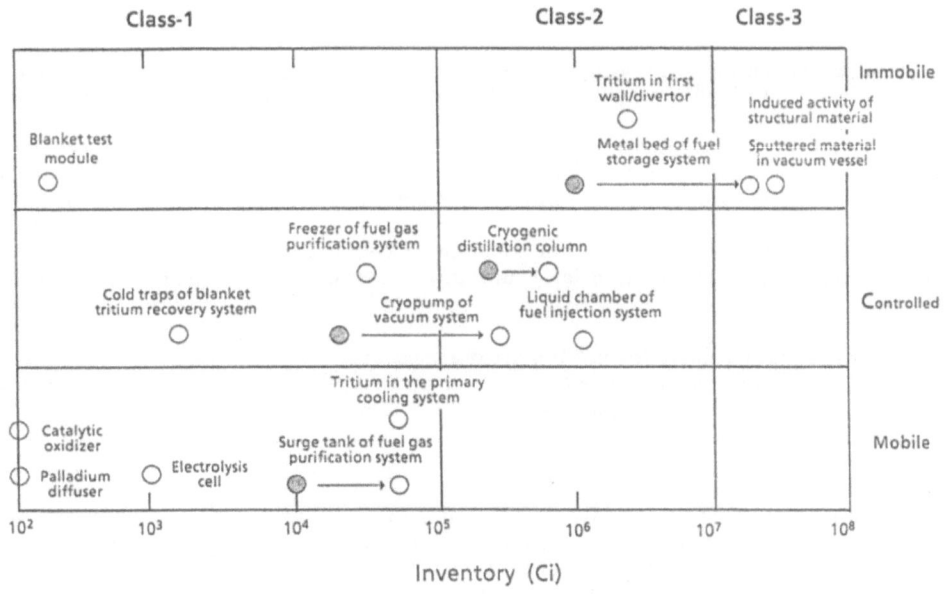

Fig.9. Classification of RI Sources in the FER

Table 10.Classification of Boundary and RI Sources of Major Components in the FER

Components		Classification of RI	Classification of Boundary
Vacuum System	Cryopump	1	1
Fuel Gas Purification System	Surge Tank	1	1
	Freezer	1	1
	Palladium Diffuzer	1	1
	Catalytic Oxidizer	1	1
	Electrolysis Cell	1	1
Isotope Separation System	Cryogenic Distillation Column	2	2
Fuel Storage System	Metal Bed	2	1
Fuel Injection System	Liquid Chamber	2	2
Blanket Tritium Recovery System	Cold Trap	1	1
Primary Cooling System	Pipings	1	1
Torus Vacuum Vessel System	Vacuum Vessel	3	1
	First wall/diverter	3	1
	Blanket Test Module	1	1

failure. The mobility of activated dust in the vacuum vessel is a key factor to decide the classification of these boundaries. Since it seems impractical to impose high reliability on these boundaries, it is necessary to clarify the mobility of activated dust and the failure probability of these boundaries in order to integrate reasonable safety requirements.

Pipings of tritium processing systems can be considered Class-1 boundary if the releasable amount of tritium is limited by the proper design of isolation systems.

Selection of Design Basis Events for FER. Figure 10 shows the event sequence initiating from the Coolant-Inleakage (AV). This typical event covers breach or rupture of coolant boundaries of in-vessel components such as first wall, divertor, blanket test module and shield. The release of the coolant into the vacuum vessel causes the pressure and thermal loads against the vacuum boundary. Chemical reaction between the coolant and the high temperature material could give an additional pressure and thermal sources.

In this situation, the following IPs are of concern:
− Vacuum boundary which serves as a barrier against RI release into the containment.
− Controllability of tritium in the first wall (Class-2 RI) ,
− Controllability of induced-radioactivety in the structural material including sputtered material (Class-3 RI).

Tritium in the cryopumps (denoted as C-RI(AV)) in Fig. 10 will be mobilized if the isolation fails. However, since the tritium in one cryopump is treated as a Class-1 RI in the FER design as shown in Fig.9 and its mobilization is considered to be within the Category-1 events, safety features to protect the mobilization, such as an isolation system, is not necessarily required in the safety design.

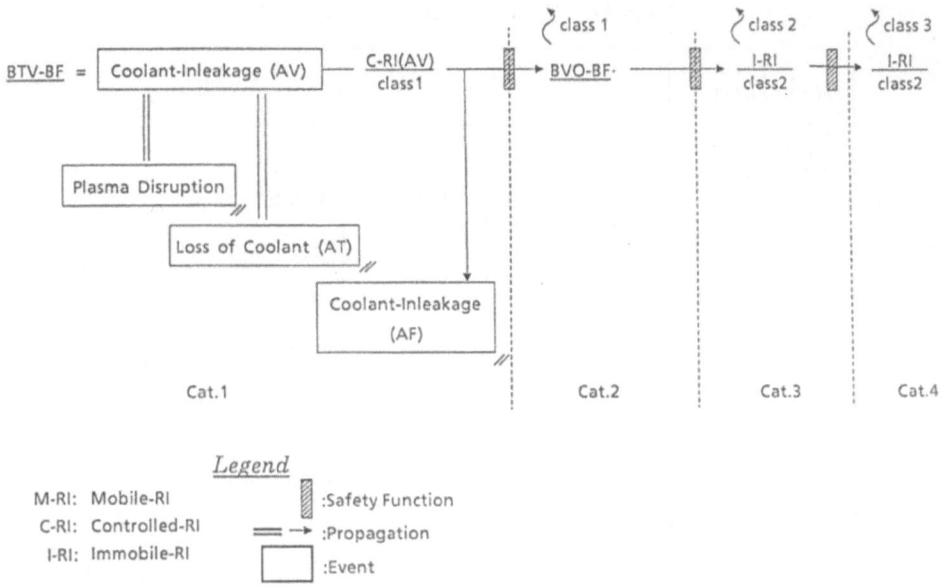

Fig.10. Event Sequence Starting from Coolant-Inleakage(AV)

If the vacuum boundary (BVO-BF) fails, most part of the mobile RI in the vacuum vessel (gaseous tritium and mobile activated dust) will be released. This situation is categorized into Category-2 events. Here we assume that most of the sputtered dust is invulnerable to release simply upon the vacuum boundary failure. If not so, we must treat this situation as events beyond Category-3 events or provide effective safety features to mitigate the release of the sputtered material.

The condition to proceed into Category-3 events is that both the controllability of the Class-2 RI and the integrity of vacuum boundary are lost. The condition to proceed into Categoty-4 is that both the controllability of the Class-3 RI and the integrity of vacuum boundary are lost. The controllability mainly depends on the temperature control of the in-vessel components. From the above discussions, the following candidates for DBEs are identified in the event sequence of the Coolant-Inleakage (AV):

Category-1 : Coolant-Inleakage
Category-2 : Coolant-Inleakage with vacuum boundary failure
Category-3 : Coolant-Inleakage with vacuum boundary failure and mobilization of the
             Class-2 RI (tritium in the firsrt wall)

Next, we incorporate the spectra of the typical initiating events, shown in Table 11, into the expression of DBEs. The "severity" of this typical event depends on the size and location of breach. Suppose a complete rupture of a large coolant pipe inside vacuum vessel, it seems impractical to assure the integrity of the vacuum boundary. Thus, in such a case, we had better assume that the boundary integrity is lost. The requirements of the event categorization is that the vacuum boundary integrity should be assumed under the severity whose occurrence frequency is more that 10^{-2}/year. For the Category-1 event, a small rupture or a leak which may occur once or more during the plant lifetime should be selected to assure the integrity of vacuum boundary. For the evaluation, it is necessary to determine the size and location to cover the spectra of piping failures resulting from various initiating events such as plasma disruption. The uncertainty in failure probability of the coolant boundary of first wall/diverter will make it difficult to set up the condition.

Table 11. Typical Events of the FER

SYSTEM ELEMENT	TYPE A	TYPE B
AV-related system element	1) Plasma Power Excursion 2) Heat Load Transient due to Plasma Heating Function 3) Plasma Disruption 4) Pressure Increase	5) Coolant- Inleakage 6) Air-Inleakage
AF-related system element	1) Pressure Increase	2) Air-Inleakage 3) Coolant-Inleakage
AB-related system element	1) Temp.-Pres. Increase	2) Coolant-Inleakage
AT-related system element	1) Thermal-hydraulic Transient	2) Loss of Coolant
AW-related system element	1) Temp.-Pres. Increase	
AC-related system element	1) Magnetic fields Transient 2) Pressure Increase	3) Coolant Release

For preventing the release of tritium in the first wall, it is crucial to keep its coolability. Supppose the cooling system has redundancy channels, the coolability will be remained even when a complete rupture of piping has occurred. For the Category-2 events, a rupture of a coolant piping which may occur more than 10^{-2}/plant-lifetime should be selected to ensure the most of the tritium in the first wall is kept immobilized.

For preventing the release of radioactivity in the structural material, the melting or evaporation of the structural material should be avoided. For the Category-3 events, we selected an event represented by combinations of a rupture of a coolant piping which is expected to occur more than 10^{-2}/plant-lifetime and a failure of one of the most effective cooling systems. As far as we assume a break in one cooling system, the coolability to assure the controllability of the Class-3 RI does not depend on the severity of the rupture. Thereby, for the evaluation, the other cooling system, which mitigates the mobilization of the Class-3RI, is postulated to be unavailable instead of assuming a severer rupture, such as breaks in multiple cooling systems. In the evaluation, it is required that most of the Class-3 RI is kept immobilized in this situation.

We examined typical events in the same manner as the above-montioned and summarized resultant candidates into the DBEs. Consequently, 19 DBEs for the FER are selected; i.e. seven Category-1 events, eight Category-2 events and four Category-3 events, as shown in Table 12. In the process of the summarization, for example, the Cate gory-3 events resulting from the Pressure Increase (AV) and Air-In-leakage (AV) are assumed to be covered by the Category-3 events from the Coolant In-leakage and excluded from the DBEs.

The failure of the metal bed casing in the fuel storage system provides an example where the integrity of the IPs and the effectiveness of the SFs are independent of the "severity". The consequence of this event depends not on the "severity" of the casing failure but on the unoperation/operation of the heating system. Thus, the casing failure

Table 12. DBES Definition and Associated Study Items for the FER

Category-1 Events

Events	Future study item
1-1 Plasma Power Excursion An off-normal plasma power excursion which is expected to occur once or more during the plant's lifetime. The integrity of coolant boundary of in-vessel components and vacuum boundary should be assured under this event. In the evaluation, the operation of safety systems such as plasma shut down system may be postulated.	Spectrum of the power excursion.
1-2 Plasma Disruption A plasma disruption which is expected to occur once or more during the plant's lifetime. The integrity of coolant boundary of in-vessel components and vacuum boundary should be assured under this event.	Spectrum of plasma disruption, feasibility of safety features to mitigate the disruption.
1-3 Plasma Heating Failure An abnormal transient in plasma heating devices. In the FER, since only RF heating is expected to use, it may be difficult to postulate the failure of the IPs such as vacuum boundary. If so, this event may be excluded from the DBEs.	Possibility of failures of related IPs.
1-4 Pressure Transient in Vacuum Vessel A pressure transient in vacuum vessel resulting from the the leakage of coolant, cryogenic material and other failures which is expected to occur once or more during the plant's lifetime. The integrity of the vacuum boundary should be assured under this event. This event envelops likely pressurizing phenomena in vacuum vessel.	Spectrum of pressure transient in vacuum vessel, feasible safety features to assure the boundary integrity.
1-5 Transient in Cooling System Abnormal transient in the cooling system such as loss of flow and loss of heat sink which is expected to occur once or more during the plant's lifetime. The integrity of the coolant boundary should be assured under this event.	Spectrum of transients
1-6 Pressure Transient in Tritium Processing System A pressure transient in tritium processing system resulting from leakage of coolant, cryogenic material and other failures which is expected to occur once or more during the plant's lifetime. The integrity of the fuel boundary should be assured under this event. This event can envelop pressurizing phenomena in fuel processing system which has no class-2 boundary.	Spectrum of pressure transients in vacuum vessel, feasible safety features to assure the boundary integrity.
1-7 Transient in Coil System Abnormal transient in coil system which is expected once or more during the plant's lifetime. The integrity of the IPs should be assured by introducing appropriate safety features.	Spectrum of transients in coil system, feasible safety features.

Continued

61

Table 12. *Continued*

Category-2 Events

Events	Future study item
2-1 Cryopump Failure The failure of vacuum boundary resulting from a rupture of cryogenic pipings in a cryopump which may occur more than 10^{-2}/plant-lifetime. Here, we assume that the integrity of vacuum boundary cannot be assured against the resulting pressure increase. This consequence should be assured within the limit of Category-2 events. This event is a similar but a severer event for Pressure Transient in Vacuum Vessel (DBE 1-4).	Frequency of cryogenic piping rupture, feasibility of safety feature to assure the vacuum boundary.
2-2 Coolant-Inleakage into Vacuum Vessel A failure of a cooling piping inside the vacuum vessel which is expected to occur more than 10^{-2} /plant-lifetime. The integrity of vacuum boundary and the isolation of the cryopumps are not assumed in the evaluation. It should be confirmed that most of tritium adsorbed on the first wall/divertor is kept unmobilized and this consequence is kept within the limit of Category-2 events. This event is selected to envelop similar accidents including vacuum boundary failure. Thus, the initial and boundary condition of this event must cover the spectra of cooling piping rupture which could be caused by plasma disruption and other initiating events.	Behavior of sputterd material, spectra of piping rupture (size, frequency, single or multiple break), chemical reaction of high temperature components with coolant and /or air.
2-3 Failure of Cooling Channel in Blanket Test Module A rupture of cooling channels in the blanket test module which is expected to occur more than 10 $^{-2}$/plant-lifetime. It should be confirmed that the released coolant into the blanket module is surged for the containment successfully to keep the integrity of the blanket module and the consequence should be kept within the limit of Category-2 events.	Feasibility of safety features to assure the blanket module integrity.
2-4 Coolant Release to the Containment A rupture of cooling pipes in the containment which is expected to occur more than 10-2/plant-lifetime. Under this event, the integrity of coolant boundaries inside the vacuum boundary should be assured assuming the operation of proper safety features such as plasma shut down system. The containment boundary must be intact against the resulting pressure increase.	Spectrum of pipe rupture
2-5 Tritium Piping Failure A failure of tritium pipes which is expected to cause a maximum tritium release into the containment taking into account the inventory and flow rate in the pipings. This event is evaluated to confirm the adequacy of the isolation system design. The consequence must be kept within the limit of Category-2 events.	Spectrum of pipe failure, consideration for a large number of tritium pipes
2-6 Failure of Metal Bed Casing A failure of the metal bed casing in the fuel storage system during the heat-up stage. In the evaluation, the heater is assumed to be turned off after the failure is detected. It should be confirmed that this consequence is kept within the limit of Category-2 events. In oreder to treat the metal bed casing as the class-1 boundary, this event is necessary to be evaluated.	Spectrum of casing failure

2-7 Fuel Boundary Failure with Pressure Increase

A failure of class-1 boundary resulted from pressure increase in the fuel processing system. A failure of the cryogenic pipings in the freezer of the fuel gas purification system is considered to be a candidate for this event. It should be confirmed the consequence is kept within the limit of Category-2 events. Here, it is assumed this frequency is 10^{-2}/plant-lifetime and there is no feasible safety features to protect the boundary integrity. If this frequency is probable to be less or the safety feature is feasible, this event may be excluded from this category.

Feasibility of safety features to protect the boundary, frequency of rupture of the cryogenic pipings.

2-8 Failure in the Waste Storage System

Abnormal transient in the waste storage system, which can challenge its boundary integrity and/or the RI controllability in the waste storage system. It should be confirmed the consequence is kept within this limit.

Category-3 Events

Events	Future study item
3-1 Coolant-Inleakage into Vacuum Vessel A rupture of a cooling pipe which may occur more than 10^{-2}/plant-lifetime with failure of a most effective cooling system. In the evaluation, it is required that most of the Class-3 RI is kept unmobilized in this situation. As far as we assume a break in one cooling system, the coolability, which assure the controllability of the Class-3 RI, does not depend on the severity of a rupture of the coolant piping. Thereby, the other cooling system, which mitigates mobilizing the Class-3 RI, is postulated to be unavailable instead of assuming severer rupture such as multiple piping break.	Spectrum of piping rupture
3-2 Large Scale Coolant Release to the Containment A rupture of a large cooling piping in the containment which is expected to occur more than 10^{-4} /plant-lifetime. Under this event, it should be confirmed that there is no abnormal progression exceeding the DBE 3-1. The containment boundary must be intact against the resulting pressure increase.	Spectrum of piping rupture
3-3 Failure of Metal Bed Casing A failure of the casing of the metal bed in fuel storage system during regeneration of tritium. In the evaluation, the maximum time delay in turning off the heater due to the failure of heater circuit or the detector is postulated. It is required to confirm that the consequence is kept within the limit of Category-3 events. The severity of the failure of the casing is postulated to be the same as that of the DBE 2-6.	Frequency of the casing rupture
3-4 Failure of Cryogenic Distillation Column or Liquid Chamber A failure of casing of cryogenic distillation column or liquid chamber. It is required to confirm that the consequence is kept within the limit of Category-3 events. In the evaluation, the effect of the pressure increase should be considered.	Frequency of the casing failure

with delay in turning off the heater is selected as the Category-3 DBEs. The severity of the failure is equivalent to that of the Category-2 DBEs. The definition of the selected DBEs and associated future study items are shown in Table 12. The development of quantitative analysis method for evaluation of these DBEs and the establishment of the judgement criteria are also the important R and D items as well as future study items listed in Table 12.

CONCLUDING REMARKS

In development of a fusion energy system, we have to intend to achieve the so-called rational safety and to harmonize it with the environment. Of course, when the fusion energy system will come to be established at the stage of the practical use, it is indispensable to view the system in the economical aspect, considering the relation with competitive energy systems. In the present stage of the system development, we have recognized that the safety analysis and evaluation methodology is one of the key issues to assess the fusion system from both view points of safety aspect and environmental one. At the same time, the safety analysis and evaluation should be performed concretely and practically since the next fusion system, which will handle a large amount of tritium and energy sources, has been planned as a near term development.

As described in this content, we have constructed the logical basis of the methodology which is consistent with the safety principle cultivated in the practical use of fusion systems. We also have prepared new approaches for the safety analysis applicable to a general fusion system. According to our methodology, one can obtain a set of the design basis events for a fusion energy system and the prospects for the siting event selection. We have tried to apply the methodology to the R-tokamak and FER designed conceptually. As a result, some design requirements at the more detailed safety designs have been furnished. Moreover, the safety research subjects necessary to evaluate quantitatively the design adequacy and the safety assurance of the fusion systems have been pointed out by checking the contents of the DBE.

One of the final targets of the safety analysis and evaluation is considered to calculate the risk of a fusion system, because the risk is one of the intelligible measure when we express the results of the safety assessment from both safety and environmental aspects. For that purpose, we have to prepare the data base concerning the relation among the energetic phenomena, the source term creation and the boundary integrity. For the safety features of a fusion system, we have to specify concretely their committed functions and characteristics, referring to the process of the DBE selection. For the risk calculation, it will be surely necessary for us to develop the computer codes for assessing the system behavior under some certain accidental conditions of a fusion system. The DBEs selected will be useful guides for such a code development.

In the discussion of the safety analysis and evaluation, it is the other important theme to arrange and explain the safety targets of a fusion system. As for the safety targets, the consideration for not only the effluence and/or the waste of radioactive-materials produced in the normal operation but also the target risk from the normal operation through the severe accidental condition is to be indispensable. Some arguments and proposals for the fusion safety targets have been started in several countries. In order to set up the safety targets, we should take much account of the safety characteristics of a fusion system and the public acceptance for a fusion systems.

REFERENCES

1. Princeton Plasma Physics Laboratory, "Final Safety Analysis Report, Tokamak Fusion Test Reactor Facilities" (1983)
2. R. Hancox and W. Redpath, Fusion Reactor-Safety and Environmental Impact, Nucl. Energy,24 (4),263/272 (1985)
3. W. Redpath, Safety Guidelines for the Design of Next-Generation Tokamak Fusion Machines (NGTEM), 7th Topical Meeting on Technology of Fusion Energy, Reno. Nevada, June (1986)

4. H. Djerassi and J. Rouillard, Fusion Reactor Blanket and First Wall Risk Analysis, Proc. of IAEA Technical Committee Meeting on Fusion Reactor Safety, Culham, UK, Nov. (1986)

5. S. J. Piet, Implication of Probabilistic Risk Assessment for Fusion Decision Making, Fusion Technology,10, 31/48 (1986)

6. Y. Seki, et al., Safety Scenario for Fusion Experimental Reactor (PER), Proc. of IAEA Technical Committee Meeting of Fusion Reactor Safety, Culham, UK, Nov. (1986)

7. Y. Fujii-e, et al., Safety Analysis and Evaluation Methodology for Fusion Systems, Res Mechanica Vol.27 Nos.1 & 2 (1989)

8. Y. Fujii-e , et al., A Role of Containment System for Fusion Safety, IAEA-TECDOC-440 (1986)

9. R-projet Desig Team, "The Second Phase Design of R-tokamak (Low Reactivity Design)", Report from IPP, Nagoya University, Dec. 1983.

10. Summary of Conceptual Design Study of Fusion Experimental Reactor (FER)(FY 1984 Report)" , Department of Large Tokamak Research, Japan Atomic Energy Research Institute, JAERI-M85-178, (in Japanese)

11. H. Iida, et al., Conceptual Design Study of Fusion Experimental Reactor (FER)(FY 1984, 85 Report) , JAERI-M86-134, 1986, (in Japanese)

12. Design Study of Plant System for the Fusion Experimental Reactor (FER), Department of Large Tokamak Research, Japan Atomic Energy Research Institute, JAERI-M86-149, 1986, (in Japanese)

13. Y. Fujii-e, et al., Development of General Methodology of Safety Analysis and Evaluation for Fusion Energy System (GEMSAFE) -(1)- Progress of the GEMSAFE, IAEA Technical Comittee Meeting on Fusion Safety, Jackson, Wyoming, April 1989.

SUMMARY OF THE U.S. SENIOR COMMITTEE ON ENVIRON-
MENTAL, SAFETY, AND ECONOMIC ASPECTS OF MAGNETIC
FUSION ENERGY (ESECOM)

B.G. Logan,[1] J.P. Holdren,[2] D.H. Berwald,[3] R.J. Budnitz,[4]
J.G. Crocker,[5] J.G. Delene,[6] R.D. Endicott,[7] M.S. Kazimi,[8]
R.A. Krakowski,[9] K.R. Schultz[10]

1. INTRODUCTION

Organized in late 1985, the ten-member, Senior Committee on Environmental, Safety, and Economic Aspects of Magnetic Fusion Energy (ESECOM) has recently completed a comprehensive assessment [1] of the potential for magnetic fusion energy (MFE) providing energy with attractive economic, environmental, and safety, characteristics compared to present and future fission energy sources. We explored the interaction of environmental, safety, and economic characteristics of a variety of fusion and fission cases listed in Section 2, using consistent economic and safety models. Our findings in Section 3 indicate that several MFE candidates have the potential to achieve costs of electricity (COE) comparable to those of present and future fission systems, and with significant safety and environmental advantages. These conclusions rest on key assumptions about plasma performance and improvements in fusion technology, which are optimistic but defensible extrapolations from current achievements. In contrast, a recent report of the Scientific, Technological Options Assessment (STOA) office of the European Parliament [3] proposes criteria for assessment of future MFE reactor safety and economics, which are generally much more restrictive than criteria used in the ESECOM study, with respect to allowing assumptions of future technology improvement. ESECOM, however, has taken the long view that the time horizon for MFE commercial application is the year 2015 at the earliest, and more probably beyond 2030. Accordingly, ESECOM chose to analyze MFE cases assuming advances of new technologies (e.g., materials) that are only in the beginning stages of development. ESECOM's work thus clarifies the promising areas for future fusion research and development. Due to lack of space, only selected portions of the ESECOM work are discussed here. For more details on all areas covered by ESECOM, the reader is referred to the published technical summary [1] of this work and to the larger main report [2].

2. COMPARATIVE ANALYSIS OF FUSION AND FISSION CASES

ESECOM selected a set of fusion, fission, and fusion-fission hybrid reactor cases for comparative analysis, listed in Table I. These cases were selected to span a wide range of technical characteristics based on reasonable extrapolation from present knowledge, permitting exploration of the impacts on safety and economics of different materials and coolant choices, power densities, energy conversion schemes, and fuel cycles.

The different cases do represent, of course, differing degrees of extrapolation from materials choices, physics parameters, and engineering features that might be considered reasonably certain to be attainable based on current knowledge. An examination that confined itself only to

Safety, Environmental Impact, and Economic Prospects of Nuclear Fusion
Edited by B. Brunelli and H. Knoepfel
Plenum Press, New York, 1990

Table I. Reference cases analyzed by ESECOM.

Fusion Cases	
1.	A "point-of-departure" D-T fusion reactor using a tokamak configuration, with vanadium-alloy structure and liquid lithium as the coolant/breeder.
2.	A helium-cooled variant of the case 1 tokamak with reduced activation ferritic steel (RAF) structure and Li_2O solid breeder.
3.	A "high-power-density," reversed-field pinch (RFP) with RAF structure, a water-cooled copper-alloy first wall and limiter, and self-cooled lithium-lead breeder.
4.	Another high-power-density RFP with a V-Li blanket minimally modified from that of the point-of-departure tokamak.
5.	A "low-activation" tokamak with silicon carbide (SiC) structure, helium coolant, and Li_2O breeder.
6.	A "pool"-type tokamak with vanadium structure and molten-salt (FLiBe) coolant/breeder.
7.	An advanced conversion variant of the point-of-departure tokamak with synchrotron-radiation-enhanced magnetohydrodynamic (MHD) conversion.
8.	An advanced fuel, water-cooled tokamak based on the D^3He fuel cycle with direct conversion of microwave synchrotron radiation.
Fusion-Fission Hybrid Cases	
9.	A "baseline" fusion-fission hybrid tokamak with RAF structure, lithium coolant, beryllium neutron multiplication, and thorium metal as the fertile material.
10.	An "advanced technology" hybrid tokamak with stainless-steel structure, helium coolant, and Li-F-Be-Th molten-salt blanket.
Fission Cases	
11.	A "best present experience" and "medium experience" pressurized water reactor (Westinghouse) (PWR-BPE), (PWR-ME).
12.	The Large-Scale Prototype Breeder (LSPB) (Electric Power Research Institute/DOE).
13.	The Power Reactor Inherently Safe Module (PRISM) Breeder design (General Electric).
14.	A modular high-temperature gas reactor (MHTGR) (GA/Gas-Cooled Reactor Associates).

conceptual designs of fusion reactors that were solidly based on existing physics and engineering data bases could not claim to have addressed fusion's full potential, nor could such a study say much about directions worth investigating in pursuit of markedly improved performance. Such cases as numbers 5 through 8—featuring (respectively) ceramic structural materials to achieve extremely low activation, a pool-type design for passive cooling under nearly any accident conditions, enhanced MHD conversion to reduce balance-of-plant complexity and cost, and a D^3He fuel cycle to reduce neutron activation and tritium problems—are currently less credible than more conventional designs. But analyzing these cases, as examples of a much larger set of "advanced" approaches, has enabled us to avoid unduly constraining our assessment of fusion's long-range possibilities. For similar reasons we included advanced fission cases (PRISM—case 13 and MHTGR—case 14) as possible future competitors to fusion.

2.1. Economic Analysis

ESECOM analyzed, in a consistent framework, the economic, environmental, and safety characteristics of the cases in Table I, including, in some cases, examining the effects of varying the plasma performance, scale, and power density within an otherwise fixed design. The fusion and hybrid breeder cases were developed and analyzed with the assistance of the Generomak magnetic fusion physics/engineering/costing model [4] modified appropriately for our purposes.

The physics/engineering part of the Generomak model accepts as input the desired values of net electric power output, plasma beta, aspect ratio and elongation of the toroidal plasma, Troyon coefficient, and maximum toroidal field at the coil. (The combination must be chosen to give an acceptable value of the edge-plasma safety factor, q.) These inputs are used together with chosen blanket/shield characteristics (materials, radial dimensions, densities, inlet and outlet temperatures), conversion-efficiency relations, and current-drive assumptions in an iterative calculation of the plasma major and minor radii R_T and a, the toroidal field in the plasma B_ϕ, and the plasma current I_ϕ, corresponding to the desired net electric power taken to be 1200 MW(electric) in all cases. Also calculated in this process are plasma volume, plasma ignition margin, fusion power, neutron wall loading, reactor thermal power, overall thermal efficiency, current-drive and other auxiliary power, "fusion island" volume, and the masses of the blanket, reflector, shield, and coils. As an example, some of the main physics and engineering parameters of the point-of-departure tokamak (case 1) are given in Table II. This reactor case assumes advances in beta and current-drive efficiency beyond those considered for the current design of the International Thermonuclear Experimental Reactor (ITER). We examined sensitivity of the case 1 capital cost and COE to variations in these and other critical parameters. Reducing the beta to 0.06, or the current-drive efficiency by a factor of five (while increasing T_e to 25 keV), for example, increased COE by 15%.

The economics part of the Generomak model uses the physics and engineering parameters to calculate the direct capital costs of the fusion island, based on unit costs supplied to the model for fabricated material (e.g., $400/kg for reactor parts fabricated from V-Cr-Ti alloy, $90/kg for superconducting coils) and for certain specific components (e.g., power supply for current drive is costed at $2.25/W). Most of these costs are based on those developed in the STARFIRE study [5], updated to the January 1986 dollars used as the cost basis throughout ESECOM's work. Some of the STARFIRE figures have been further modified based on the Committee's judgment that more recent information warranted changes.

Costs of the blanket, limiter, coolant, and other major items that turn over on a short time scale compared to the plant lifetime are treated analogously to fuel costs in the fission fuel cycle, following the methodology embodied in the Nuclear Energy Cost Data Base (NECDB) at ORNL [6]. Calculation of other operation and maintenance costs also follows the NECDB model. Following standard engineering-economics techniques, as embodied in the NECDB, we then obtain a levelized constant dollar COE, in units of 10^{-3} U.S. dollars (1986) per kilowatt hour, or mills/KW·h.

The results of the basic economic calculations are shown in Table III. Here, the "overnight" costs include the application of indirect and contingency factors but not interest during construction; they are the costs that would result if construction were instantaneous. The total capital costs are obtained by accounting for interest during the assumed 6-yr construction period (adjusted to 1986 dollars). The additional fission case (11' PWR-ME) in Table III is the "median experience" PWR and provides a second reference point for the U.S. (The design and construction lead time for this case is 12 yr, and the indirect costs are 100% instead of 37.5%.) Particularly noteworthy in these results is that the COEs for the best experience and median experience PWRs bracket the range of costs estimated for the various fusion, hybrid breeder, and advanced fission cases.

2.2. Safety/Environment Analysis

ESECOM's analysis of environmental and safety characteristics included qualitative and, where possible, quantitative assessment of (a) possibilities and consequences of major releases of radioactivity from reactor accidents, (b) magnitude of the radioactive waste burden, (c) oc-

Table II. Parameters of the ESECOM point-of-departure reference fusion reactor.

	V-Li/TOK
Aspect ratio, A	4.0
Plasma elongation, κ	2.5
Total plasma beta, β	0.1
Safety factor, q_ψ	2.3
Maximum field at coil, $B_{\phi c}$ (T)	10.0
Toroidal field in plasma, B_{phi} (T)	4.29
Major radius, R_T (m)	5.89
Plasma current, I_ϕ (MA)	15.8
Neutron wall loading (MW/m^2)	3.20
Fusion power (MW)	2862.
Blanket thickness (m)	0.71
Blanket/shield gap (m)	0.10
Shield thickness (m)	0.83
Neutron energy multiplication	1.27
Tritium breeding ratio	1.28
Total thermal power (MW)	3563.
Primary coolant inlet, T_i (°C)	300.
Primary coolant outlet, T_o (°C)	550.
Thermal conversion efficiency	0.404
Recirculating power fraction	0.12
Net electric power [MW(electric)]	1200.
Volume of fusion power core (m^3)	2669.
Mass of fusion power core (tonne)	11482.
Mass power density [kW(electric)/tonne]	105.

cupational and public exposures to radiation in routine operation, and (d) unwanted links to nuclear weaponry.

ESECOM's calculations of activation product inventories were carried out at LLNL using the TART, ORLIB, and FORIG computer codes and their associated data bases [7–10]. These codes operated on cylindrical approximations to our toroidal blanket configurations.

The Monte Carlo calculations employed by the TART code to determine the neutron and gamma spectra in the various layers of the blanket, manifold/reflector, and shield (and, in one case, magnets) used 20 samples with 5000 particles per sample. These spectrum calculations accounted for materials compositions down to the level of 0.1 wt%. The activation calculations performed by the ORLIB averaging code using the ACTL cross-section library accounted for impurities to levels below 1 ppm by weight. The constituent and impurity compositions used in these calculations came mainly from the BCSS [11] and, in a few instances, from the design groups working on particular blankets. Based on a neutron fluence limit of 20 MW·yr/m^2 at the first wall, it was assumed that solid blanket components in reactors with first-wall fluxes in the range of 3 MW/m^2 were changed after each 6 full-power years (FPY) of operation, while those in reactors with first-wall fluxes around 15 MW/m^2 were changed after each full-power year of operation. Shields, magnets, and liquid constituents of blankets were assumed in most cases to be irradiated for 30 FPY, as was the entire blanket of the D^{3}He case.

For purposes of assessing accident potential and occupational hazards, reactor radioactivity

Table III. Comparative costs without safety assurance credits (1986 U.S. dollars).

	Unit capital costs [$/kW(electric)]			COE (mill/kW·h)			
Case	Direct	Overnight	Total	Capital	Fuel and other O&M	Fission fuel sales	Total
1. V-Li/TOK	1378	2178	2365	35.1	18.1	0.0	53.1
2. RAF-He/TOK	1387	2193	2380	35.3	13.2	0.0	43.5
3. RAF-LiPb/RFP	949	1501	1630	24.2	13.5	0.0	37.7
4. V-Li/RFP	963	1523	1655	24.5	12.8	0.0	37.3
5. SiC-He/TOK	1621	2563	2785	41.3	13.4	0.0	54.6
6. V-Flibe/TOK	1184	1873	2035	30.1	17.8	0.0	47.9
7. V-MHD/TOK	873	1380	1500	19.2	16.1	0.0	35.4
8. V-D^{3}He/TOK	1763	2787	3025	38.9	8.9	0.0	47.8
9. RAF-Li/HYB	1649	2608	2830	41.9	21.7	−23.2[a]	40.3
10. SS-He/HYB	1343	2123	2305	34.1	21.7	−16.0[a]	39.8
11. PWR-BPE	740	1170	1270	18.8	14.6	0.0	33.4
11.′ PWR-ME	980	2260	2620	41.0	15.6	0.0	56.6
12. LSPB	1040	1645	1785	26.5	16.7[b]	16.7[b]	43.2
13. PRISM[c]	996	1575	1710	25.3	18.5[b]	18.5[b]	43.8
14. MHTGR[c]	885	1400	1520	22.6	19.4	0.0	42.0

[a]These figures for hybrid fissile fuel sales are based on MHTGR clients.

[b]Fuel sales credits for LSPB and PRISM are based on costs of reprocessing at central facilities (see Table II) and sale of resulting plutonium at $50/g. Reprocessing costs may be higher for on-site processing proposed by PRISM designers.

[c]Some safety assurance credits were embedded in the vendor/designer estimates of PRISM and MHTGR capital costs and remain in the cost figures shown here.

inventories were evaluated at their maximum levels, that is, those attained just before blanket change-out. Radioactive waste calculations were based on "life cycle" waste quantities for 30 FPY of operation, including all changed-out components.

Estimates of tritium inventories in the fusion cases were based on the BCSS [11] and on subsequent design studies, and included tritium in structure, coolant, breeder, and neutron multiplier materials.

To facilitate analysis of accident hazards associated with radioactive materials of different degrees of inherent mobility, we divided the radioactive inventories of fusion and fission reactors alike into five mobility categories:

1. Elements gaseous or extremely volatile under thermochemical conditions of normal operation.

2. Elements somewhat volatile under thermochemical conditions of normal operation.

3. Elements somewhat to highly volatile under conditions likely to be encountered in an accident.

4. Elements somewhat volatile under conditions that may be encountered in severe accidents.

Table IV. Dose-threshold release fractions by component and mobility category.

| Case and mobility categories | Inventories (MCi) | | Release fraction that would produce | | | |
| | | | 200-rem critical dose from plume at 1 km | | 25-rem 50-yr ground dose at 10 km | |
	First wall	BOFC	First wall	BOFC	First wall	BOFC
Case 1: V-Li/TOK (Fusion)						
I	5	0.077	52.	7100.	15.	260.
I-II	10	6.0	6.3	5.0	0.78	0.82
I-III	10	560.	5.1	0.027	0.55	0.00011
I-IV	95	670.	3.7	0.027	0.021	0.00010
I-V	540	2400.	0.036	0.015	0.0016	0.00009
Case 11: PWR-BPE (Fission)						
I	380		0.38		28	
I-II	1300		0.017		0.00013	
I-III	1500		0.011		0.00012	
I-IV	2600		0.0058		0.000086	
I-V	5600		0.0025		0.000048	

BOFC = Balance of fusion core (other than first wall).

5. Elements resistant to volatilization even under extreme accident conditions.

Given the radioactive inventories and the mobility-based classification scheme just described, we can calculate the off-site doses that would result from release of 100% of the radioactive inventory in each mobility category for each design. Then we can deduce how large the actual release fractions of these materials would have to be to produce any particular dose of interest. We calculated for each design, for example, what fractions of the radioactive inventories in each mobility category and each reactor component would have to be released in order to generate, under adverse weather conditions, an acute whole-body dose of 200 rem at a distance of 1 km from the reactor (corresponding approximately to the threshold below which no early fatalities would be expected); we also calculated such "dose threshold release fractions" corresponding to a 50-year dose of 25 rem from ground contamination at a distance of 10 km from the reactor. (For conversion of dose units, use 1 rem = 10^{-2} Sv.) The higher these dose-threshold release fractions are, the better, since a large figure indicates that the threshold dose will not be exceeded unless a large fraction of the inventory escapes. (A dose-threshold release fraction exceeding unity means that not even a 100% release of the inventory would suffice to produce the threshold dose.) Table IV gives dose-threshold release fractions by mobility category for the point-of-departure tokamak (case 1) and the PWR fission (case 11). The dose-threshold-release fractions for the fusion case 1 are larger than those for the PWR fission case in all mobility categories. The isotope contributing most of the mobility category I dose for fusion is tritium.

When we take the next step of comparing the release fractions needed to generate threshold doses with the fractions that may be physically plausible for isotopes in the different mobility categories in the fusion and fission cases, the advantage of fusion widens. Table V gives estimates of the maximum plausible release fractions based on analysis at MIT by Kazimi, combined with test data from INEL on volatilization from the vanadium and steel alloys used in the fusion cases.

ESECOM found it useful to work with a classification of safety levels that defines four levels of "safety assurance," given in Table VI, based in substantial part on the work of Piet [12]. These level are based on differences in the extent and nature of dependence on passive versus

Table V. Estimates of maximum plausible release fractions.

Mobility category	Fusion			Fission
	V-Li/TOK	RAF-He/TOK	V-Li/RFP	LWR, LMFBR
I	1.0	1.0	1.0	1.0
II	0.1	0.7	0.7	0.2 to 0.7
III	0.05	0.1 to 0.7	0.2	0.2 to 0.4
IV	5×10^{-4}	0.01 to 0.1	2×10^{-3}	0.05 to 0.1
V	5×10^{-4}	1×10^{-4} to 1×10^{-3}	2×10^{-4}	0.003 to 0.05

Note: The time-temperature scenarios assumed in estimating the fusion release fractions are as follows:

V-Li/TOK: Li-air fire + decay heat produce 1300°C for 10 h followed by 40 h at 900°C.

RAF-He/TOK: Decay heat produces 900°C for 50 h.

V-Li/RFP: Li-air fire + decay heat produce 1500°C for 10 h followed by 40 h at 1200°C.

active design features needed in a given design to provide assurance of public safety—more specifically, to preclude any off-site early fatalities from release of radioactivity.

By "passive design features" we mean combinations of materials properties and configurations of structural components such that natural processes of energy removal (conduction, natural convection, radiation) suffice to limit accident sequences and the resulting radioactivity releases. Relevant materials properties include inventories of radioactivity, masses, heat capacities, strength versus temperature, melting points, vapor pressures (as functions of temperature), and susceptibility to formation of volatile oxides. By "active design features" we mean pumps, valves, switches, sensors, and the like, as well as containment buildings with many doors and controlled penetrations.

ESECOM concluded that high "levels of safety assurance," (low LSA numbers) as described above, could be the basis for reduced requirements for expensive "nuclear-grade" materials and components compared to the requirements imposed in fission power plant construction in the U.S. today. To give cost credit to low LSA-numbered ESECOM fusion reactor cases, we used a set of cost-reduction factors for various reactor subsystems developed by Perkins in the MINIMARS study [13] to estimate the potential cost savings associated with the use of non-nuclear grade materials and components. Reactor cases with LSA = 1 received 100% of maximum credits of reference 12, those with LSA = 2 received 50%, those with LSA = 3 received 25% and those with LSA = 4 receive no safety assurance cost credits. The maximum cost credits amount to approximately 30% reduction in COE. Table VII lists the LSA rating and corresponding COE of the ten fusion cases for optimistic, nominal, and conservative evaluations. All fission cases were rated LSA = 4, except for the PRISM (case 13) and MHTGR (case 14), which designs included passive safety features, for LOCA accidents, and included cost reductions for the passive safety features estimated by their designers.

3.   ESECOM FINDINGS AND CONCLUSIONS

Some of the most important findings of the ESECOM analysis are summarized in the following sections. For brevity, we have not included all of the findings or supporting analysis, for which we refer the reader to our longer reports [1,2].

Table VI. Definition of ESECOM's levels of safety assurance (LSA).

LSA	Concise description	Accident class		
		Large-scale reconfiguration	Small-scale violation of geometry [e.g., loss-of-coolant accident (LOCA)]	Transient without violation of geometry [e.g., loss-of-flow accident (LOFA)]
I	Inherent safety	If event occurs, material properties suffice to prevent fatal release.	If event occurs, material properties suffice to prevent fatal release.	If event occurs, material properties suffice to prevent fatal release.
II	Large-scale passive protection	Reconfiguration severe enough to lead to off-site fatality is made incredible using passive design features.	If event occurs, material properties and passive mechanisms suffice to prevent fatal release or escalation to next class.	If event occurs, material properties and passive mechanisms suffice to prevent fatal release.
III	Small-scale passive protection	Reconfiguration severe enough to lead to off-site fatality is made incredible using passive design features.	Violation severe enough to lead to off-site fatality is made incredible using passive design features.	If event occurs, material properties and passive mechanism suffice to prevent fatal release or escalation to next class.
IV	Active protection	There are events in one or more of these categories that, if they occur, require active systems to preclude an off-site fatality, and that cannot be made incredible by passive design measures alone.		

3.1. The Potential of Magnetic Fusion Energy

Our analysis indicates that MFE systems have the potential to achieve costs of electricity comparable to those of present and future fission systems, coupled with significant safety and environmental advantages. The most important potential advantages of fusion with respect to safety and environment are as follows:

1. high demonstrability of adequate public protection from reactor accidents (no early fatalities off-site), based entirely or largely on low radioactivity inventories and passive barriers to release rather than an active safety systems and the performance of containment buildings.

2. substantial amelioration of the radioactive waste problem by eliminating or greatly reducing the inventories of radioactive isotopes with long half-lives. Under current U.S. waste-management regulations, fusion could greatly reduce or eliminate high-level wastes that require deep geologic disposal.

3. diminution of some important links with nuclear weaponry (easier safeguards against clandestine use of energy facilities to produce fissile materials, no inherent production or circulation of fissile materials subject to diversion or theft).

Table VII. Levels of safety assurance (LSA) and COE with safety assurance cost credits.

	Case	COE (mill/kW·h) and (LSA in parenthesis)			
		Optimistic concept evaluation	Nominal design estimate	Conservative concept evaluation	No safety assurance credits
1.	V-Li/TOK	46.2 (2)	49.7 (3)	53.1 (4)	53.1
2.	RAF-He/TOK	42.6 (2)	42.6 (2)	45.6 (3)	48.5
3.	RAF-PbLi/RFP	35.7 (3)	37.7 (4)	37.7 (4)	37.7
4.	V-Li/RFP	35.2 (3)	37.3 (4)	37.3 (4)	37.3
5.	SiC-He/TOK	40.3 (1)	40.3 (1)	47.5 (2)	54.6
6.	V-FLiBe/TOK	38.0 (1)	42.9 (2)	42.9 (2)	47.9
7.	V-MHD/TOK	31.0 (3)	35.4 (4)	35.4 (4)	35.4
8.	V-D^{3}He/TOK	34.9 (1)	41.3 (2)	41.3 (2)	47.8
9.	RAF-Li/HYB				
	With LWR clients	39.1 (3)	39.4 (4)	39.4 (4)	39.4
	With MHTGR clients	40.1 (3)	40.3 (4)	40.3 (4)	40.3
10.	SS-He/HYB				
	With LWR clients	38.4 (3)	38.8 (4)	38.8 (4)	38.8
	With MHTGR clients	39.4 (3)	39.8 (4)	39.8 (4)	39.8
11.	PWR-BPE	33.4 (4)	33.4 (4)	33.4 (4)	33.4
12.	LSPB	43.2 (4)	43.2 (4)	43.2 (4)	43.2
13.	PRISM	43.8 (3)	43.8 (3)	(4)	
14.	MHTGR	42.0 (3)	42.0 (3)	(4)	

Neither the economic competitiveness nor the environmental and safety advantages of fusion will materialize automatically. Economic competitiveness depends on attaining plasma and engineering performance, such as high beta, efficient current drive, and ease of maintenance consistent with high capacity factor, that are not yet assured. Achieving the potential environmental and safety advantages depends in large measure on designs specifically tailored to do so and on the use of low-activation materials whose practicality for fusion applications remains to be demonstrated.

It is essential, in this connection, that sufficient R&D be devoted early to determining which of a variety of confinement schemes, structural materials, blanket types, and fuel cycle/energy conversion combinations can actually be made practical.

3.2. Technology and Economics

The design characteristics offering the most important potential benefits for reducing fusion costs are as follows:

1. compactness (including but not limited to high power output per unit mass), which reduces the capital cost of the fusion power core; which reduces, as a result, the sensitivity of COE to plasma performance; and which also may ease maintenance.

2. high level of safety assurance, meaning demonstrability of public safety based on low radioactive inventories and passive mechanisms for preventing releases, which should reduce costs for active safety systems and nuclear-grade components as well as facilitating siting and licensing.

3. advanced energy conversion systems, which should be able to reduce balance-of-plant (BOP) costs and may increase capacity factors.

Each of these features has the potential to generate COE reductions in the range of 20 to 30%. If two or more of them can be combined in one design, the resulting COE reduction could be even larger.

The fusion cost estimates we have derived necessarily embody many uncertainities. The magnitudes of these cost estimates relative to one another are more informative than their absolute values, and serve to indicate promising areas of research to improve fusion.

3.3. Environment, Safety, and Economics

We believe the categorization of different designs into four levels of safety assurance, based on the extent to which assurance of public safety depends only on low inventories of radioactivity and passive mechanisms to prevent releases, is an informative way to characterize differences relevant to the interaction of safety and economics.

There is a potential conflict between pursuing higher neutron wall loading to reduce cost through higher power density, and pursuing the economic, regulatory, and public acceptance benefits of high levels of safety assurance.

With suitable choice of structural materials and blanket design, even a large lithium fire would not produce any prompt fatalities off-site. The potential destructiveness of lithium fires in terms of plant investment and public acceptability nonetheless dictates the use of special design features against such fires in plants that use liquid lithium as the primary coolant breeder.

Active inventories of tritium in current reactor designs are small enough that even complete release under adverse meteorological conditions would not produce any prompt fatalities off-site.

Proper choice of fusion reactor structural materials can reduce or eliminate formation of the most troublesome long-lived activation products, and therefore can significantly reduce radioactive waste hazards, compared to fission.

An electricity supply system based on MFE would be less likely than a fission energy system to contribute to the acquisition of nuclear weapons capabilities by subnational groups, and would also be easier to safeguard against clandestine use for fissile material production by governments. Except for hybrid breeders, fusion reactors need not produce or contain any fissile material, and a fusion-based electricity supply system would not circulate any. Because fusion reactors could be modified to produce fissile material, however, they will need to be subjected to international safeguards.

3.4. Implications for MFE R&D

Far more system level design and analysis work than has been conducted so far is needed to better define the economics and safety characteristics of fusion. Emphases in these systems studies should include:

1. improved characterization of accident pathways and radioactivity release mechanisms;

2. development of reactor designs **combining** high levels of safety assurance, high mass power density, direct conversion, and design simplicity for reliability and ease of maintenance.

The ultimate viability and attractiveness of MFE depends so strongly on materials issues that a strong, sustained materials development and testing program must be considered second only to confinement studies as a prerequisite for fusion's success. The materials program should be closely integrated with the systems studies called for above, as well as responding to the materials issues posed by current fusion devices.

Notwithstanding the difficulty of the physics and engineering challenges that must be addressed in the next generation of fusion facilities, such as the compact ignition torus and the International Thermonuclear Experiment Reactor (ITER), it is important that these facilities also be used to develop and demonstrate the kinds of safety features that will be needed for commercial reactors.

4. ACKNOWLEDGMENTS

The committee members wish to recognize the extraordinary work of Steve Fetter (at LLNL through much of the study, now at Harvard and MIT), Steve Piet (INEL), Carroll Maninger (LLNL), and John Massida (MIT) in computing the environmental and safety indices central to our effort.

Work was performed under the auspices of the U.S. Department of Energy by the Lawrence Livermore National Laboratory under contract number W-7405-ENG-48.

Addresses of the authors

[1] *Lawrence Livermore National Laboratory, Livermore, CA 94550 (Presenter at IAEA)*

[2] *University of California, Berkeley, Energy and Resources Program, Berkeley, CA 94720 (Chairman of ESECOM)*

[3] *Grumman Aerospace Corporation, MS C47-05, Bethpage, NY 11714*

[4] *Future Resources Associates, Berkeley, CA 94720*

[5] *Idaho National Engineering Laboratory, EG&G Idaho, Idaho Falls, ID 83401*

[6] *Oak Ridge National Laboratory, Oak Ridge, TN 37831*

[7] *Public Service Electric and Gas Company, Newark, NJ 07101*

[8] *Massachusetts Institute of Technology, Department of Nuclear Engineering, Cambridge, MA 02139*

[9] *Los Alamos National Laboratory, Los Alamos, NM 87545*

[10] *General Atomics, P.O. Box 86508, San Diego, CA 92138-5608*

References

[1] HOLDREN, J.P., BERWALD, D.H., BUDNITZ, R.J., CROCKER, J.G., DELENE, J.G., ENDICOTT, R.D., KAZIMI, M.S., KRAKOWSKI, R.A., LOGAN, B.G., and SCHULTZ, K.R., "Exploring the Competitive Potential of Magnetic Fusion Energy: The Interaction of Economics with Safety and Environmental Characteristics," *Fusion Technology* **13**, 7 (1988).

[2] HOLDREN, J.P., BERWALD, D.H., BUDNITZ, R.J., CROCKER, J.G., DELENE, J.G., ENDICOTT, R.D., KAZIMI, M.S., KRAKOWSKI, R.A., LOGAN, B.G., and SCHULTZ, K.R., *Report of the Senior Committee on Environmental, Safety, and Economic Aspects of Magnetic Fusion Energy*, UCRL-53766, Lawrence Livermore National Laboratory (1988).

[3] Report of the Office of Scientific, Technological Options Assessment (STOA) "Criteria for the Assessment of European Fusion Research," Vol. I and II, The European Parliament, Luxembourg, May, 1988.

[4] SHEFFIELD, J., DORY, R.A., COHN, S.M., DELENE, J.G., PARSLY, L.F. ASHBY D.E.T.F., and REIERSEN, W.T., "Cost Assessment of a Generic Magnetic Fusion Reactor," *Fusion Technology* **9**, 199 (1986).

[5] Argonne National Laboratory, McDonnell-Douglass Aeronautic Co., General Atomic Co., and The Ralph M. Parsons Co. "STARFIRE: A Commercial Tokamak Fusion Power Study," Argonne National Laboratory, Report ANL/FPP-80-1 (1980).

[6] "Nuclear Energy Cost Data Base," DOE/NE-0044/3, U.S. Department of Energy, Office of Nuclear Energy (1985).

[7] FETTER, S.A., "Radiological Hazards of Fusion Reactors: Models and Comparisons," Ph.D. Thesis, University of California, Berkeley (1985).

[8] PLECHATY, E.F. and KIMLINGER, J.R., "TARTNP: A Coupled Neutron-Photon Monte Carlo Transport Code," UCRL-50400, Vol. 14, Lawrence Livermore National Laboratory (1976).

[9] BLINK, J.A., DYE, R.E., and KIMLINGER, J.R., "ORLIB: A Computer Code that Produces One-Energy-Group, Time- and Spatially Averaged Neutron Cross Sections," UCRL-53262, Lawrence Livermore National Laboratory (1981).

[10] BLINK, J.A., "FORIG: A Computer Code for Calculating Radionuclide Generation and Depletion in Fusion and Fission Reactors," UCRL-53633, Lawrence Livermore National Laboratory (1985).

[11] SMITH, D.L., BAKER, C.C., SZE, D.K., MORGAN, G.D., ABDOU, M.A., PIET, S.J., SCHULTZ, K.R., MOIR, R.W., and GORDON, J.D., "Overview of the Blanket Comparison and Selection Study," *Fusion Technology* **8**, 1, Part 1, 10 (1985).

[12] PIET, S.J., "Approaches to Achieving Inherently Safe Fusion Power Plants," *Fusion Technology* **10** 1, 7 (1986).

[13] PERKINS, J., Lawrence Livermore National Laboratory, Personal Communication (Aug. 1985). See also LLNL report UCID-20773 "MINIMARS Conceptual Design: Final Report," Sept. 1986.

II. GENERAL PROGRAM EVALUATION

THE ENERGY SCENE IN THE MID-21st CENTURY

L. Gouni

Electricité de France, Direction Générale

32 rue de Monceau, Paris, France

INTRODUCTION

Nuclear fusion technology will not enter the energy market before several decades and so must be placed in the context of the energy scene over the next half century or hundred years. The economists' classical backward look and forecasts consider ten to twenty year spells; however, the position at the time of analysis influences the outcome too greatly not to be accepted without due caution. Therefore, by contrast, the basic and permanent, - one may speak of secular trends - must be distinguished. The paper examines the following points:

- energy demand and supply equilibrium mechanisms (para. 1),
- energy demand and supply scenarios (para. 2 and 3),
- estimate of a future price of energy and "base-load" electricity (para. 4).

1. THE LESSONS OF THE PAST - MARKET EQUILIBRIUM MECHANISMS

1.1 Teachings of the 20th century as regards energy economics

Looking at the "commercial" energy market since the beginning of the century, the following basic trends - which, according to experts, will still mark future decades - can be observed:

- Energy is indispensable for the economic and social development of mankind, whether for the heat applications entering the matter transformation processes and the improvement of comfort, or its mechanical uses which replace animal strength, multiplying it considerably. World energy consumption has thus increased greatly as population has grown and living standards risen (see Fig. 1).

- A geographical analysis of consumption nevertheless reveals major differences which reflect the profound disequilibria in economic conditions: OECD countries consume ten times more per person than developing countries, the differences being still greater in extreme cases (the ratio is 90 for commercial energy sources between Canada and Vietnam). Thirty years ago, this North-South disequilibrium was still hardly perceived: the modest volume of trade and the difficulty of

Safety, Environmental Impact, and Economic Prospects of Nuclear Fusion
Edited by B. Brunelli and H. Knoepfel
Plenum Press, New York, 1990

81

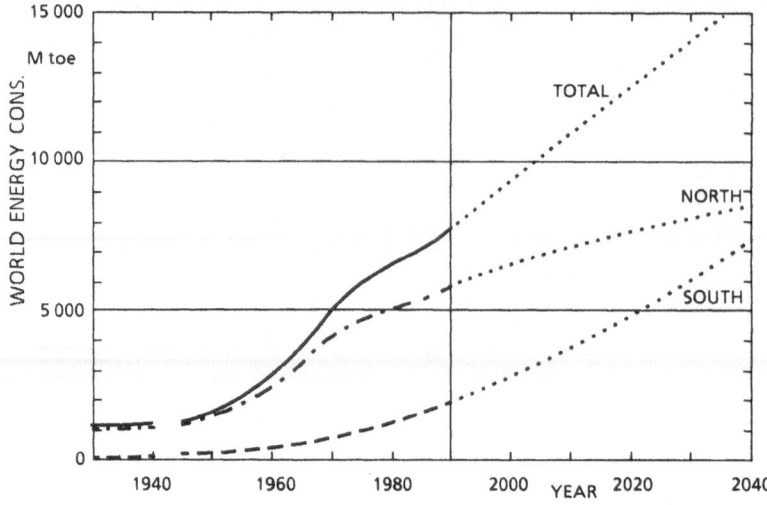

Fig. 1. World energy consumption (WEC)

transmitting information went hand in hand with this ignorance. Now, however, with the opening of borders and progress in communications, the world is well aware that the anomalies of these affairs cannot persist for long. This theme, it is reasonable to suggest, will dominate the energy scene of the coming decades. The developing countries' population is mostly rural; it must be fed and its ability to act multiplied. Although commercial energy sources today are beyond its means, the development of the Third World's energy consumption will nevertheless have to be organised: the basic character of future development lies here (see para. 2.3). This disparity can also be observed in the world distribution of mineral resources. Industrialised countries have already slipped largely into their own energy resources and have been importing large quantities of conventional energy sources for a long time. This flow can only increase. Where will North-South co-operation stand for these exchanges (see para. 3.3.)?

- This last phenomenon, intensified by the progress achieved in transport, has given an international dimension to the energy market since World War II. Although before that, setting aside oil for motors, most industrially advanced areas secured their own supplies, this has no longer been true for many years. In taking over from European coal since 1955, world oil has become Western Europe's staple energy source (see Fig. 2). This explosion of energy boundaries is irreversible. Long distance transport of LNG developed in the last two decades. The question for tomorrow is when and how major electricity exchanges will take place over similar distances.

- The continuation of growth of the share of electricity in total energy consumption is a recognised phenomenon. Consumption of this energy source in OECD countries has increased from 7% before the war to 35% today (see Fig. 3). The quality of this energy, by allowing varied and economical applications suitable to automatic systems, guarantees that this trend will continue.

1.2 Practical equilibrium mechanisms of energy supply and demand

a) The data of economic theory. In the theoretical model, energy, like other products useful to or desired by man, is subject to the laws of

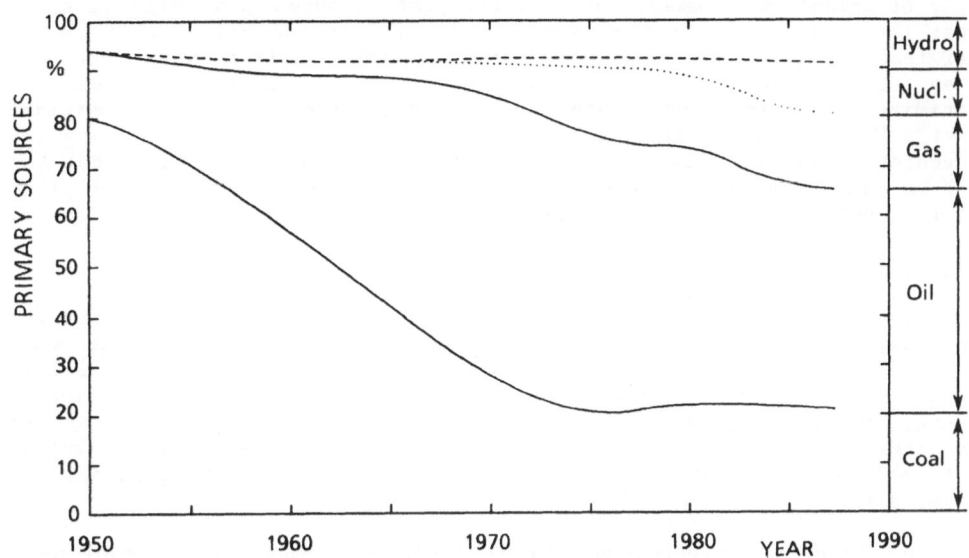

Fig. 2. Share of different primary energy sources for Western Europe
(incl. Scandinavian countries) (ENERDATA)

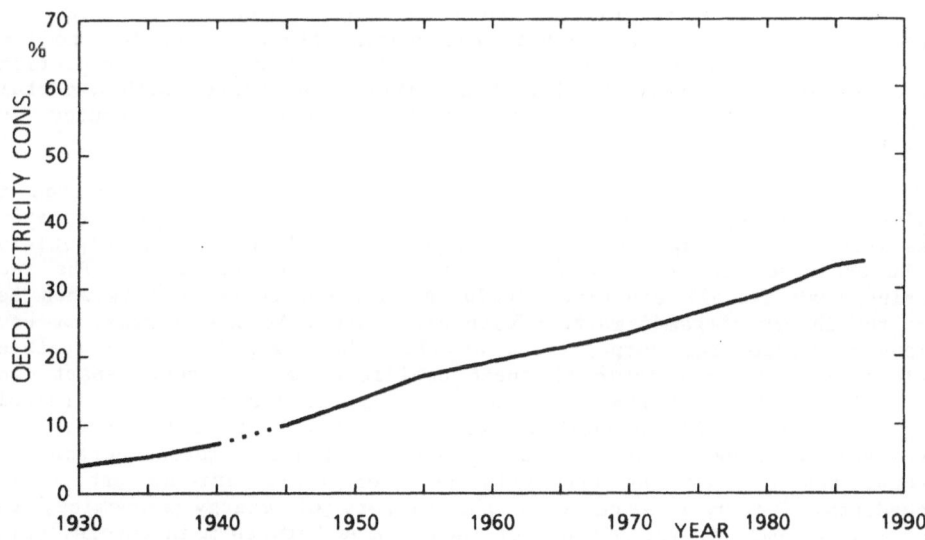

Fig. 3. Electricity share in final consumption for OECD countries
(ENERDATA)

the market and the rules of competition. Supply should be adequate to satisfy a given demand. The resources for this purpose are created and exploited so as to reduce the necessary expenditure to the minimum. This rule of optimization means that the cheapest resources are used first and that the cost of the last resource used will set the price: this is the marginal cost. As the infra-marginal production facilities are cheaper, they yield a profit often referred to as a "rent". In this balanced and optimized scenario, supply matches demand and price is equal to "marginal cost". When demand increases, a new equilibrium is created, inducing the recourse to dearer resources. Marginal cost and price increase, but the growth of demand is reduced by the rise in price. Finally, the price settles both supply and demand at equilibrium. Using a broad brush image, energy abundance is associated with a low price, energy shortage with a high price.

Theory reaches far beyond in the quest for optimization. At the optimum, the differences between supply techniques disappear as does rent. Similarly, taking account of the limits of resource deposits, it has been shown (Hotelling) that under quite restrictive hypotheses (1) the price of energy increases at the same pace as the rate of interest. (One of the hypotheses is that the level of deposits in reserve be perfectly known, but this level depends on technical progress and research spending, see para. 3.1).

b) Practical mechanisms. In reality, the theoretical model pre-supposes that economic operators respect the laws of perfect competition, i.e., on the one hand, that they are kept informed of market data at all times and have the means to react immediately when they receive them and on the other hand, that in making the best choice they are willing to play the game fairly.

With today's communication facilities, it might be thought that the decision-maker has near perfect knowledge of the state of the market. Unfortunately, this information is in general insufficient. He must also have an idea of the relatively long-term future. However, the future is affected by uncertainty as to both the general trend and deviations from it. Any forecaster knows that reality will not obey his prediction, especially if it becomes "official" and therefore tainted with a certain purposefulness; tension may then erupt, engendering divergences and reactions.

Futhermore, the laws of fair competition cannot be respected in reality. It may be that the existence of a large number of small decision-makers creates an image of a framework generally allowing fair competition to be achieved, masking the real behaviour of the individual. The fact remains that overall behaviour includes many confrontations between men that the theory of games masters with difficulty. At one extreme, we have friendly or sporting competition; at the other, armed conflict between nations. The characteristic of these conflicts is that they depart from rationality and, since the actors are no longer solidary, it is difficult to predict what will happen: Who will win? How long will the conflict last? What will be its effects and repercussions? The more important the actors, the greater the uncertainties, for there are no arbitration procedures. For over a century, the history of energy economics, es-pecially for oil, is the history of these games between main actors, games which go beyond the framework of the energy sector to encompass taxation, the sharing of oil rents, using as reference unit an international curren-cy - currently the dollar - which is not defined today (see para. 3.3).

These modulations of the laws of perfect competition are not   neutral regarding trends. On the one hand, they influence decisions and long-term orientations. On the other hand, they engender relatively long periods of tension which in turn disturb market trends by distorting some decisions; in reality, the market seeks an equilibrium without ever reaching it on a durable basis: disturbing events always occur. The sharp fluctuations in energy prices about the average therefore convey a meaning that cannot be neglected: they correspond to periods of re-equilibration born of strategic changes, many of which involve conflicts, sometimes referred to as shocks.

2. FUTURE ENERGY DEMAND AND ITS CONTROL

2.1 A basic lesson of the last oil shock: consumption can participate to a large extent in the equilibrium of supply and demand

Until 1978, probably through lack of experience, energy economists considered that total energy demand was not very sensitive to price; of course, there was strong competition between different types of energy, but overall consumption was held to depend virtually on the rate of economic growth alone. By multiplying oil prices by 3 in 1974 and then by 2 in 1979 (at constant prices), the 1973 oil crisis triggered action in all countries to reduce consumption, supported by governments. Whereas it was then thought that the new equilibrium could only be secured by increasing supply, i.e., the use of more expensive fossil energy deposits or recourse to new technologies (nuclear, solar, biomass, etc.) and that prices would thenceforth be higher, experience has shown that considerable savings in consumption could be derived from the improved operation of consuming appliances and the design of new appliances or even new techniques. In the light of this, a major effort was made to innovate, which, just as it started to materialise in 1985, was thwarted by the reduction in energy prices. Consider for instance, the energy-saving automobile: the first improvements appeared in 1978; but it is known that the manufacturers conducted studies with a view to halving consumption by comparison with the vehicles of 1979. Another example is the generalisation of heat insulation of premises, replacing heating energy by an investment in insulating materials; energy consumption was more than halved.

In the upshot, control of demand was one of the basic factors which allowed the return to energy abundance in 1984. Between 1975 and 1985, world energy intensity (excluding Eastern countries and China), i.e., the ratio of energy consumption to GNP, viewed as stable before 1974, fell by 17% overall, and by 23% in OECD countries (see Fig. 4).

By the way, this difference between industrialised and developing countries warrants attention, as it points out a phenomenon that will probably continue in the future. The major reduction in energy intensity in industrialised countries is partly explained by a modification of economic structures. Large energy consuming basic industries, such as the steel and iron industry or the chemistry of raw materials, moved to industrialising countries endowed with abundant low cost energy, while the industrialised countries focused their activities on less energy-intensive high technology industries. It will be interesting to see how far this transfer will go in the light of broad industrial strategy and commercial and political risks.

2.2 A factor taking over from energy control: the environmental constraint

Owing to the steep drop in energy prices since 1985, energy consumption control has become less strict, which is a source of concern

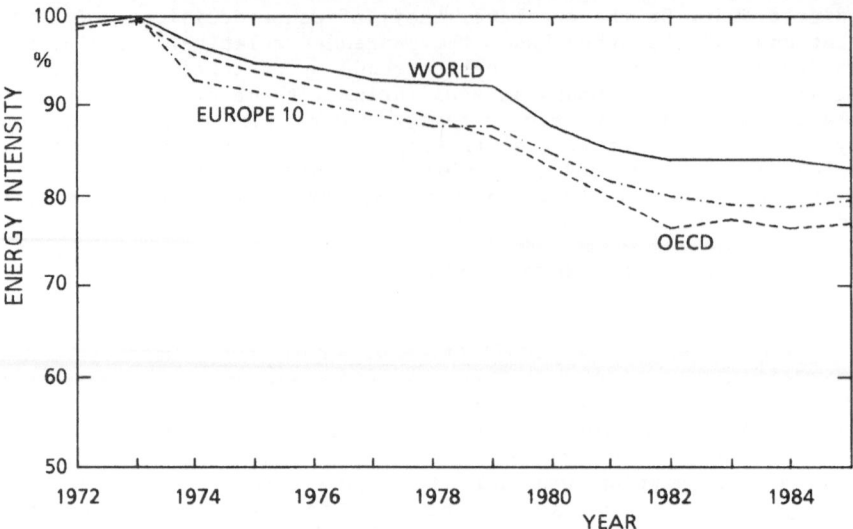

Fig. 4. Energy intensity (with index 100 in 1973) for OECD, Europe 10, and World (Eastern countries and China not included) (SHELL)

for many governments. However, it also appears that an underlying tendency has emerged favouring the development of a completely new range of utilisation technologies which will be of all the more interest to the market in that, even if the price of energy remains quite low, the constraint of environmental protection will growingly become an obligation for energy suppliers and users.

From a technical point of view, the problem is quite complicated in that the mechanisms that alter the natural environment involve space and time dimensions which are far from known and a fortiori far from mastered. It is easy to influence public opinion by rumours whose accuracy cannot be checked; these games are played the more willingly as the serious hazards are of concern to the population at large. The political decision-maker is therefore compelled to take account not only of scientific data, but also of how the risk is perceived by the population. This will probably entail major, but unjustified economic consequences, such as, for example, the hasty measures taken recently to replace the polychlorinated biphenyls (pyralenes) of electric transformers. The fact is that, without scientific data, the media have a free hand.

Considering the opinions expressed by scientific organisations, it can be stated, simplifying to the extreme, that two environmental factors will dominate energy economics of the next few decades:

- Firstly the factor relating to <u>fossil fuels</u>, coal , oil, gas and other fuels. The nuisances are at transformation (electrical plants) and final energy consumption level. These fuels are responsible for the major increase in the content of CO_2 in the atmosphere, which has risen from 270 to 335 ppm over a century; it is predicted more than 500 ppm for the middle of the next century, with a disquieting impact on the temperature rise of the earth, climatic change and the sea level. This phenomenon is much more worrying than acid rains (SO_2 and NOx), on which the media have tended to concentrate; from this last point of view, the cost of desulphurisation of the fluegas released by a conventional electrical plant is now known. It amounts to approximately

10% of cost per electrical kWh.

- Secondly, the factor relating to <u>nuclear technologies</u> and consequently, with little connection to final energy consumption. Radioactive emissions in normal operating conditions of reactors and low radioactivity waste are under control and the disposal of highly radioactive waste will also be mastered within the next fifteen years. The most critical problem is the safety of the nuclear reactors themselves in the event of an incident, as was shown by Three Mile Island and Chernobyl. Future technologies will thus have to improve greatly the safety characteristics of these reactors. The Chernobyl accident showed the world what a nuclear accident can mean: although the death toll was relatively low (31), the evacuation of 135.000 inhabitants from a zone which will remain prohibited for a long time is an unacceptable phenomenon. In addition, the use of nuclear power for military purposes cannot fail to affect the development of nuclear generation.

Nevertheless, in the coming decades, given the new awareness of the deterioration of the environment, the environment constraint will be, directly or indirectly, a major factor in limiting energy consumption.

2.3 <u>Some ideas on future energy consumption</u>

The major disturbances affecting the world economical and energy scene have led forecasters to alter their methods to take account of the important factors listed earlier. There is a certain number of predicative studies today that are much more cautious than those of the mid-1970s. The main trends are as follows (see Fig. 1):

- Although world commercial energy consumption increased six-fold between 1930 and 1985 (i.e., over 55 years), this factor will only be 2.5 over the next 55 years (from 1985 to 2040). Consumption per inhabitant which was 0.55 toe in 1930 and 1.4 toe in 1985 will be 1.75 toe in 2040.

- Consumption growth will take place at a moderate pace in what is referred to as the North (the Western industrialised countries and the Eastern bloc) and at a much higher rate in the South (developing countries). Between 1985 and 2040 growth will only be 25% or 30% in the North as opposed to 350% in the South where consumption will thus be 4.5 times that of 1985. In 2040, the North will consume 5.2 toe per inhabitant and the South 1 toe: this is very far from equality, but the South will be closing the gap. This means that the new large energy markets will no longer be in advanced countries and that the structure of international energy trade in the mid-21st century will differ substantially from today's.

2.4 <u>Future electricity consumption and the electrical network</u>

The share of electricity consumption in total energy demand will keep on rising, as in the past (see para. 1.1), although the constraints referred to above will also lead to a reduction in its consumption per unit of GDP.

Under these circumstances, although some prefer to define electricity consumption on the basis of total energy consumption, it is easier to predict electricity than energy consumption. In 1985, electricity consumption in the 12 European Community countries was approximately 1500 TWh; the EC Departments have now made forecasts for 2010 at 2400 TWh, with a probable range of + or - 15%. Exponential extrapolation, classical in the case of electricity consumption, should not be undertaken over a 50 year period; preference must be given to linear extrapolation. Electricity consumption will be between 3000 TWh and 4000 TWh around 2050.

It is not very likely, though not impossible, that by this time new techniques will allow economical electricity storage to be achieved. Consequently, the present structure of the grid which, through an interconnected system, pools large generating units, thus affording the indispensable compromise between the number of power stations and their size, will probably not be modified. It can then be estimated that the EC will need to build roughly fifteen 1000 MW power stations per year around 2050. This leaves room for nuclear fusion power stations built in series in satisfactory economic conditions.

3. FUTURE LONG-TERM ENERGY SUPPLY AND ITS COSTS

3.1 Broad lines. The increasingly surprising capacity of techniques

Three main categories of energy enter the formation of future energy prices, each with very different profiles: coal, hydrocarbons, and new energies (first nuclear and then solar energy). A word must be said about this.

Mankind's ability to react to untoward events is another basic teaching of the 1973 crisis. Twelve years after the first oil shock, energy abundance returned and energy prices fell to the average levels experienced in the past (at constant prices). What happened? After a period of doubt (two or three years), a fantastic programme of technical innovation and re-structuring was implemented in every energy field. New energy sources, especially nuclear, emerged, and considerable progress in conventional energy sources led to a new evaluation of economically viable reserves at the same time upsetting the geographical distribution of international trade in energy as new producers appeared. The most striking examples are probably in the hydrocarbon industry which progressively controlled geological uncertainty through a rational approach based on high technology, each year broadening its field of exploration by implementing very efficient underwater techniques. The humorous remark currently made by oil geologists is worth the forecaster's attention: "Beware of statements relating to the exhaustion of hydrocarbon reserves in the next century; similar statements were made for coal a hundred years ago; coal reserves are ever-growing today". (P. Oppenheimer, Shell, WEC Cannes, Oct. 1986).

3.2 Coal

The logic behind the effect of coal cost in the formation of regional or international energy prices is quite clear in the sense that technical costs are accessible with reasonable accuracy, while for certain reasons, the burden of the distribution of mine rents, with the irrational human behaviour this implies, has not been fundamental and durable.

Initially, when closed borders protected activities, the coal industry was a regional industry. Correspondingly, there were no major agreements amongst coal suppliers; prices were formed in each region on the basis of a reasonable cost, and the concept of a rent did not greatly distort these prices. It is true that risks in exploration are limited because the discovery of coal is simpler and cheaper than that of hydrocarbons. In parallel, the concept of reserves is expressed in terms of centuries of consumption, which shows that this is not a real economic factor. Competition on world coal markets only appeared after World War II, incidentally just as Europe was opening up its borders and preparing itself to close down some of its mines.

As a result, coal is not a great world traveller. In 1985, despite a substantial increase since 1978, international maritime trade accounted for only 275 million tons out of world consumption of 3200 million tons (9%); in addition, half of the volume traded was coking coal. Maritime coal transport has therefore not made much progress, in contrast to oil which has long been a great voyager: approximately half the oil consumed in the world travels 8000 km on average and yet remains competitive.

In a nutshell, with large deposits scattered around the world and a system of prices immune - at least for now - to disorientating geopolitical developments, coal over the last few decades has been and will remain the energy source used to calm tension on the energy market and to limit changes in oil prices, albeit oil has progressively taken over from coal on the world market (see Fig. 5).

3.3 Oil and natural gas

Over the past three decades, oil has been the main hydrocarbon fuel (natural gas was thought of by oil prospectors as a by-product). However, if an oil price is to be defined, it turns out that, from the outset and for various reasons, some technical such as the risk and the cost of oil exploration, others geo-strategical, the world price of oil is not rigidly tied to a classical technical cost. (Consequently, for oil, the concept of reserves has an economic meaning. Since 1960, the ratio reserves/annual production is roughly the same, i.e., 30 to 35 years).

A logic of costs could probably be determined, but it is easier to describe it a posteriori than to predict it. Before the second World War, the United States supplied the Western world, and prices were quoted FOB from the Gulf of Mexico. After the war, the Middle East deposits were

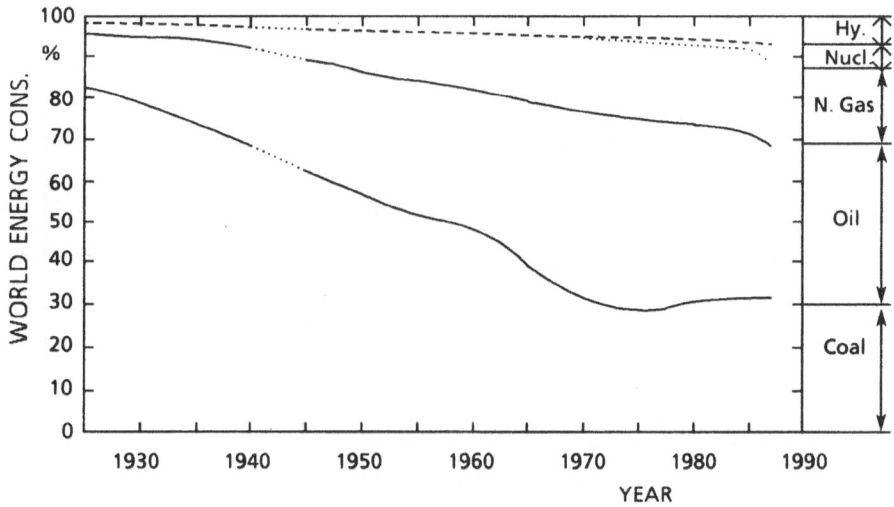

Fig. 5. World energy consumption: primary sources (WEC)

developed and an equilibrium between United States and Middle East prices was created in Europe. This equilibrium changed again when the United States became net oil importers. Similarly, in theory, oil whose marginal cost is equal to the world market price could be found on the international scene at this point.

However, in a context in which politico-oligopolistic structures define scarcity and dominate technical factors in the formation of prices, the nature of costs remains to be defined. The complex history of oil, the repeated confrontations of the biggest economic operators with no great concern for the laws of the free market must then be borne in mind: the large international concerns, the consumer countries, the Third World countries endowed with deposits and, now, the Third World countries without oil resources (see para. 1.2).

The relative strength of social groups in the determination of world oil prices is a basic element; it may indeed be the image of what will happen in the future in international transactions for strategic products such as uranium or even coal. Be it as it may, the effects of these relationships on trends are obviously different from those of a free market. Consider for example, the change in oil structures after the 1973 crisis (see Fig. 6). OPEC's output fell by 45% between 1973 and 1985; its share was taken up by non-OPEC oil; notwithstanding, this only became evident in 1979, at the time of the second oil shock, six years after the first.

Some characteristics explain why the history of natural gas is not as troubled as that of oil, even though it has close economic links to it. For a long time, natural gas was not the prime objective of oil exploration, while its transport costs are much higher than for oil, the profit from deposits much lower. Quite to the contrary: as the possibilities of long-distance transport improved, it enjoyed regular growth, and its share of world supply has now reached 20% (see Fig. 5). With larger reserves than oil, located in zones viewed as politically more stable than the OPEC countries (share of OPEC in proven reserves: 34% for gas and 66% for oil), natural gas appears today as an element regulating the tension on the energy market.

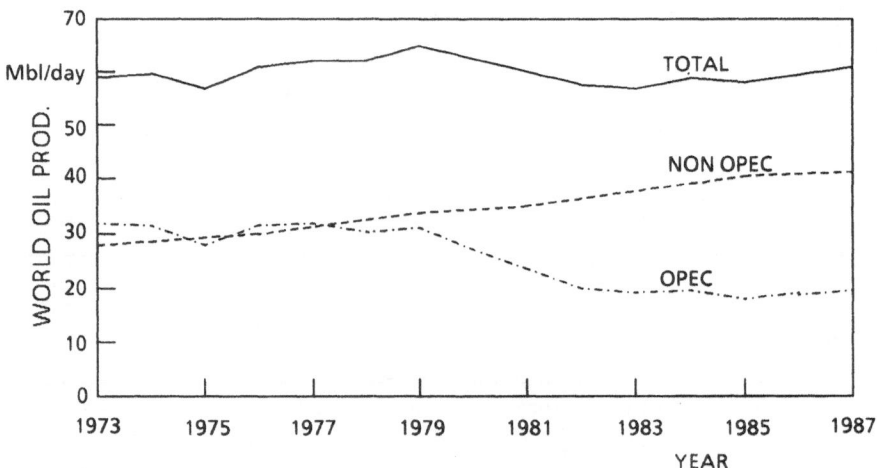

Fig. 6. World oil production (Shell)

3.4 The main new energy sources

Of the energy sources which have appeared since 1974 and whose use could grow in the coming decades, two seem destined to play a role in international trade and the setting of world energy prices: nuclear and solar energy by photovoltaic cells. Their economic breakthrough owes much to the 1973-1979 energy crises.

Nuclear power, with its different variants, has proven its potential, although its limits cannot be ascertained today. From 0.8% of world supply in 1973, its share has increased to 5% today; this increase has greatly contributed to the drop in the share of oil (see Fig. 5). Experience shows that industrialised countries can substantially reduce their dependence on countries endowed with abundant raw materials. There are still two major areas of concern: the public's fear of nuclear hazards has not diminished, quite to the contrary; in addition, the strategic aspects of the market in fissile substances should not be under-estimated (see para. 2.2).

Photovoltaic energy is probably the most economical form of commercial solar energy. It will certainly be much used in remote regions where there is no competition from other energy sources. It will suffer the drawbacks of the irregularity of solar radiation and the cell area needed to produce large quantities of energy.

4. ENERGY AND ELECTRICITY GUIDING PRICES IN EUROPE

4.1 Energy sources in competition: disequilibria and balancing forces

The previous paragraphs have described the mechanisms and factors contributing to permanent market equilibrium and the formation of prices, underlining the differences between theory and practice. The two most important teachings drawn from the 1973 crisis are that:

- There is no long-term risk of energy shortage given human ability to act on both supply and demand. It is true that there is a limit to non-renewable energy deposits, but this limit is always being pushed back, and other reserves are sure to be found, providing a major ongoing research and development effort is implemented. Physicists, in particular nuclear physicists, believe that energy deposits are practically limitless, but this view is not held by economists, who also take cost into account.
- A disequilibrium in as inflexible a domain as this cannot be quickly remedied, the more so as the balance of the forces entering the field increase uncertainty and delay decisions. It would be naive to think that these external phenomena will disappear with increasing international trade. In the energy field, it takes 5 to 10 years to correct a disequilibrium, others will follow. For this reason, forecasters always speak in terms of diversification and flexibility, even in periods of abundance.

The battle of energy prices on the international scene, similar to that of coal and oil over the last 35 years, will repeat itself, of course in different forms. Electrical power stations will be the key focus, since excluding the "captive" energy sources of energy consumption (raw materials for chemistry and the steel and iron industry, fuel supplied for explosion - driven machines), the electricity producer is the biggest consumer of primary energy and will be increasingly so. On examination of the imported oil and coal prices at constant terms feeding electrical power stations in France and plotting the mean price coal curve, it can be

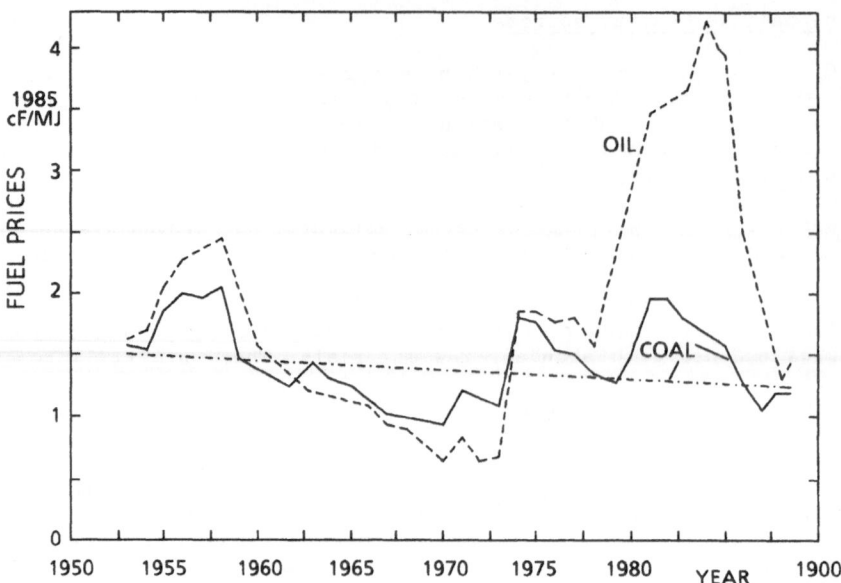

Fig. 7. Imported fuel prices for electricity generation in France (in cen-
times of French franc)

seen that average coal prices have remained roughly stable over the period
considered, although with major oscillations (see Fig. 7):

- in 1956-1958, the price of imported coal - the cheapest fuel - was
  approximately 35% higher than the average price of the period; it was
  par excellence the fuel used in European electrical power stations.

- After 1958, the oil price dropped markedly to become lower than the
  coal price in 1963. Owing to its superior technical properties, since
  1965, oil largely replaced coal in the supply of power stations.

- By 1970, the oil price runs at approximately 50% of the average coal
  price over the period.

- In 1975, the prices of oil and coal were at a par, both approximately
  30% higher than the average price of the period; given the uncertainty
  of oil supply, power stations had returned to coal since 1976 for their
  supplies.

- In 1982, coal was 50% dearer than the average price; oil was 170%
  dearer than the average price; the price of oil was no longer the
  benchmark.

- In 1988, the price of imported coal was quite near the average price;
  that of oil was only 5% higher.

Despite the strong competition levied by oil before 1973, its price
was finally kept in check by that of coal. It is also evident that the
decision-maker choosing equipment must view the prices of the last few
years and their short-term trend with a critical eye. He will make fewer
mistakes by extending the past long period trend in the absence of factors
suggesting a real change in world equilibrium over the coming decades. For
this reason, I am proposing the following benchmark price scenarios for

coal for European electrical power stations for 2050:

- the mean price of the last 35 years,

- this average price increased by 50% maximum, applicable especially over periods of disequilibrium which could last roughly ten years.

4.2 The price of electricity in Europe at the base of the load diagram by 2050

For electricity generation at the base of the load diagram in Europe, the two reference competitors will be classical fuel-fired power stations (imported coal), and fission power plants. Taking as the benchmark the scenario of the average price of coal over the last thirty-five years equal to 100, Table 1 (giving the probable costs of electricity by 2050) can be drawn up. In the second scenario (stresses on fossil fuel prices), the coal-fired stations are not competitive (cost 127) and the guiding price is the cost of the fission nuclear station (120).

These two benchmark prices (100 and 120) can be used to ascertain the conditions for the competitivity of new technologies for base-load electricity generation (fast breeder reactors and nuclear fusion), providing of course that the costs are calculated following the same conventions. The gap between these two long run estimates is strikingly narrow (100 to 120).

CONCLUSIONS

Any new energy production technology must be competitive with the energy sources which dominate the market. The analysis of the conditions of equilibrium of supply and demand yields an appraisal of the long-term price guideline for energy, a benchmark for the appraisal of this competitivity.

The experience of the last fifty years and reflection about the future provide the basic elements listed below:

- Man's capacity for innovation, when confronted with serious difficulties, whether they stem from natural events or social conflicts, is such that there is no need to fear an energy shortage in the long term.

- Nevertheless, periods of major tension induce price rises which can last up to roughly ten years; the effects of alternative sources of adaptation are not felt immediately.

- While energy is essential for the economy, the control of its utilisation is a major factor that henceforth has a strong influence on the balance between demand and supply. Concern for respect of the environment will also act in the same direction.

- In these circumstances, as in the last fifty years, there is no fear of witnessing in the long run factors justifying a major rise in benchmark energy prices. This paper therefore, sets out fossil fuel (coal) price scenarios up to 2050 which are based on either the stability of the average price over the last 35 years, or a 50% rise. The benchmark price of base-load electricity would be in a narrower range (100 to 120).

Table 1. Cost per base-load kWhe (discount rate: 5%)

Scenario	Average price of coal over the last 35 years		Average price of coal over the last 35 years increased by 50%	
	Coal	Nuclear PWR	Coal	Nuclear PWR
Capital charges	29	43	29	45
Operation-maintenance	17	20	17	20
Fuel – transformation	–	24	–	24
– raw material*	54	13	81	31
Total:	100	100	127	120

* i.e., uranium and plutonium for nuclear energy.

– However, this outcome can only be obtained if research and development provide re-assuring intermediate relief from the inevitable exhaustion of traditional mineral reserves. This has been the role of nuclear fission which is now in competition with the conventional electricity generation technologies.

REFERENCES

1. Frish J.R. – Abondance énergétique: mythe ou réalité 1986. Technip – Paris.
2. Giraud A. et Boy de la Tour X. – Géopolitique du pétrole et du gaz 1987. Technip – Paris.
3. Institut d'Economie et de Politique de l'Energie (Grenoble) Energie Internationale 1987 - 88. Economica Paris.
4. Institut d'Economie et de Politique de l'Energie (Grenoble) Energie Internationale 1988 - 89. Economica Paris.
5. International Association for Energy Economics – Proceedings of the 10th International Conference. "Energy and Economic Growth Revisited" Luxemburg 1988.
6. International Institute for Applied Systems Analysis (W. Häfele). Energy in a finite world 1981. Ballinger Publishing Co., London.
7. Jones P.M.S. – Nuclear Power: Policy and Prospects 1987. John Wiley and Sons Ltd., Chichester.
8. MITI – The twenty-first century energy vision 1987. Japan Co-operation Center for Petroleum Industry Development, Tokyo.
9. Percebois J. – Economie de l' Energie 1989. Economica – Paris.
10. World Energy Conference (WEC) – Proceedings of the 13th Congress. Cannes, France 1986.
11. World Energy Conference (WEC) – Technical papers – 14th Congress. Montreal, Canada 1989.

REVIEW OF PLASMA PHYSICS CONSTRAINTS

R.S. Pease

Newbury, Berkshire, RG16 OAW, G.B.

INTRODUCTION

The review covers the recent advances in the plasma physics of the large tokamak experiments, of reverse field pinches, and of stellarators, relevant to their potential to form the basis of a fusion reactor. All these systems are, of course, toroidal magnetic confinement systems. The first two are toroidal pinch discharges and are axisymmetric, whereas the stellarators have no confining toroidal pinch current, but have non-axisymmetric magnetic fields to perform the plasma-confining function.

As we shall see, the reverse field pinches and the stellarators are at an earlier stage of research than are the tokamaks. The paper looks specifically at the tokamak physics for reactors, summarizing the constraints, the improvements needed and the uncertainties.

LARGE TOKAMAK EXPERIMENTS

Physical Characterisation

The principal engineering features of these experiments are listed in Table I. A general discussion of tokamaks is given in Ref. 1. Simple signposts for the principal physics needed are as follows:

a) The pressure balance in the minor cross-section is described by

$$\beta_p I^2 = 2\pi a^2 Kk(\bar{n}_e \bar{T}_e + \bar{n}_i \bar{T}_i) \tag{1}$$

where I is the toroidal current (e.m.u.), a is the minor radius, 2Ka is the height of the minor cross-section parallel to the major axis of the torus, k is the Boltzmann constant and $\bar{n}$ and $\bar{T}$ are the mean number densities and temperatures of the electrons (e) and ions (i) respectively. This equation defines the poloidal beta β_p. For steady state ohmic heating, in the long mean-free-path case, neo-classical theory gives $\beta_p \simeq 0.56$. Additional heating is required if β_p is to exceed this ideal figure, or to reach it if the diffusion is higher than neo-classical.

b) Ohm's law parallel to the magnetic field line is, on neo-classical theory, given approximately by

$$j_{\shortparallel} = E_{\shortparallel}\sigma_{sp}[1 - \alpha(r/R)^{\frac{1}{2}}] + [p/B_\theta(r/R)^{\frac{1}{2}}](2.44/L_n + 0.135/L_T) \tag{2}$$

where $j_{\shortparallel}$, $E_{\shortparallel}$ is the current density and electric field parallel to the

Safety, Environmental Impact, and Economic Prospects of Nuclear Fusion
Edited by B. Brunelli and H. Knoepfel
Plenum Press, New York, 1990

95

TABLE I. Tokamak Experiments

	JET	TFTR	D-III-D	JT60
SITE	CEC CULHAM	USA PRINCETON	USA SAN DIEGO	JAPAN NAKA
FIRST OPERATION	1983	1982	1986	1985
MAJOR RADIUS (m)	2.96	2.48	1.67	3
MINOR RADIUS (m)	2.1/1.25	0.85	1.3/0.67	0.95
TOROIDAL FIELD (T)	3.5	5	2.2	4.8
PLASMA CURRENT (MA)	7	3	3.5	3.2
PULSE LENGTH (s)	35	2	6	10
ADDITIONAL PLASMA HEATING: NEUTRAL BEAM (MW) RF (MW)	10 18	30 4	12 2	20 1
SPECIAL FEATURES	CAPABLE OF EXTENDED D-T OPERATION	LIMITED D-T CAPABILITY	HIGH β	POLOIDAL DIVERTOR

field, $1/\sigma_{sp}$ is the plasma resistivity,[2] α is a coefficient of order unity, R, r_{sp} are the major and minor radii, p is the total plasma pressure, L_n and L_T are the scale lengths of the gradients of number density and temperature respectively.[3]

c) The thermal insulation expected on neo-classical theory is given by

$$n_e \, \tau_E \propto I^2 c_s \, (R/a)^{\frac{1}{2}}/Z_{eff},$$

$$\cong 1.5 \times 10^{20} \text{ m}^{-3}\text{s} \quad \text{for} \quad I = 1.5 \text{ MA}, \ Z_{eff} = 1.$$

Here τ_E is the global energy confinement, c_s is the ion thermal speed, and Z_{eff} is the effective ion charge number. To reach reactor conditions (Lawson's criterion), I > 1.5 MA (see Ref. 4). Magnetic confinement of the 3.5 MeV α-particles from the D(T,n)α reaction requires I > 3 MA.

d) The ohmic-bremsstrahlung radiation limit is given by

$$I \simeq 1.4/\beta_p \quad \text{MA}$$

When the currents exceed this, radiation collapse will occur unless the radiation losses are countered by additional heating.

Duration and Control of the Current

Figure 1 shows a 35 s duration of a 3 MA current discharge in JET, driven by induction.[5] Large currents, 2.6 MA, have also been driven by lower hybrid current drive in JT-60. There are apparently two potential obstacles to the duration and magnitude of the currents. First, emission of impurities at the walls (a build-up of impurities must be avoided); and secondly, disruptions of the current flow need to be avoided. This latter, it appears, is mainly a matter of avoiding radiation collapse. The discharge in Fig. 1 avoids both these pitfalls, but at the expense of a low value of β_p. Figure 2 shows that generally ohmic heated discharges have β_p values well below those expected from the neo-classical value, due presumably to anomalous cross-field outward diffusion of the plasma. Figure 3 illustrates attempts to control the disruption at low currents. Such feedback may be useful for the detection of precursor conditions, but is unlikely to rectify a major power imbalance from radiation cooling.

Ohms Law Parallel to B.

Experimental evidence for the neo-classical Ohm's law parallel to the magnetic field (Eq. 2) is now becoming convincing.[8,9] Figure 4 illustrates agreement between neo-classical theory and experiment for the bootstrap current, i.e., for the pressure-gradient-driven term of Eq. 2; and Fig. 5 illustrates the agreement with theory for the trapped-particle (toroidal correction), i.e., the first term of Eq. 2. The individual effects of L_n and L_T have not been established experimentally.

MHD Stability and β-values

Values of the mean ratio β of plasma pressure to the total magnetic pressure $B^2/8\pi$ have reached 8% experimentally, using additional heating.[10] The values obtained experimentally are related to the Troyon-Gruber formula, with a favourable value (3.5 in the usual units) of the coefficient (Fig. 6). This formula is obtained by assuming that the plasma pressure and current profile is optimum, and that unstable modes of sufficiently low growth-rate can be ignored; kink modes provide the theoretical limit to mean β values.

The theory of both these and of the localized flute-type ballooning instabilities is authoritatively reviewed in Ref. 11. Comparison between theory and experiment of the localized modes is shown in Fig. 7. It would appear that in TFTR the pressure gradients know about this theory. The figure also shows the so-called 1st and 2nd stability regions. As shown in Refs 9,11, there is no impenetrable boundary between the two regions: there is always stability at low or zero shear whatever the value of the pressure gradient. However, in the vacuum region outside the plasma the shear parameter S has the value 2. Consequently, the plasma profile has to get from the zero shear region (characteristic of the centre) to the S=2 region outside, without traversing the danger area of high α and high S shown in Fig. 7. Obviously, a specific calculation for each hypothetical profile is necessary.

Confinement Time

The energy confinement time has exceeded 1 s in JET, but is disappointing in two respects: it is about 10-100 times less than neo-classical prediction (Eq. 3); and it decreases with increasing heating power. As shown in Fig. 8, the confinement follows approximately the Goldstone scaling law, namely,

$$\tau_E \propto I P_o^{-\frac{1}{2}} a^{-0.37} R^{1.75} K^{\frac{1}{2}} \tag{4}$$

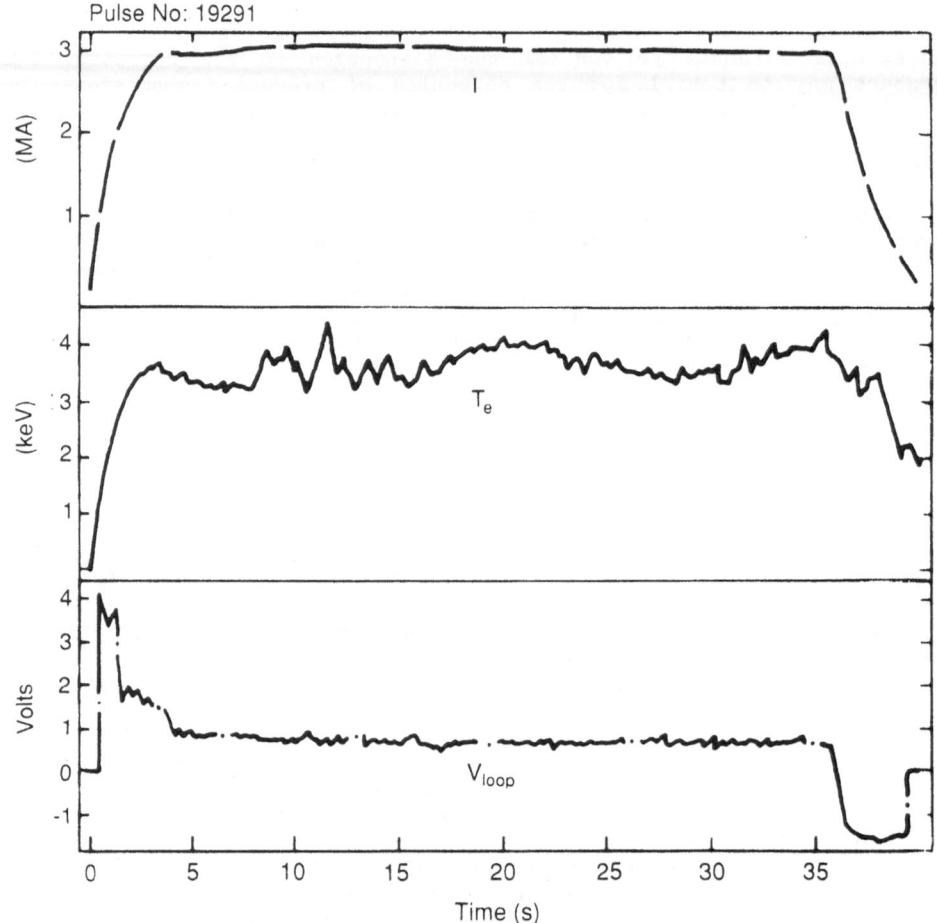

Fig. 1. Long-duration JET pulse, May 1989,[5] showing plasma current I, electron temperature T_e, and loop voltage V_{loop}.

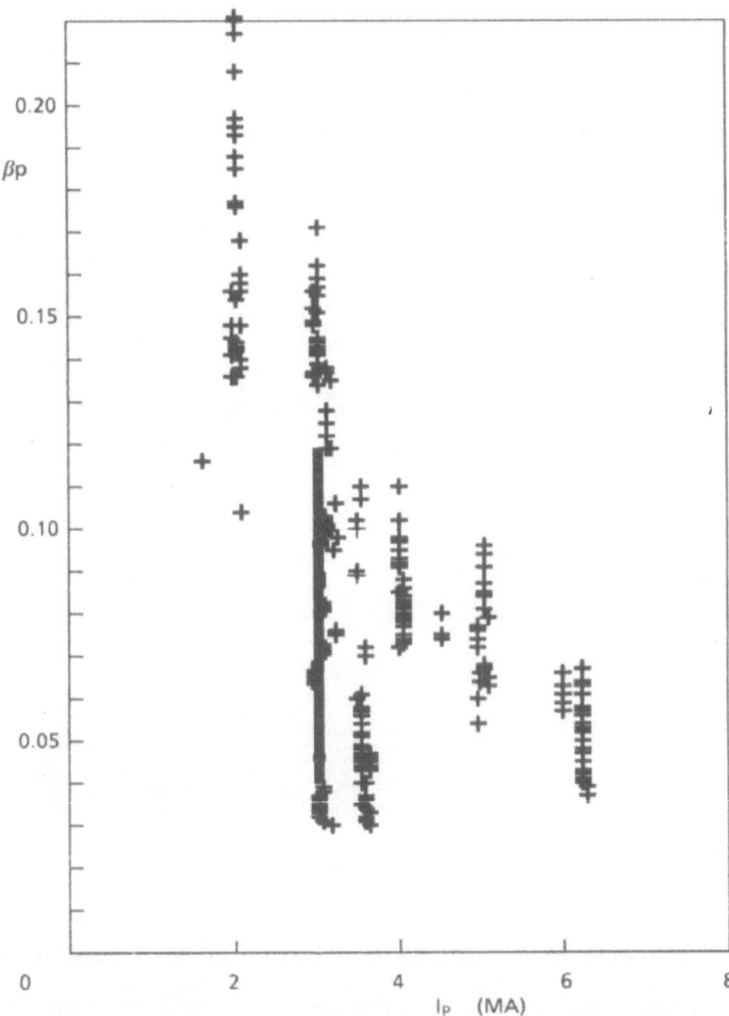

Fig. 2. Steady state, ohmic heating alone; values of β_p vs plasma current I_p observed in JET by ECE and x-ray diagnostics.[7]

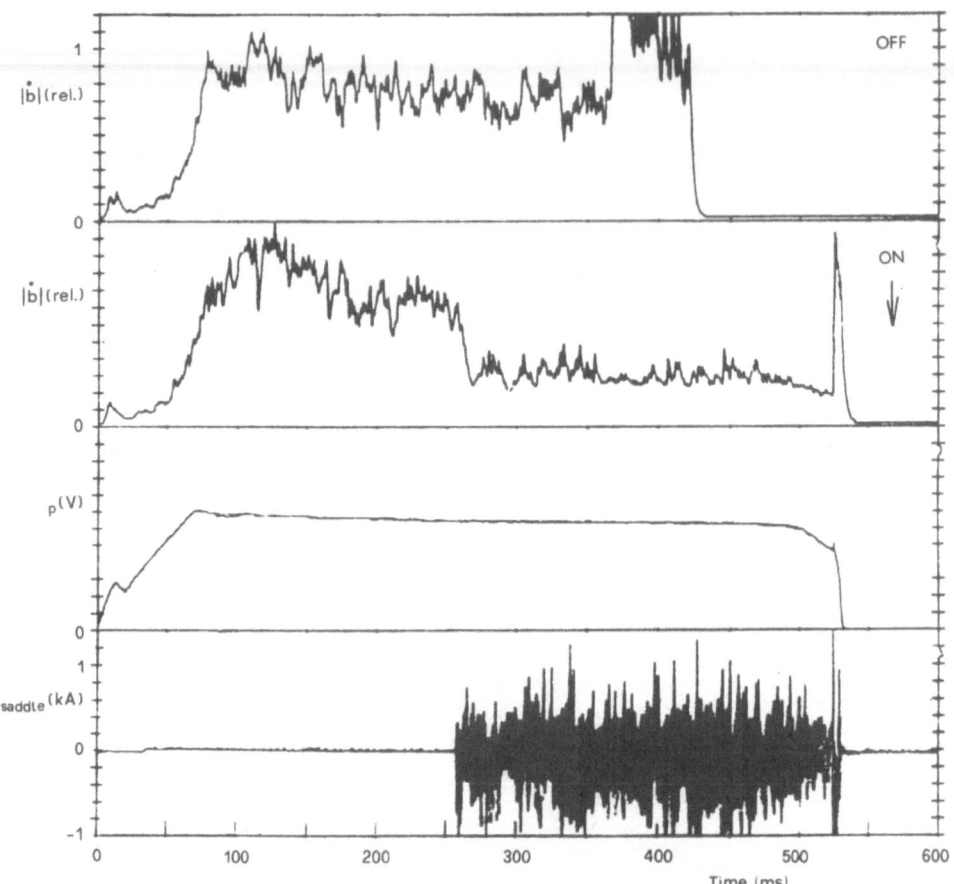

Fig. 3.   Disruptions of the plasma current in DITE showing postponement effects of feedback control currents (I_{saddle}) on $\dot{b}$ signal and plasma current (I_p) (discharge DMI 356.72); top $\dot{b}$ signal is without control, for comparison.[6]

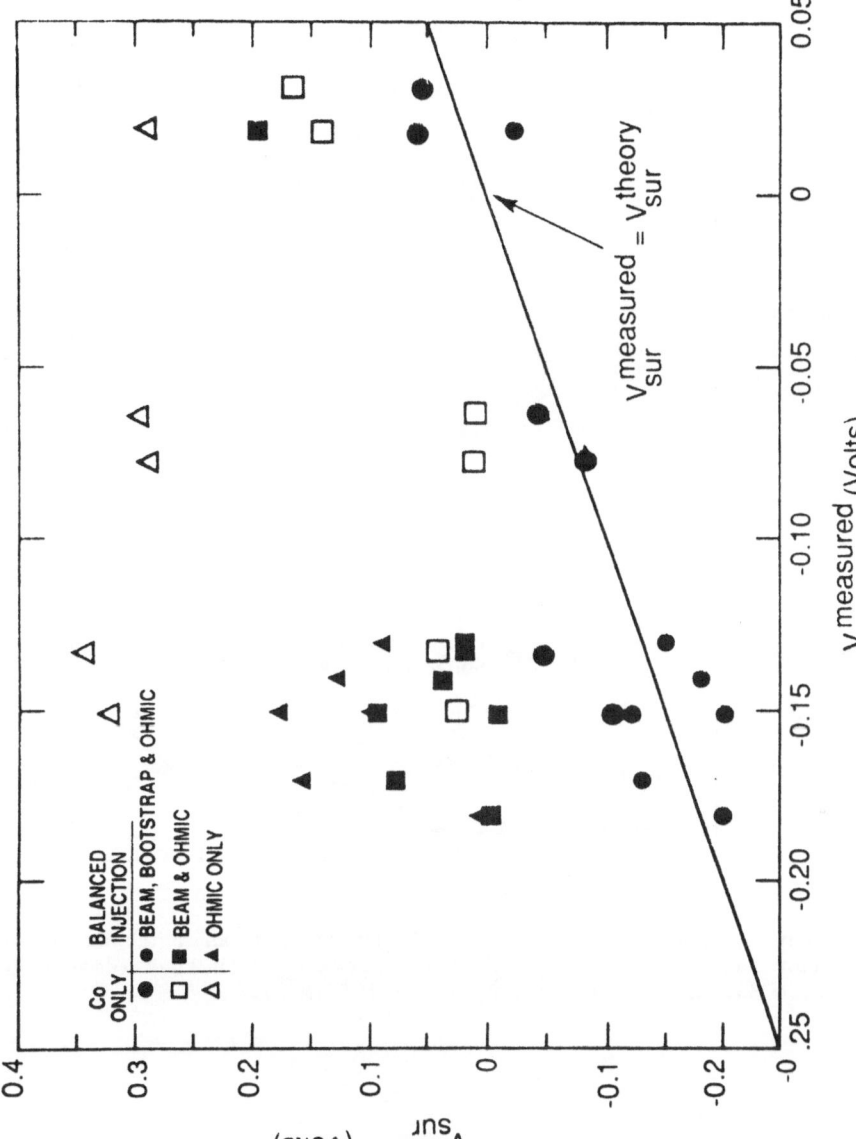

Fig. 4. Comparison of observed surface EMF (volts per turn) in TFTR with neoclassical theory, at high β_p. Bootstrap current term is essential for agreement.[8]

101

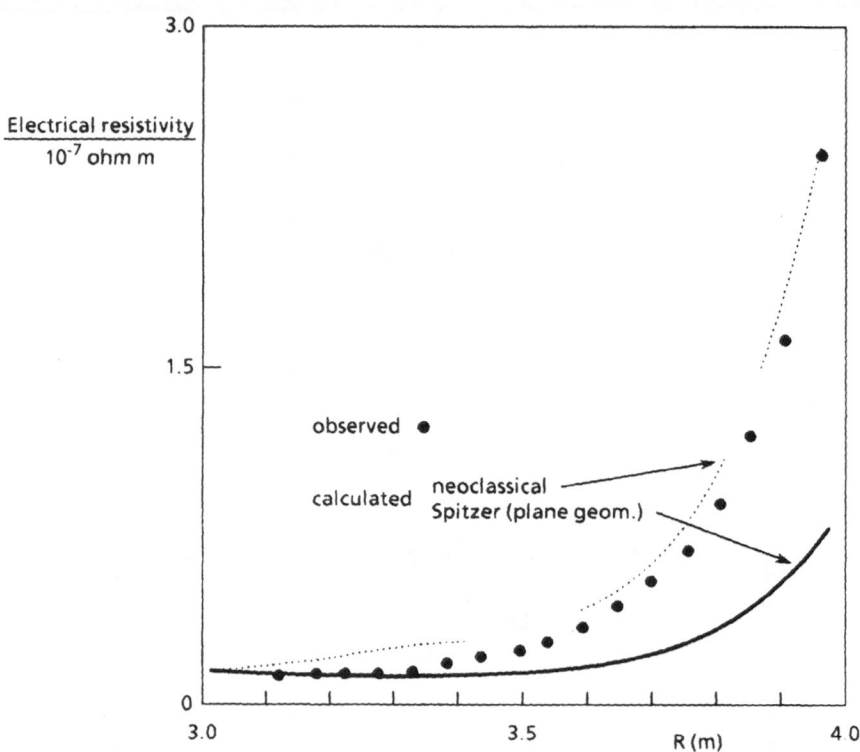

Fig. 5.   Observed and calculated resistivity in JET as a function of major radius, at low β_p, showing agreement with Eq.2 (discharge nr. 4783, with I_p = 4 MA, $\bar{n}_e$ = 2.7 10^{19} m^{-3}).[9]

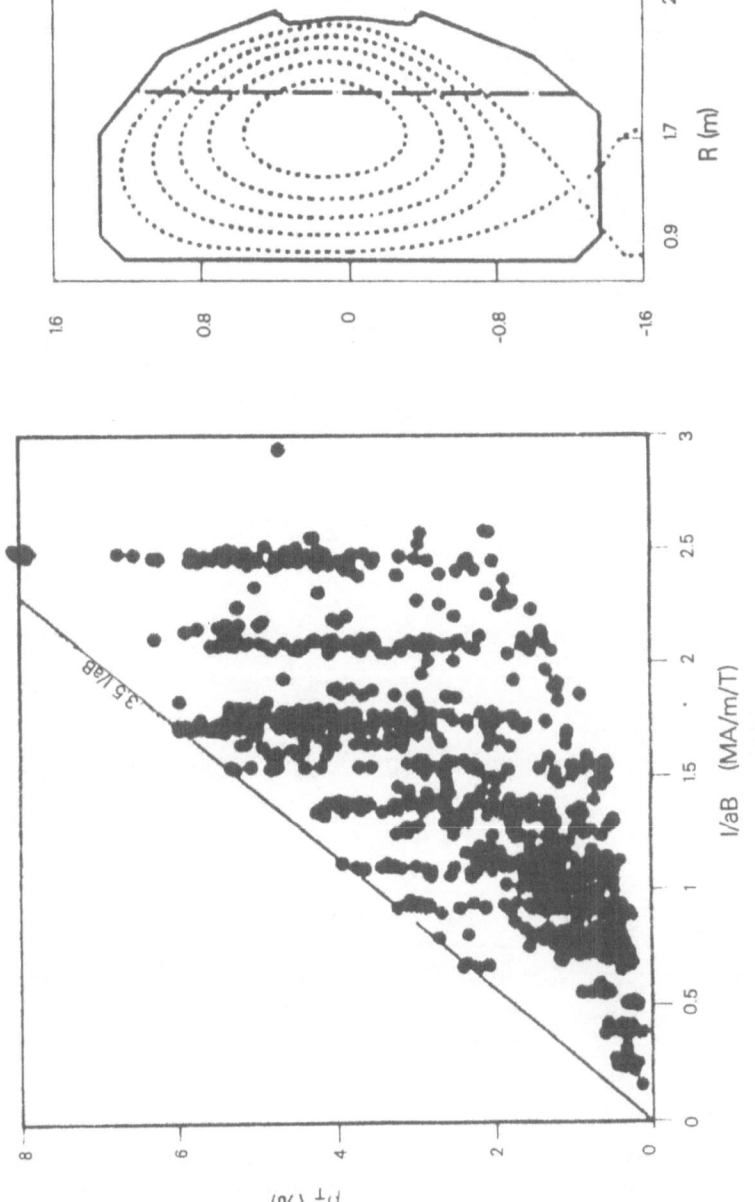

Fig. 6.   Average of total beta β_t observed in DIII-D vs normalized plasma current, compared with Troyon-Gruber Scaling;[10] also shown is an equilibrium flux plot at β_t = 6.8% (typical shape for most high beta discharges).

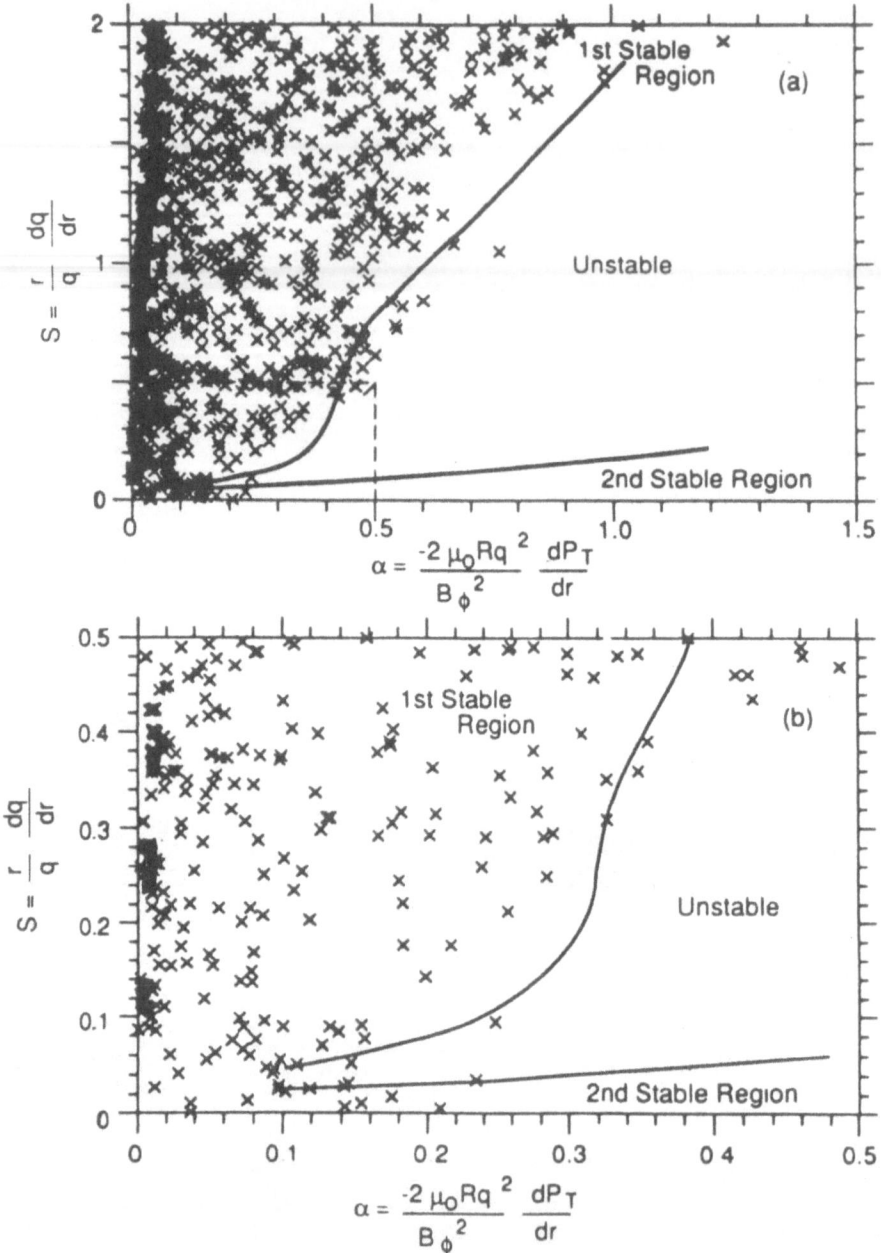

Fig. 7. Observed local pressure gradients, α, in TFTR compared with
stability theory of ideal MHD ballooning, as a function of
normalized shear (S is shear parameter).[8] The unstable region
contains essentially no observed points. The low-S, low-α
section of Fig. (a), which corresponds to the central plasma
region, is shown enlarged in (b).

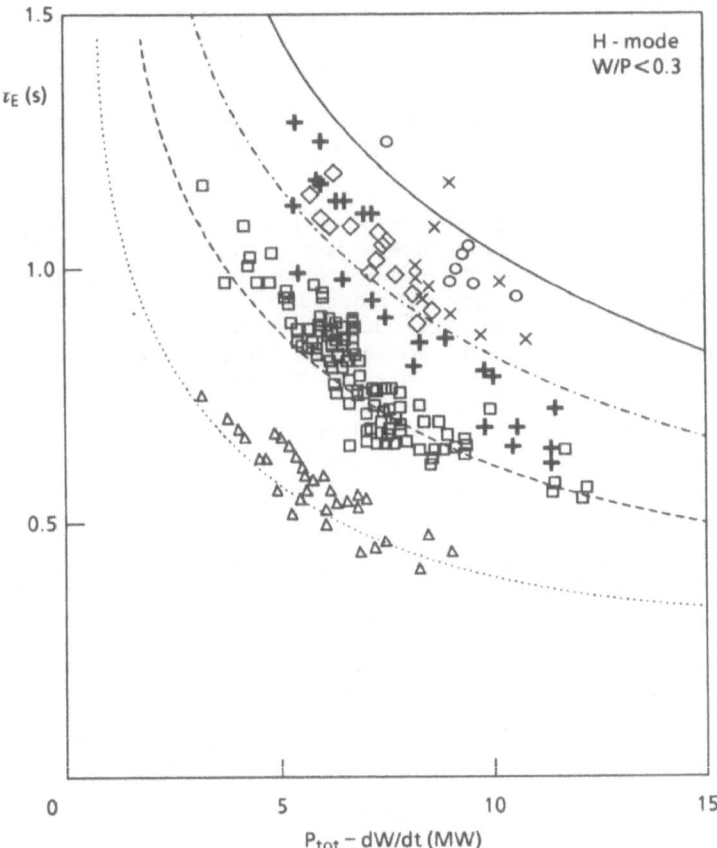

Fig. 8. Observed global confinement time in JET as a function of heating power and current. The dotted curves are 2 x Goldstone scaling (equation 4).[13] The experimental points refer to: "triangles" 2 MA, B < 2.2 T; "squares" 3 MA, B < 2.7 T; "rhombii" 4 MA, B < 2.7 T; "o" 5 MA, B > 2.5 T; "+" 3 MA B > 2.7 T; "x" 4 MA, B > 2.7 T.

where P_o is the total heating power. The important feature of this equation for reactor design is the increase of τ_E with current and major radius. Evidently, the plasma knows about the confining current and its magnitude. Combining equation (4) with the definition of τ_E,

$$\tau_E \equiv 6\pi a^2 \; K \; R \; \bar{n} \; \bar{T}/P_o, \tag{5}$$

yields

$$\tau_E \propto (I^2/\bar{n} \; \bar{T}) \; (R/a)^{2 \cdot 5} \; a^{-0 \cdot 24} \tag{6}$$

Thus it would appear that the quantities governing the confinement are the poloidal beta and the ratio R/a. However, the scaling with the aspect-ratio is not well-tested, and needs to be checked at the expected reactor values.

It is also known that the values of τ_E obtained experimentally can be improved by profile control and central refuelling (H-mode, supershots, etc.) by about a factor of 3 over the normal (L-mode) values given by Ref. 4. The corresponding variations found in JET in the thermal diffusivity are illustrated in Fig. 9. The sensitivity of χ in the outer region is particularly interesting; the change from L to H mode is accompanied by a striking reduction in magnetic fluctuations (Fig. 10). However, there is as yet no satisfactory theory of the observed variation of τ_E with these various tokamak parameters, so that extrapolations, for example with heating power P_o to the 500 MW α-particle heating power foreseen in reactors, cannot be regarded as secure. The present triple product $n_D T_i \tau_E$ for JET and TFTR at end 1988 reached about 2×10^{20} keV s/m^3. Recently, JET experiments have restarted, and have used evaporated beryllium to reduce impurities. Although preliminary results only are available,[15] they indicate that:

- The effect of the 100 Å thick Be layers lasts about 20 shots.
- Oxygen is strongly gettered, and Z_{eff} is reduced.
- Density control is improved.
- The value of $n_D \tau_E T_i$ is doubled to a new record value of 4×10^{20} keV s/m^3.
- Q_{DD} has exceeded 10^{-3} which converts to a Q_{DT} value of 0.35.
- H-mode is achieved with ICRH.

Finally, ion cyclotron heating, which produces a large number of fast moving charged particles (up to 1 MeV), has been used to simulate some features of α-particle heating.[16] It produces L-mode scaling. It can be argued that true α-particle heating has prospects of being less disturbing than ICRH to the confinement: α-particle heating is in the centre of the plasma; it has no wall interaction; and it is symmetric and uniform in real and velocity space. However, in a reactor, the α-particle heating at ~ 500 MW is so much larger than any additional heating, or profile control that can be imagined, that it leads to a profile which cannot easily be altered. No one yet knows if these profiles of n_e, T_e, T_i and $j_{\shortparallel}$ are going to be favourable or not for all the other attributes needed.

REVERSE FIELD PINCHES

Reverse field pinch experiments yield plasma parameters well below those for reactors, except that of β, which is generally in the range 1-10%. The density control problem is different from that in tokamaks. It is strongly determined by wall recycling. Figure 11 shows that even with pellet refuelling the density returns to a preferred value, in times of about 1 ms. The plasma and energy confinement time is about 0.5 ms, much less than that achieved in the large tokamaks. Although this shortfall is

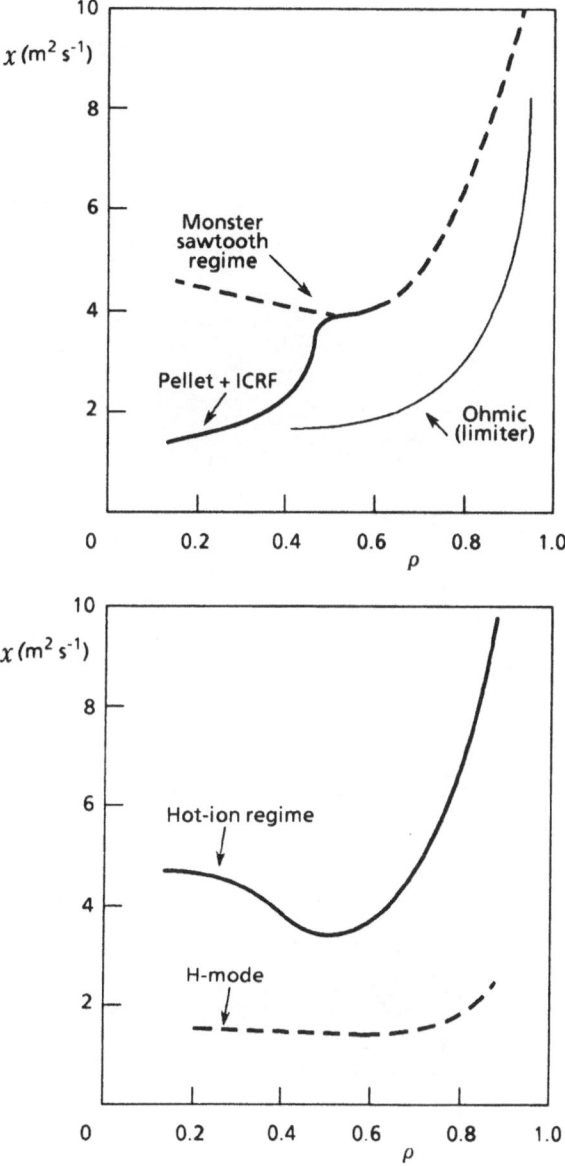

Fig. 9.   Observed local thermal diffusivity, χ, as a function of normalized minor radius ρ, for various regimes in JET.[9]

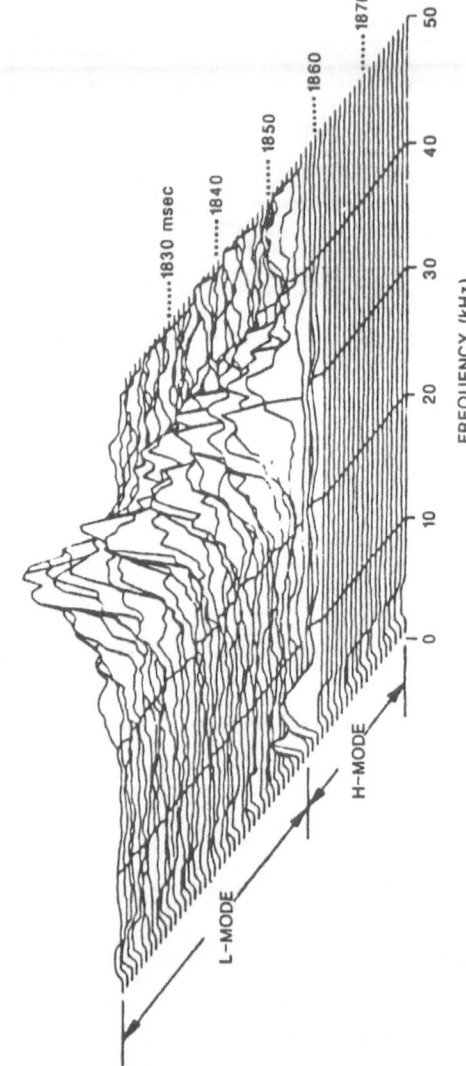

Fig. 10. Magnetic field fluctuations observed at the edge of DIII-D plasma. Note sharp reduction with onset of H-mode.[10]

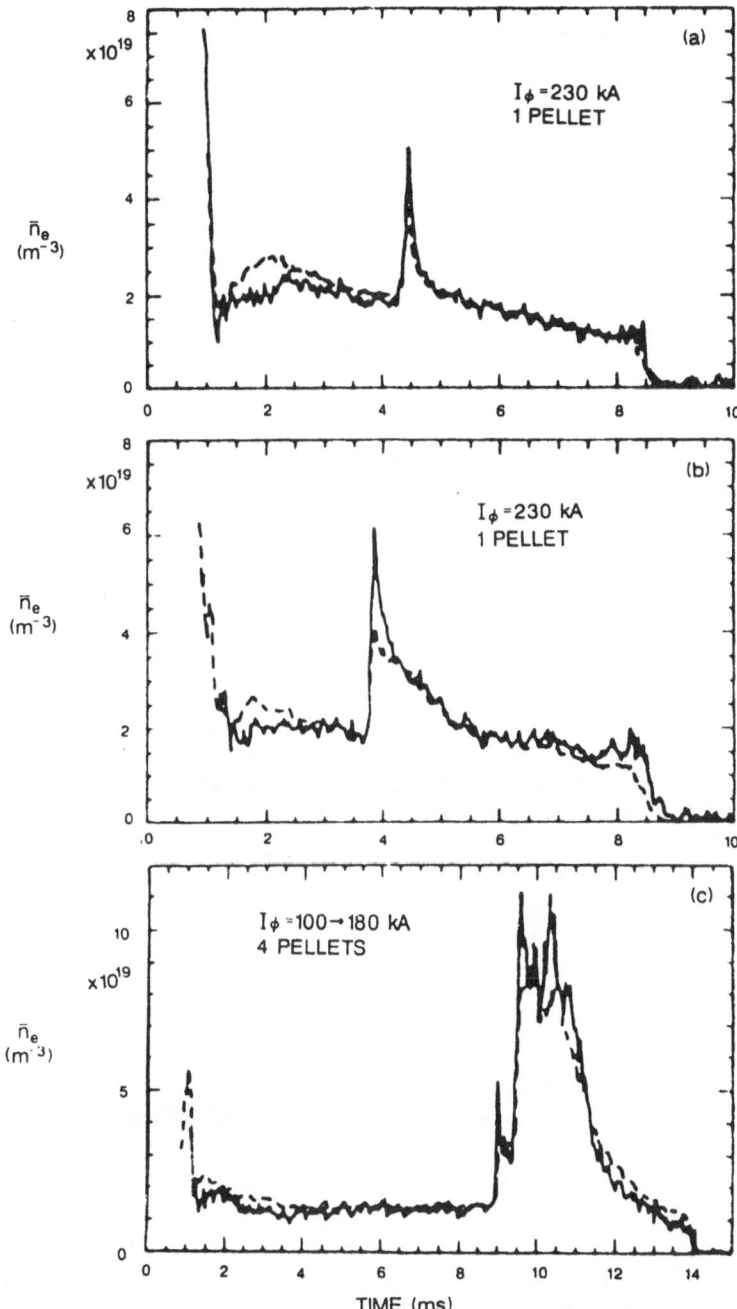

Fig. 11. Electron density as a function of time in ZT-40 reverse field
pinch. Density is measured along the chords r=o (continuous
curve), and r=0.12 m (broken curve). Pellet injection: (a) Deep
penetration; (b) shallow penetration; (c) multiple pellets.

TABLE II.   Stellarator Experiments

	HELIOTRON-E	W.VII AS	ATF
R/a (m)	2.2/0.2	2.0/0.2	2.1/0.27
B_T (T)	2.0	3	2
τ, ℓ	2.5, $\ell=2$	0.4	1.0, $\ell=2$
I_{eq} (kA)	430	120	340
T_e,T_i (keV)	$\sim$ 1,1	$\sim$ 1	$\sim$ 1.2
n_e ($10^{19}m^{-3}$)	2-20	2	3
τ_E (ms)	$\sim$ 10	$\sim$ 5	10
HEATING (MW)	ECRH/0.72 ICRH/2 NBI/2	ECRH/0.4 ICRH/- NBI/- OH/-	NB/0.6 ECRH/0.2
OBS. SCALING	$\tau_e \propto P_{tot}^{-0.64} n^{+0.54}$		$\tau_e \propto P^{-0.65} n^{+0.57}$

partly due to the much smaller size and currents in present experiments, there is little doubt that a model akin to the tangled discharge model – in which the lines of force are chaotic – applies. Plasma and its energy are lost by flow along the lines of force; the energy confinement time is characterized by a variation

$$\tau_E \propto \pi a^2 \bar{n}/I \tag{7}$$

(see Ref. 4). Moreover, unlike the tokamak, the interception of the lines of force with the torus wall provides an "excess resistance" over and above the resistivity $1/\sigma_{sp}$ given in Eq. 2 (see Fig. 12). The major instability responsible is MHD in origin, and is a consequence of sustaining the currents and the reverse field with an induced electric field drive.

New apparatus, RFX and ZT-H, are under construction, and hopefully will tell us whether a favourable scaling can be found at higher (megamp range) currents.

STELLARATORS

The potential for DC operation, and the absence of large plasma currents, are well-known advantages of stellarators. The parameters of two new experiments, W.VII AS (Garching) and ATF (Oak Ridge), are shown in Table II together with those of Heliotron-E, the largest of the earlier generation of experiments. From the value of the "tokamak-equivalent current" we can infer that their confining potential is that of substantial medium-sized tokamaks. W.VII AS is particularly important for the reactor designer, since it is exploring a modular coil configuration,

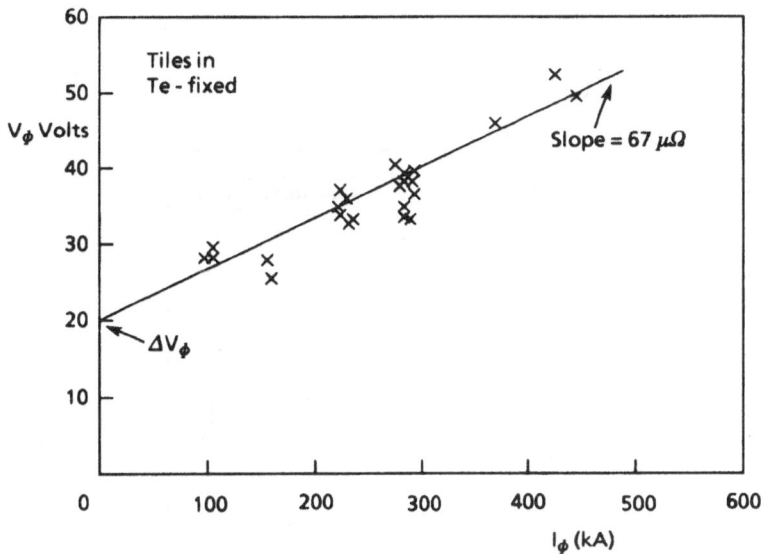

Fig. 12. Observed surface voltage in the HBTX IB reverse field pinch, as a function of current at constant electron temperature (250 + 25 eV). The slope of the dotted line corresponds to Spitzer resistivity with Z_{eff}=2. The finite intercept at I=0 is evidence of non-Spitzer impedance of the discharge.

which avoids the interlinked coil configuration of earlier stellarators and which could therefore be used in a reactor.

Like the reverse field pinch, these experiments (Fig. 13) also show parameters which are rather far from the reactor regime. For our discussion here, the most important of the initial results is the appearance of a bootstrap current in rough accord with theoretical prediction, and which is opposed and backed-off (e.g., by an applied toroidal electric field) to zero net current in the W.VII AS experiments. The observed heat fluxes in all three experiments indicate a non-classical component which, with strong additional heating, follows a degradation of confinement scaling law reminiscent of that found in tokamaks.[19,20,21] (see Fig. 14) Unlike the reverse field pinch, the beta value is low, < 1%, in these experiments. (The highest recorded value in stellarators is 3%) Particularly with the new experiments, it is too early to draw any firm conclusions in these initial stages of work.

Therefore, if we wish to consider real prospects for reactors depending on the achievement of ignition, such as NET and ITER, there is no present choice but that of basing designs and calculations on the tokamak system.

TOKAMAK ADVANCES EXPECTED AND NEEDED

a) Confinement times

Present D-T reactor designs have currents in the range 10 - 30 MA. To achieve ignition, a scaling of 2 - 3 × L-mode is sufficient. This could come (for example) from using H-mode discharges with a margin provided by the ion-mass dependence, the superiority of α-particle heating, or by reduced trapped-particle instability.[22] The extrapolation of a

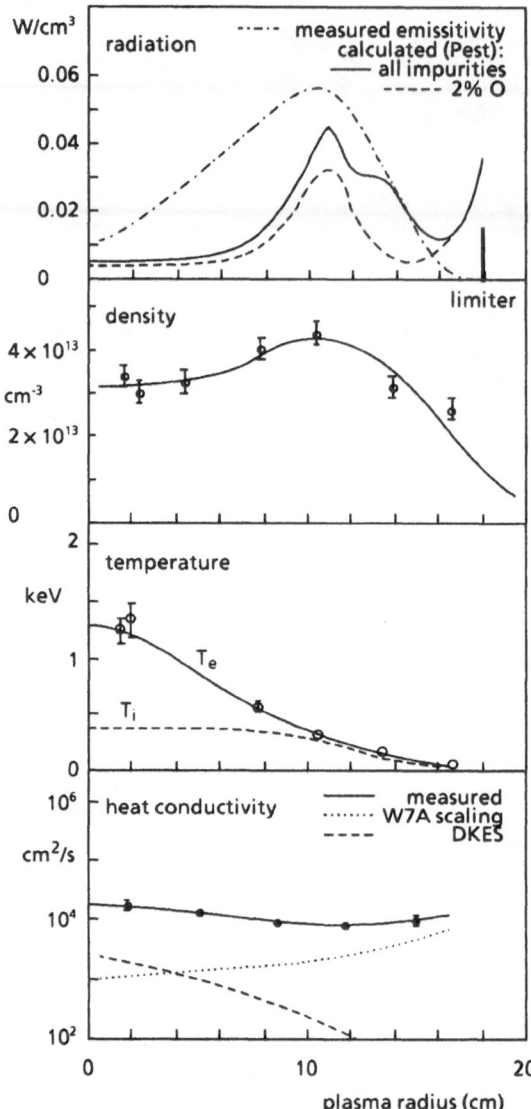

Fig. 13. Temperature, density and radiation from the W.VII AS stellarator
plasma. The observed heat diffusivity is shown as a function of
minor radius as well as the one calculated by the DKES code
(drift kinetic equation solver). Also shown is the curve related
to the old W.VII A stellarator. Bootstrap current of about 2 kA
is suppressed. ECRH heating, 400 kW.[21]

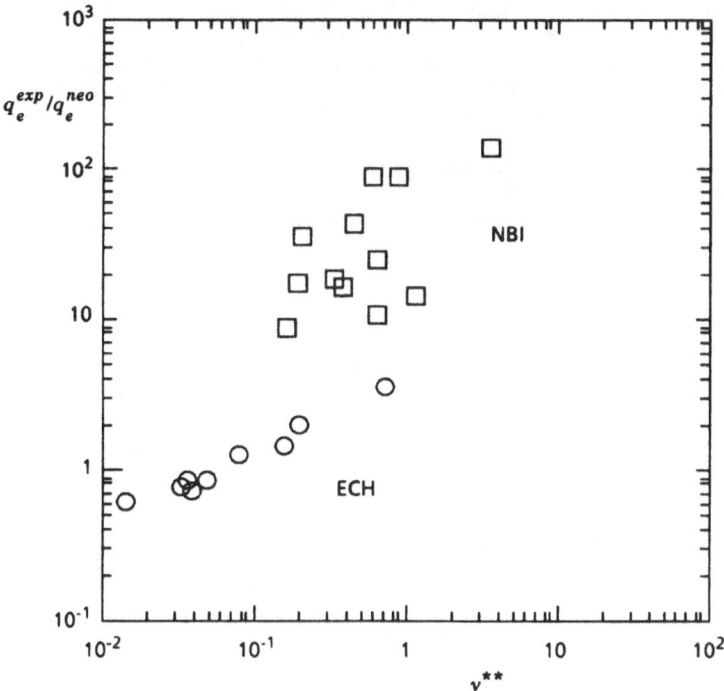

Fig. 14. Observed electron heat flux q_e^{exp}, divided by neoclassical theoretical value q_e^{neo}, vs collisionality parameter, $\nu^{**} = 2\pi\nu/\varepsilon_h^{32} = 2\pi R_o/mV_{th}$ for ECRH and NBI heated plasmas at r = a/2 in Heliotron E.[19]

factor of ten in power and five in current requires underpinning with understanding. There is a very large margin for improving the confinement time towards the neo-classical value. If some modest fraction of this improvement could be achieved, then the reactor designer might be able to rely on ohmic heating alone to reach ignition - a significant simplification.

b) Current Drive

Good efficiency is essential if we are to avoid uneconomic circulating powers in the reactor (>10%). Present efficiency of 0.35 (10^{20} A/W m^{-2}) can be extrapolated to 0.6 in some cases, but 0.7 is needed;[23] ideas exist for efficiencies > 1. But why not go strongly for near 100% bootstrap current drive? Recent theoretical papers[24] indicate that this can be achieved, and be compatible with ideal MHD stability; this again would lead to simplification of the ultimate reactor.

c) β-values

The value of the coefficient in the Troyon-Gruber formula needs to be 3.5, and this has been achieved. With tight aspect ratios, β > 10% used in the ESECOM analyses can be achieved. The limitation is the ballooning-kink-mode, which has a soft (i.e., growth-rate-dependent) theoretical limit. Obviously, we would prefer to have a better margin in hand than the present barely reached reactor values of 8%. The designer can

however rely on ideal MHD theory; but at low growth rates, there is a competition between β and τ_E. The chief need is to have a self-consistent model of the reacting plasma which has simultaneously:
- an energy balance determined mainly by the α-particle heating;
- a pressure and currents profile determined mainly by the α-particle heating and the bootstrap current;
- MHD stability at the required β;
- the appropriate confinement time;
- a high confinement time for OH conditions;
- means of refuelling, and for exhausting the reactor products.

We have been encouraged to hear from Dr. Salpietro (see these proceedings) that he can now see a way towards solving the thicket of interlinked problems associated with the divertor design, which also requires certain limitation on the edge plasma. Finally, it remains to add that 50% enhanced reaction rates could be obtained by nuclear-spin "polarized" D-T fuel, and this could ease the problem of β.

CONCLUSIONS

General

a) The large tokamak experiments give reactor-like conditions: $T_e < 12$ keV, $n_e < 2 \cdot 10^{20} m^{-3}$, $\beta < 8\%$, $\tau_E < 1.3$ s; $n_D \tau_E T_i \cong 4 \cdot 10^{20}$, compared to $5 \cdot 10^{21}$ keVs/m^3 required in a reactor. No other magnetic system has such experiments or can offer such significant near-term prospects.

b) The new reverse field pinch and stellarator experiments (ATF, W.VII AS and RFX) will give important new results for these alternate systems; but the conditions in them, will be still rather far from reactors.

Tokamak Physics

c) There is no apparent limit on current magnitude and duration in a tokamak. But it is necessary to have $q > 2.5$, radiation collapse has to be prevented by additional heating, and for D-T reactors accumulation of impurities must be avoided.

d) Ohm's law parallel to B is neo-classical, at least at $T_e \simeq 2-3$ keV. Bootstrap current and toroidal corrections exist.

e) Plasma pressure gradients achieved are related to ideal MHD stability. Troyon multipliers of 3.5 are achieved. Edge-conditions and ballooning kink affect results sensitively. Better margins between achieved and needed values of β require demonstration.

f) Confinement: L-mode scaling is pessimistic. Both experimentally and theoretically one can do better. But there is no good theory, so that extrapolation with respect to I, R/a, is uncertain.

g) α-particle heating has been simulated by ICRH. In practice, true α-particle heating may be less degrading because it is central, isotropic, and symmetric. In a reactor, α-particle heating will be large and probably inflexible; the profiles it will provide remain to be established.

h) Impurities: low Z-walls and divertor work at present power levels. The extrapolations needed are a matter of engineering calculation and testing.

i) For a reactor, improvements (over present design) to develop are:
- Bootstrap current drive near 100%;
- high β (aspect ratio, second stability regime);
- ohmic heating ignition by improved confinement and spin-polarized fuel;
- α-particle heating to give required profiles.

ACKNOWLEDGMENTS

I am most grateful to Dr. A. Gibson and Dr. D. Düchs of the JET Team for letting me use material they have provided; to Prof. H.P. Furth, Prof. K. Burrell and Dr. H. Renner who have allowed me to draw on material presented at the Venice EPS Conference prior to publication. Dr. D.C. Robinson and J. Hugill have kindly let me use material from their work prior to publication.

REFERENCES

1.   J. Wesson et al., Tokamaks (Oxford University Press, Oxford 1987)
2.   L. Spitzer, Physics of Fully Ionized Bases (Interscience, New York 1962)
3.   J.W. Connor, in Ref. 1, p. 95. (1987)
4.   R.S. Pease, Plasma Physics and Controlled Fusion, 29, 1171 (1987)
5.   A. Gibson, paper given at the Institute of Physics, Plasma Physics Conference, York, July 1989, and Culham News Sheet, June 1989.
6.   J. Hugill, 1989, to be published.
7.   D. Düchs, private communication (1989).
8   H.P. Furth, invited paper, 16th EPS Conference on Controlled Fusion and Plasma Physics, Venice, March 1989.
9.   R.J. Bickerton et al., paper AI-3, 12th IAEA Conference on Plasma Physics and Controlled Nuclear Fusion Research, Nice 1988, and D.V. Bartlett et al., Nuclear Fusion 28, 73 (1988).
10. E.J. Strait et al., paper AII-1, 12th IAEA Conference on Plasma Physics and Controlled Nuclear Fusion Research, Nice 1988, and K.V. Burrell, invited paper, 16th EPS Conference on Controlled Fusion and Plasma Physics, Venice, March 1989.
11. O. Pogutse and E.I. Yurchenko, Reviews of Plasma Physics 11, 80 (1986) (Consultants Bureau/Plenum).
12. See Ref. 8.
13. P.J. Lomas and the Jet Team, invited paper, 16th EPS Conference on Controlled Fusion and Plasma Physics, Venice, March 1989, and M. Keilhacker et al., paper AIII-2, 12th IAEA Conference on Plasma Physics and Controlled Fusion, Nice 1988.
14. See Ref. 9.
15. A. Gibson, private communication.
16. P.R. Thomas et al., paper AIV-1, 12th IAEA Conference on Plasma Physics and Controlled Nuclear Fusion Research, Nice 1988.
17. K.F. Schoenberg et al., 11th IAEA Conference on Plasma Physics and Controlled Nuclear Fusion Research, Vol. 2, 423 (1987).
18. B. Alper et al., Plasma Physics and Controlled Fusion, 30.2, 843 (1988).
19. T. Obiki et al., paper CI-1, 12th IAEA Conference on Plasma Physics and Controlled Nuclear Fusion Research, Nice 1988.
20. M. Murakami et al., 16th EPS Conference on Controlled Fusion and Plasma Physics, Vol. 2, p. 591 (1989).
21. H. Massberg et al., ibid. Vol. 2, p. 631 and p. 635 (1989).
22. T. Okawa et al., post deadline paper, 12th IAEA Conference on Plasma

Physics and Controlled Nuclear Fusion Research, Nice 1988.

23. D.C. Robinson, unpublished report: Tokamak advance, needed and expected; June 1989.

24. Y.I. Kolesnichenko and V.P. Nagornyi, 16th EPS Conference on Controlled Fusion and Plasma Physics, Vol. 4, p. 1279 (1989); J.M. Ané et al., ibid. Vol. 4, p. 1323 (1989).

25. K.V. Burrell, 1989, see Ref. 10.

THE STATUS OF INERTIAL CONFINED FUSION RESEARCH IN THE US [☆]

R.W. Conn

Fusion Engineering and Physics Program,
UCLA, Los Angeles (USA)

Introduction

I thought I would give you essentially a status report of what is happening in inertial confinement fusion (ICF) in the US. First of all, I hope you are all aware that ICF involves the use of a powerful, pulsed driver, with a pulse length that is typically 0.01 ns to 0.1 ns, and driving a very small target (which contains DT fuel) to compressions that are the order of 1000 times liquid density. The driver must do this job with a pulse shaped such that the shocks converge at the target centre, the dissipation of the shock energy heats the core, and the core ignites; the alpha particle range is short compared to the pellet size, allowing the burn to propagate, inertially, through the fuel. The burn-up is typically about a 30%. And that is essentially the physics of how a target should work.

There are, at present, at least four types of drivers being pursued (Table I) and two types of targets. One type of target is referred to as an *indirect drive target*. This means whatever the drive energy, it is used to convert that energy into X-rays, which in turn somehow drive the target. The second target type is a direct-drive target in which the driver power directly illuminates the target. Illumination uniformity is a key requirement.

ICF Drivers

The major program in the US is centred on glass lasers. The largest glass laser is NOVA, which is at the Lawrence Livermore National Laboratory. It was designed to deliver 100 kJ at 1.06 microns, which is the frequency of lasing of a neodynium glass laser. At LLNL they can frequency-quadruple, which means that they can reduce the wavelength to 1/4 μm, with an efficiency approaching 80%. If you were to ask ten years ago what

Safety, Environmental Impact, and Economic Prospects of Nuclear Fusion
Edited by B. Brunelli and H. Knoepfel
Plenum Press, New York, 1990

117

would be the efficiency of a glass laser to deliver 1/4 micron light, people probably would have said, "You will lose at least 75% of the energy in going from 1 micron to 1/4 micron". In ten years of development in glass laser technology, that number has been reduced by a factor of 5. It is really a remarkable technological achievement.

At the moment NOVA delivers 30 kJ at 2500 Å. They had problems with the glass. That glass has been replaced, and they now hope in the next year to get to 100 kJ. What has been achieved so far are compressions of the order 200 times liquid density. Remember that the density of liquid DT is 0.2 g/cm^3. However, there is not simultaneously compression and heating, so in these experiments there is very little neutron yield and essentially no central temperature rise.

TABLE I

ICF DRIVERS

1. **Glass Lasers**

a) Indirect Drive Targets & Glass Lasers (LLNL)

   NOVA $- E_{laser} \cong 30\,kJ$ at a wavelength of

   $1/4$ of $1.06\mu m$

   1. $\rho_{compression}/\rho_{liquid\ DT} \cong 200$ $\qquad \rho_{DT} \cong 0.2\,G/cm^3$

      but no central temperature rise.
   2. No fast electron problems at the scale lengths of the targets employed.
   3. Much work on cryogenic targets.
   4. Relevant underground tests have been carried out.

b) Direct Drive (Univ. of Rochester/NRL)

   $E_{laser} \cong 10\,kJ$ at $\frac{1}{2} \div \frac{1}{4} 1.06\mu m$

   Symmetric compressions are coming along reasonably well - NRL technique of scale-length smoothing.

2. **Pulsed Power Systems (Sandia)**

a) Development of diode to deliver Li^+ ions at several MeV to targets.
b) Target experiments are late by at least 2 years. Four key experiments planned for next 9 months.

3. **KrF Lasers (Los Alamos)**

a) Optical multiplexing for pulse compression and shaping has been demonstrated at ~ 1 kJ level.
b) Wavelength is very good (~2000 Å)
c) Gas phase laser - can be rep rated for reactor applications (still to be demonstrated).
d) Very large physical system.

4. **Heavy Ion Drivers (LBL)**

a) Recent demonstration of multiply-charged (+3) heavy ions.
b) Accelerators remain expensive.

By now, one has been able, by going to short wavelengths, to essentially eliminate all of the problems that had been associated some years ago with fast electrons. The problem was that if the wavelength is too long, the electromagnetic waves couple to the electrons at low density on the outside of the target. This creates fast electrons by various processes. These fast electrons then preheat the core and prevent it from being compressed efficiently. This was essentially the Achilles heal of the CO_2 laser program and is the reason CO_2 lasers are no longer pursued for laser fusion research. The core preheat has been stopped (for the scale lengths of the targets presently being examined) by going to short wavelengths. However, problems may come back again when reactor-size targets are used.

There has been a lot of work on cryogenic targets in the US, and absolutely remarkable progress is being made in technology to produce very complicated, multilayer cryogenic targets for the ICF program. Basically, one is able to make the targets which are needed, though not cheaply.

Finally, there has been a press release indicating that underground tests have been carried out that are relevant to ICF research and that have been encouraging.

The direct drive work is unclassified. It mainly takes place at the University of Rochester and in the Naval Research Laboratory. At Rochester, there is a 10 kJ glass laser. They are working again at half and quarter frequencies. They are running most of the time between 1 and a few kJ on target. Their work is, of course, about symmetric illumination of targets. There have been very important developments at NRL in the last 5 years in this field, as well. If you look back at the late 70's, one was talking about 1 kJ lasers and delivering maybe 1/2 kJ to a target. Now they are delivering 10 kJ.

There are at least three other driver programs under research in the US. There is a whole program at Sandia on pulsed power. The pulsed power program derives its technology entirely from defense applications. We have had some discussions here about how much military and defense applications can help in the development of fusion technology; inertial confinement is almost entirely being developed as a result of military and defense interests. This technology, which is required for the production of X-rays for hardness studies, has been simply taken over and applied for inertial confinement.

The technology of the pulsed power driver was applied first to accelerate electrons and because the electrons did not work as efficiently as drivers, they then started to accelerate ions. Sandia found that hydrogen ions will not work and so have shifted to producing lithium ions from a specially designed diode. The diode provides several MeV lithium+ ions. This program is probably two years late with respect to target experiments. We are still waiting for the first real target experiments to see whether or not this driver technique can properly pulse shape the power to the target in a way that is adequate for achieving reasonable target compressions. Such experiments are planned within the next year and a half.

Krypton fluoride lasers are being developed at Los Alamos. The KrF is a gas phase laser which has an intrinsic efficiency of about 5%. It has the technological features needed for rep-ratability, which is what is needed for a reactor. By the nature of the way it lases (it does not have long-life upper state), it decays right away. Therefore, it is necessary to apply optical multiplexing. Multiplexing is one way to compress a relatively long pulse of laser light into the pulse length and shape needed to drive targets. From a reactor point of view this means a very large laser building because multiplexing is obtained by making the beams travel different distances and then recombining them at the end. The wavelength of KrF is excellent for driving purposes; it is something in excess of 2000 microns. Los Alamos has recently demonstrated pulse compression multiplexing operation of a laser in excess of 1 kJ. There have been no target experiments yet.

Heavy ion beam drivers are being studied primarily in the US at the Lawrence Berkely Laboratory. Heavy ion beam drivers based on large accelerators are going to be very expensive. The higher the charge you can make on the heavy ion, the smaller the accelerator can be. LBL has recently demonstrated that they can produce the heavy ions needed in charge state +3 without much +2.

Let me state clearly that the only real ICF target physics experiments at high power have been done on NOVA with glass lasers.

In conclusion, the status of ICF R&D is as follows: about five driver systems are being developed (which is analogous to five alternative concepts from a magnetic fusion point of view). Good compression ratios and reaction rates in targets are obtained, target fabrication is well advanced, and some underground tests have been done that are encouraging.

ICF Program Plans

Program plans, which are not yet approved but are under serious discussion in the US (they may also have an impact on the plans in the magnetic fusion program), are shown in Table II: Livermore is essentially suggesting but not yet formally proposing to build a laser microfusion facility called LMF. DOE has recently been suggesting they might like to make a decision to start construction as early as 1994. The characteristics of the laser microfusion facility, as conceived by Livermore, are that it would have a laser driver energy of 10 MJ. (The largest laser built up to now is 100 kJ.) The expectation is that a glass laser driver

TABLE II

ICF PLANS

(not yet approved by DOE)

a) Plan suggested by LLNL.

Build Laser Microfusion Facility (LMF) with decision date ~ 1994.
LMF Characteristics

$E_{laser} \approx 10 MJ$

Target gain expected ~ 100
Cost (depends on laser):
(For a glass laser, project cost will be ≥ 1 billion dollars.)

b) Plan suggested by Los Alamos

Wait on 10 MJ LMF - build 1 MJ KrF facility as intermediate step
from NOVA ($\leq$ 100 kJ).
Commit to 10 MJ system only around 2005 if intermediate 1 MJ facility is successful in pellet performance.

will cost at least 1 billion dollars. LLNL believes that with some development they might get this cost down to $750 million. (A glass laser,
Athena, that might do the job is estimated to cost 1.5 billion dollars.)
From physics experiments which can be done at 100 kJ over the next two
years, combined with other information, Livermore feels that they would
have a sufficient data to design the LMF facility with the objective of
achieving a target gain of 100. This means the thermonuclear energy output divided by the laser energy delivered to the target (which is
different from the absorbed energy) is 100.

Los Alamos, on the other hand, has a very different idea. They argue
one needs an intermediate step at about 1 MJ (possibly KrF lasers) to be
close enough to the performance one is actually asking for in this LMF in
order to have a reasonable probability that the LMF will be successful.
The 1 MJ facility could be built between now and the year 2000, and the
LMF facility after that. So these two plans are being argued about at the
moment.

Reactor Issues

Because ICF has been primarily a program funded for defense purposes, there has only been a modest activity (almost entirely at
Livermore and Los Alamos) on the matter of what is a reactor and what are
its characteristics and technological requirements. There is very little
work up to now on the technologies that are needed to make a reactor
practical. There is a recent proposal by the DOE to provide extra money
to the ICF program for the development of reactor relevant drivers, probably particularly for KrF. They might be willing to add as much as 100
million dollars over the next two years under the energy program, as

opposed to under the defense program. Reactor studies were performed in the early 70's when laser fusion was first let out of the bag (Table III). I was involved in leading a study called SOLASE which was published in 1978. It proposed the concept of using a gas filled reactor chamber at a pressure of the order of 0.1 mtorr to control the thermonuclear explosion, i.e., to lengthen the time of the explosion shock. The assumptions we made at the time were that the driver energy should be 1 MJ and that the gain could therefore be 100. We were not specific about the laser - we simply said it had to be a gas phase laser such that it could ensure repeatability, and it needed an efficiency on the order of 10%. The KrF laser begins to fulfill some of these requirements, but probably a gain of 100 will require a laser energy of at least 5 MJ. So one of the things that has continued to happen in the ICF program is that the minimum energy required to ahieve high gain has been going up as a function of time.

In the period 1980-1985, people began to realize that you do not need very low pressure within the reactor chamber. They began to design reactors whose chambers were filled with falling lithium walls (*waterfalls*). Inside the chamber you had flowing liquid metal streams that would be relatively thick (mainly 5-6 neutron mean free paths), where most of the neutrons are stopped and their energy is deposited. The liquid walls themselves absorb essentially all the shock energy, so there is no shock transmission to the chamber walls. The neutron flux at the chamber wall is very low and the radioactivity generated in the chamber is dramatically reduced.

TABLE III

ICF REACTOR TRENDS

ICF is not funded now as an energy program - work on reactor concepts is much less mature than in magnetic fusion.

1978 - SOLASE (graphite cavity and gas fill to 0.1 mtorr to control shocks from fireball)

$E_{laser} \approx 1 MJ$; $G \approx 100$;

1980 - 1985 Lithium fall designs - use ~0.6 m thick falling lithium to protect structure - achieve low activation - provide breeding and energy capture.

1985 - Present - CASCADE: solid SiC granules falling in a rotating reactor chamber - to achieve characteristics of Li-fall designs without safety concerns associated with the liquid lithium.

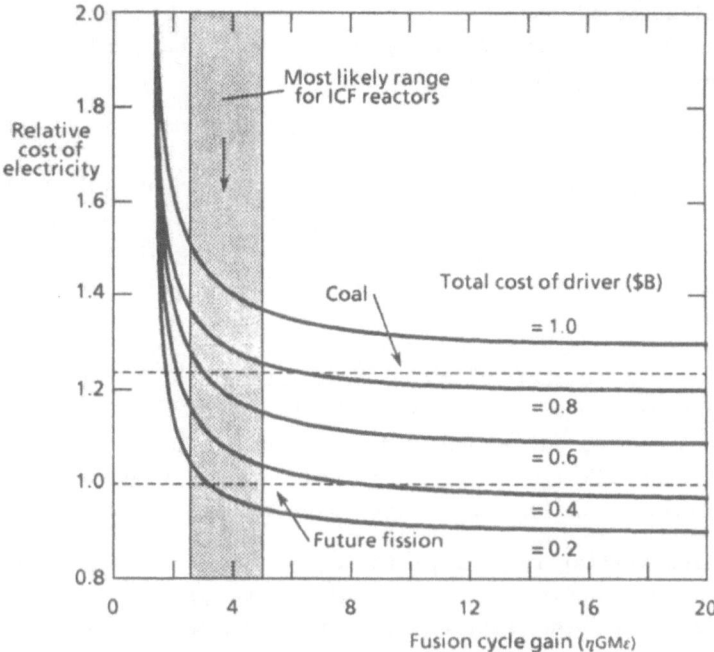

Fig.1 The relative cost of electricity for an ICF reactor as a function of the fusion cycle gain parameter.

However, many important questions remain open - how to really deliver the target, do the tracking, fire the laser at the appropriate time, ensure that the liquid wall always reestablished itself properly on the time scale between pulses, and so on. Remember that in a reactor, to get 1000 MW electric with 1 GJ per pulse, one typically has 200 to 250 ms between shots to reestablish the inside of the reactor chamber to the conditions required to propagate the laser. Lower yields mean short terms between shots.

In 1985, having become concerned about the liquid metal safety problems, the ICF reactor designers began to develop a new concept called CASCADE, which is based on the use of solid particles or granules. My image of it is a *cement mixer*. You have basically a funnel and it mechanically moves; the granules come in from the top and are kept against the rotating wall, then come out at the bottom. The explosion of course takes place in the middle and the granules absorb the energy, protecting the wall.

The present status of reactor design for ICF is primitive relative to where we are in magnetic fusion, primarily because the program has not been energy oriented. It has provided limited funding.

An indication of what has to be achieved in ICF follows from Fig. 1, where the relative cost of the electricity to be produced is plotted against the fusion cycle gain

$$\eta GM\epsilon$$

where η is the driver efficiency, G is the target gain, M is the blanket energy multiplication, and ϵ is the power cycle conversion (*thermal*) efficiency. A realistic, maximum set of parameters at present could be $\eta \simeq 0.05\text{-}0.1$, $G \leq 100$, $M \leq 1.3$, $\epsilon \simeq 0.4$. The likely total driver energy and cost is 5-10 MJ and 0.5-1.5 billion dollars (Table IV). On the graph this results in electricity costs at or above those obtainable with coal. Clearly, such considerations are at best indicative, but they can provide some insight into the sensitivity of the various parameters and processes. For example, pulsed power drivers can be 25-35% efficient. On the other hand, the likelihood of being able to get high gain with pulsed power is lower, so the product of ηG is probably the same.

Following similar reasoning, it follows that one can afford to spend something on the order of 2 to 3 dollars to produce each target. The smaller the yield, the smaller the amount of money one can spend per target.

Considering that each shot (with a rep rate of 4 Hz) yields the equivalent of 1/8 to 1/4 of a ton of TNT explosive, the isolation of the driver facility from the reactor chamber and the reactor building is a key safety issue, as well as a practical problem. If the driver is a KrF-type laser, that laser building is going to be very large (100's of meters by 100's of meters). If this building had to be contained, for example, the cost implications would be phenomenal. So it is really imperative to isolate the reactor and laser buildings, yet that may not be easy because the isolation of the optical path has to be done with optically transparent material.

TABLE IV

KEY ICF REACTOR ISSUES

a) Target fabrication $\leq$ 2 dollars/target.

b) Target injection, targeting, tracking, and orientation control.

c) Rep rate of driver $\simeq$4 Hz.

d) Driven energy $\leq$5 MJ.

e) Target gains of 100 $\Rightarrow$
Yields are 1/8 to 1/4 ton of TNT equivalent.

f) Practical approach to rep-rated chamber design.

g) Isolation of driver facility from reactor chamber and reactor building. A safety as well as practical problem.

h) Isolation and protection of last optical element for laser drivers.

TABLE V

SAFETY, ENVIRONMENT AND LICENSING ISSUES

a)   Targets must be unclassified - not yet clear this will be possible.

b)   Relationship to nuclear weapons may impact licensing and public acceptance.

c)   Cascade-type designs can be very low activation - waste disposal could be dominated by materials used in targets.

d)   Separation of driver building and reactor building is possible - maybe. It is probably required and the issue will be if holes can be allowed from reactor building to driver building.

e)   Very low afterheat levels in chamber in CASCADE designs ⇒ safety features may be enhanced (0.24 G Ci at shutdown, 0.008 G Ci one day after shutdown. These are ≈3 to 1 x less than MFE.

f)   Fractional burnup in targets is ≈30% ⇒ smaller tritium in recirculation system and lower inventories than in MFE.

Safety and Environmental Issues

Now some considerations on the safety, environmental and licensing issues (Table V). The relationship to nuclear weapons may impact licensing and public acceptance. That is a nonnegligible point. There is a much closer relationship between inertial confinement fusion and weapons than there is between magnetic fusion and weapons. Cascade-type designs, if they were to prove feasible, could lead to very low activation, and waste disposal could be a minor problem, with the possible exception of target material activation. The separation of the driver building and the reactor building can pose a safety and environment licensing issue because of the connections between the buildings. Probably this can be solved, but it is an issue. In the Cascade designs there is very low afterheat in the chamber so safety is enhanced. Finally, the fractional burnups in the targets, if these designs work at all, will be intrinsically higher, typically 30%. This means that the tritium inventory problems and the tritium recycling will be lower relative to what we are considering in magnetic fusion.

Concluding Remarks

Summarizing, here are the main points. Very little work has been done in ICF as far as reactor design and safety (and classification) issues are concerned by comparison to magnetic fusion. As is usual at the very early stage of a project, some very optimistic prospects have appeared and the ideas are attractive. We have to continue to see how feasible those ideas are. What is needed is a better characterization of the reactor concept to be able to begin safety and environmental studies. If Livermore is right and the data base is sufficient to design a Laser

TABLE VI

MAIN CONCLUSIONS

- Very little work has been done so far on ICF reactor safety, environment, licensing and classification issues.

- As usual at an early state, some very optimistic prospects appear possible.

- Needed is a better characterized reactor concept around which to begin safety and environmental studies.

- Classification is essentially a defense/political issue, despite desires of energy enthusiasts.

- Military/defense applications may provide money to develop ICF as an energy source through first demonstration phase (as in LWR's).

- Defense may drive technology for low activation materials.

Microfusion Facility, then it is important to begin the reactor activities seriously. Classification is essentially a political and defense issue, despite the desires of the energy enthusiasts, and will remain a major difficulty for ICF energy applications.

Finally, I wanted to comment about a point which may have a big impact on the development of fusion in general, although it may be clearer in ICF than in magnetic fusion. Defense provides the money for the development of ICF technology, not unlike the situation with fission reactors developed for the submarines and aircraft carriers. The initial monies are being spent by the military, but later there can be a civilian application or spin-off. This is an important point. In addition, there can be other technologies of use to fusion which are being developed for defense purposes in other fields. An example is isotopic and elemental tailoring of materials to allow the use in fusion of well-characterized materials such as steels, but with much lower activation levels. In the last 10 years, technological advances have been dramatic enough that the prospect of being able to isotopically tailor metals is probably a reality. These materials could be used in targets and in the blanket structure. The US program will spend billions to build a laser isotope separation facility for military applications in the next several years, and this same technology can be used (in fact it is easier) to isotopically tailor metals for use in fusion. In short, defense developments will push forward both ICF and technologies such as isotope separation which will aid the development of fusion for energy applications, without requiring money from the fusion energy program.

EDITORS' NOTE [*]

During the seminar, Prof. R.W. Conn very kindly agreed to prepare (on one day's notice) and give a lecture before the participants on the ICF program in the US. This paper was prepared by the editors from the taped lecture and is reprinted with the permission of the author.

III. FEASIBILITY PROBLEMS

MAIN ISSUES IN FUSION REACTOR SYSTEM ENGINEERING

E. Salpietro

The NET Team
c/o Max-Planck-Institut für Plasmaphysik
D-8046 Garching bei München

INTRODUCTION

The Tokamak reactor will be one of the most complex machines man has ever built. The research and development needed to demonstrate its feasibility encompasses the areas of physics, technology and engineering and the solution will be a good compromise between the requirements of these three different fields. It is not advisable to give priority to the needs of any one of them because good performance in all three areas is essential for the realization of a successful fusion reactor.

The main issues to be considered in the development of the fusion reactor are cost, safety and environmental impact. Safety has become the main concern for the production of energy from nuclear fission. Environmental impact has started being an issue of great importance not only for nuclear energy sources, but also for conventional energy sources (e.g. green-house effect). Cost is usually considered the most important parameter, but it might not remain so in the twenty first century. Today we can already see a concern for safety and environmental issues becoming more important than cost savings, especially in the case of nuclear fission, which, even though cheap, is not welcomed because of concern about its radioactive waste and safety. Because of this increasing environmental awareness, fission reactors are now being designed with more emphasis on safety rather than on cost savings.

Therefore, engineering solutions for fusion power generation and its technological development must be focused on the development of a safe, environmentally acceptable and competitively priced reactor _in this order of priority_. For example, although high power density in the reactor seems to lead to lower costs, it is likely to require more difficult and less reliable technical solutions, which can have a strong impact on plant safety. Moving towards high power density therefore violates the above priorities. Other considerations are the plant unit size and the operating conditions. These are mainly determined by plasma physics constraints. However, clever engineering and good technological solutions can reduce the unit size and mitigate unattractive operating conditions (e.g. by means of efficient current drive and heating systems).

Safety, Environmental Impact, and Economic Prospects of Nuclear Fusion   129
Edited by B. Brunelli and H. Knoepfel
Plenum Press, New York, 1990

DESIGN CONCEPT DRIVERS

Whilst the concept of the fusion reactor should allow all systems to work at optimum conditions, there are a number of characteristics which have a major influence on the design of a reactor.

The maintenance procedure for replacing plasma facing components has a big impact on the reactor concept. The vertical maintenance scheme allows one to have a more compact machine with less magnetic energy especially in the poloidal system. The requirements for fully remote maintenance have a significant influence on the machine's design. Components are generally classified in two divisions: 'Semi-permanent' and 'replaceable' components. The semi-permanent components are not expected to be removed during the machine's lifetime. Their failure would imply a replacement time of about 1-2 years, and their removal should be made as easy as possible. The components that will be replaced during the lifetime of the machine, including all the nuclear components (divertor, first wall, blanket), must be designed for rapid removal and replacement. Removal of the larger segments is by a vertical/oblique lift, whereas smaller components such as divertor plates can be accessed through the equatorial ports.

With plasma elongation, the more elongated the plasma, the more compact the reactor. However, a vertically elongated plasma is vertically unstable, and the more elongated the plasma, the more unstable it is. To operate with as elongated a plasma as possible, active control coils have to be installed inside the vacuum vessel as close as possible to the plasma in order to be most effective and to reduce power supply requirements and AC losses in the magnetic system.

Another essential parameter, more so for the power reactor than for the next-step device, is the blanket tritium breeding ratio. The local breeding ratio can be well above one, but an overall breeding ratio of one could be difficult to achieve, due to the tritium recovery efficiency and the limited blanket coverage. In order to reduce the breeding ratio required, it is possible to work with a tritium depleted fuel mixture in the plasma without penalising the power density too much.

The unit size of the reactor is determined by the confinement, the beta limit and the impurity content. These parameters determine how easy it is for the plasma to ignite. At present the most uncertain parameter is the impurity content and typically an increase of one percentage point in low-Z impurities requires an increase of about 1 MA in the plasma current needed to achieve ignition.

DESCRIPTION OF A TYPICAL FUSION REACTOR

Prototypical of a fusion reactor are the ITER and NET machines whose parameters are shown in Table I[1,2]. ITER/NET is proposed as a machine typically producing about 1 GW of fusion power, but if operating at the Troyon beta limit it is able to produce about 2 GW of fusion power. Taking into account the energy multiplication in the blanket, a total thermal power of about 2.5 GW would be produced. The electric power required by the plant during the burn is about 0.2 - 0.5 GW electric (depending on the chosen operating conditions). However, during the transients the peak power is about 1 GW. Thus ITER/NET could produce in principle about 200-600 MW electric power if 2 GW of fusion power can be handled by the plasma facing components.

Table 1

Major Parameters of Next-Step Machines

	NET	ITER
Major radius (m)	6.3	5.8
Minor radius (m)	2.05	2.2
Aspect ratio	3.1	2.6
Plasma current (MA)	25	22
Toroidal field on axis (T)	6.0	5.0
Plasma volume (m^3)	1100	1030
Fusion power (MW)	1100	1000
Neutron wall load (MW/m^2)	1.0	1.0

The overall layout of ITER/NET is shown in Fig. 1. The heart of the machine is formed by the plasma, shown here with a "double null" configuration (although a "single null" is also possible). The plasma is surrounded by the first wall, typically 15 cm from the plasma surface. Initially, this will be largely protected by graphite tiles against thermal radiation and particle bombardment from the plasma. Above and below the plasma are the divertor regions, where the majority of the charged particle loading occurs. The lower one is connected to 16 pumping ducts (situated between each of the toroidal field coils) and the heat loading and erosion by sputtering is taken on the divertor plates, which have either graphite or refractory metal protection.

Magnetic confinement of the plasma is provided by two sets of coils. The inner set, consisting of 16 identical D-shaped coils, generates the toroidal field (TF) in the plasma region (Fig. 2a, 2b). The outer set consists of 7 up-down pairs of circular poloidal field (PF) coils coaxial with the machine's vertical axis (Fig. 2c). These generate magnetic fields that keep the plasma in position and also act as the primary circuit of a transformer that drives current in the plasma's 'secondary' circuit. Each of the TF coils can be removed without disturbing the others and only need the removal of one PF coil (the upper "P5 coil"). All the PF coils can be independently removed. The solenoid in the machine centre is free to be withdrawn vertically and the coil under the machine can be lowered between the main support legs.

The space between coils and the first wall is occupied by the primary vacuum barrier that provides the plasma vacuum, (and also acts as the main tritium containment) and two levels of shielding for the coils (Fig. 3). The innermost shield extracts most of the heat from plasma-generated neutrons and can also serve the function of breeding tritium (from lithium compounds) for use in the plasma. In a power reactor all the tritium must be generated in this breeding blanket therefore more space will be required for it than in ITER/NET. The outermost shield is primarily a steel/water mixture and exceptionally has lead or tungsten in critical areas. The combined shield is needed to keep the nuclear radiation dose to about 10^9 rads at the end of the machine lifetime to avoid damage to the organic (epoxy) insulation/bonding between the conductors in the coils. The coils themselves are contained in a secondary (cryostat) vacuum, which has as its outer boundary the main biological shield for the machine. This consists of a thick concrete wall with a metal inner lining to provide the vacuum seal.

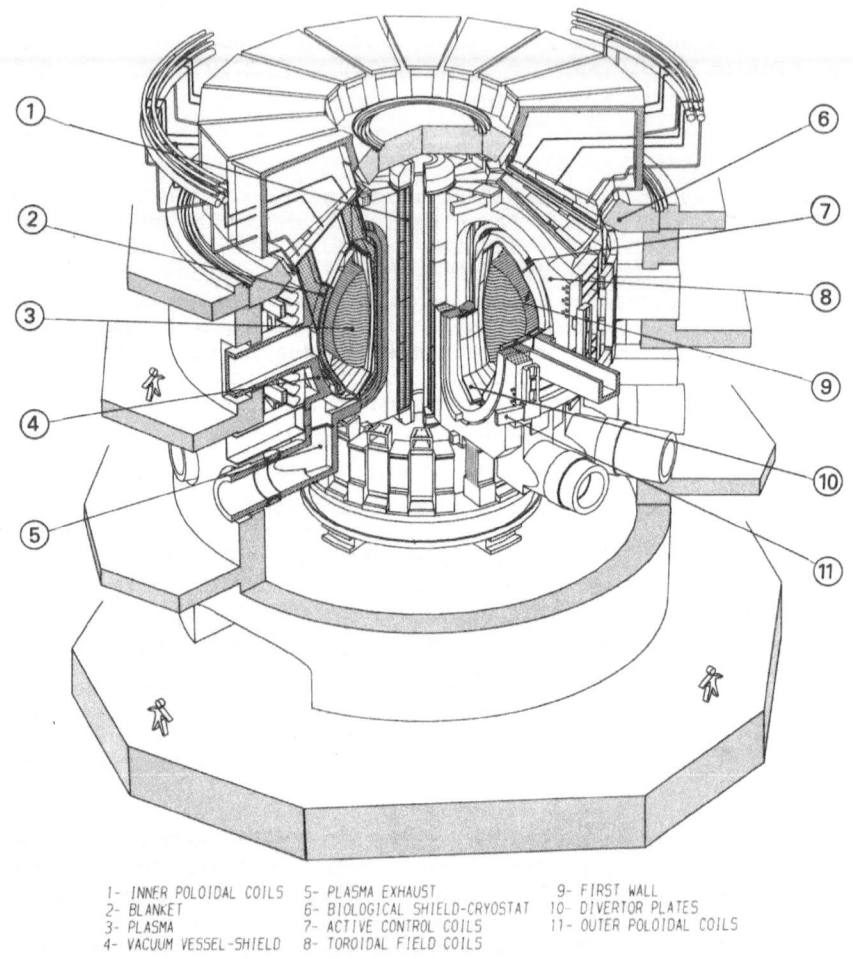

1- INNER POLOIDAL COILS 5- PLASMA EXHAUST 9- FIRST WALL
2- BLANKET 6- BIOLOGICAL SHIELD-CRYOSTAT 10- DIVERTOR PLATES
3- PLASMA 7- ACTIVE CONTROL COILS 11- OUTER POLOIDAL COILS
4- VACUUM VESSEL-SHIELD 8- TOROIDAL FIELD COILS

Fig. 1. ITER/NET overall layout

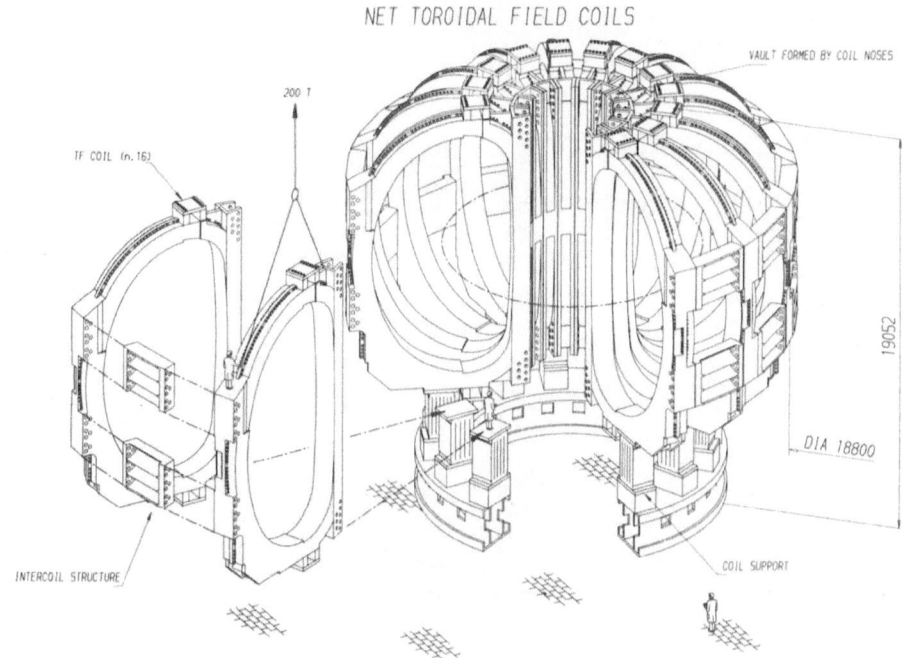

NET TOROIDAL FIELD COILS

Fig. 2a. Toroidal field coils assembly

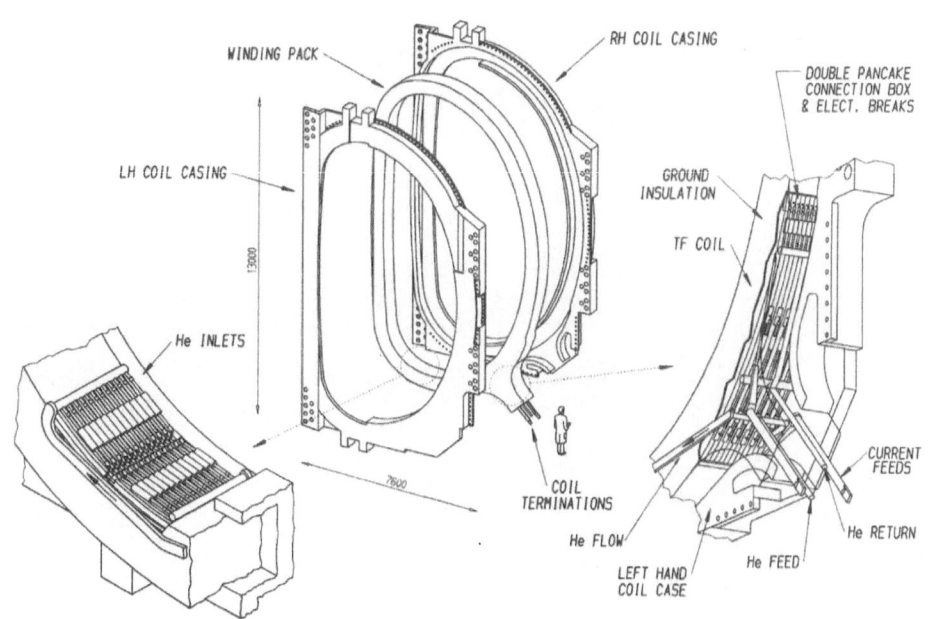

Fig. 2b. Toroidal field coil parts

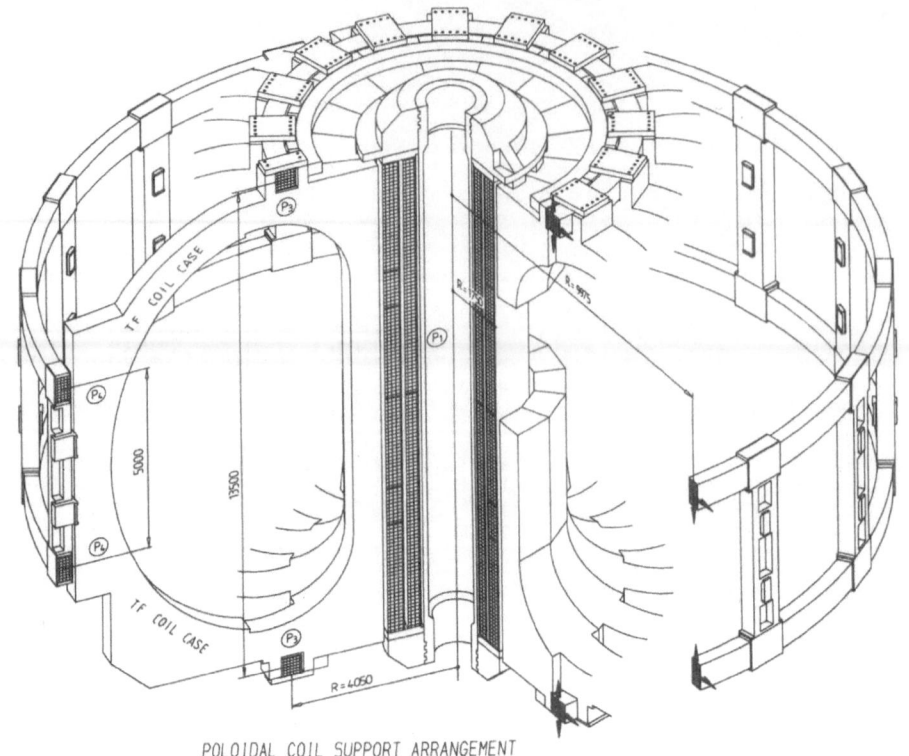

POLOIDAL COIL SUPPORT ARRANGEMENT

Fig. 2c. Poloidal field coils

NET VACUUM VESSEL ASSEMBLY

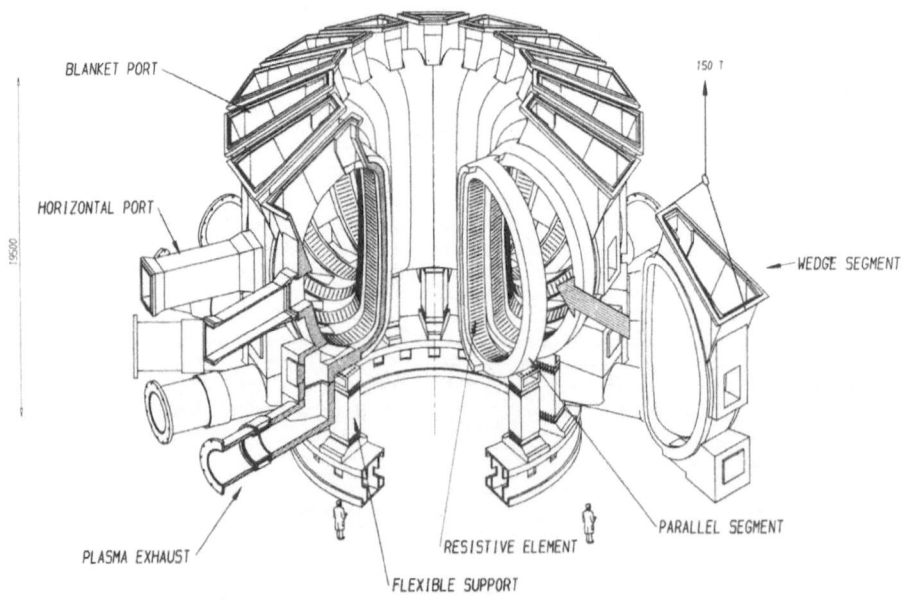

Fig. 3. Two levels of shielding (blanket, vacuum vessel)

The internal components, such as the divertor plates, first wall and blankets, have to be removed through the enclosing outer shielding, coils and cryostat. These components are the most heavily loaded (from a thermal and nuclear viewpoint) and will require regular replacement. To allow this removal, large vertical access ports have been left between each of the toroidal field coils (Fig. 4). In addition there is an equatorial access port through which an articulated boom or vehicle can be inserted for the replacement of smaller components (such as protective tiles, divertor plates). These equatorial ports are also used as access ducts for plasma auxiliary heating and non-inductive current drive systems (RF antennae and neutral beam injection ducts).

MAGNETIC SYSTEM

Superconducting coils offer substantial advantages over conventional copper coils. The vast electrical and cooling requirements and thermal expansion problems are eliminated, and one can take advantage of the high current density of the superconductor and of the high material strength obtainable at cryogenic temperatures.

The optimum peak magnetic field required for a reactor lies in the range of 12 tesla. At this level of magnetic field only the A15 superconductors can be used. Despite the fact that it was discovered before NbTi, the use of A15 compounds is less common, the reason being the strain-sensitivity of the current density and the brittleness of the A15 compounds. Two main techniques can be used to manufacture the coils in A15:

Wind and React (W/R): This consists of producing the conductor in the unreacted state, winding it with the insulation matrix and treating it at about 700°C. Finally the coil is epoxy vacuum impregnated.

React and Wind (R/W): This is similar to the above but the heat treatment to form the A15 compound is carried out before the winding of the coil.

The major problems to be resolved in the W/R process are that all coil materials (e.g. glass, steel) must resist the heat treatment, and the sintering of strands must be avoided. The advantage of the W/R technique is to avoid handling the reacted A15 compound, thereby limiting the risk of damage.

The major risk during operation of the coils is the transition from superconducting to normal conducting status. This is caused by temperature rise of the conductor due to the Joule effect or friction. The tokamak environment is very noisy from this point of view (e.g. plasma disruptions, high stress) therefore the conductor must be designed with an ample thermal stability margin. The technology of the superconducting coils is well established, although not at the scale required for NET/ITER. However, the coil systems constraints significantly affect the machine's cost and it is desirable to obtain maximum performance from them without compromising their reliability.

The coils are wound from a 40 kA, 12 T (operating field level) $(NbTi)_3Sn$ cable-in-conduit type superconductor, (Fig. 5) where liquid helium at 4.2 K flows through the interstices of a cable made up of the individual superconducting strands. This cable is enclosed in a steel jacket and the winding is bonded and insulated by vacuum-pressure-impregnated glass-fibre-reinforced epoxy resin, in the same way as a normal-conducting coil. Ceramic insulation could possibly offer better radiation resistance but has much lower mechanical and electrical strength and is a novel and high risk technique in this application.

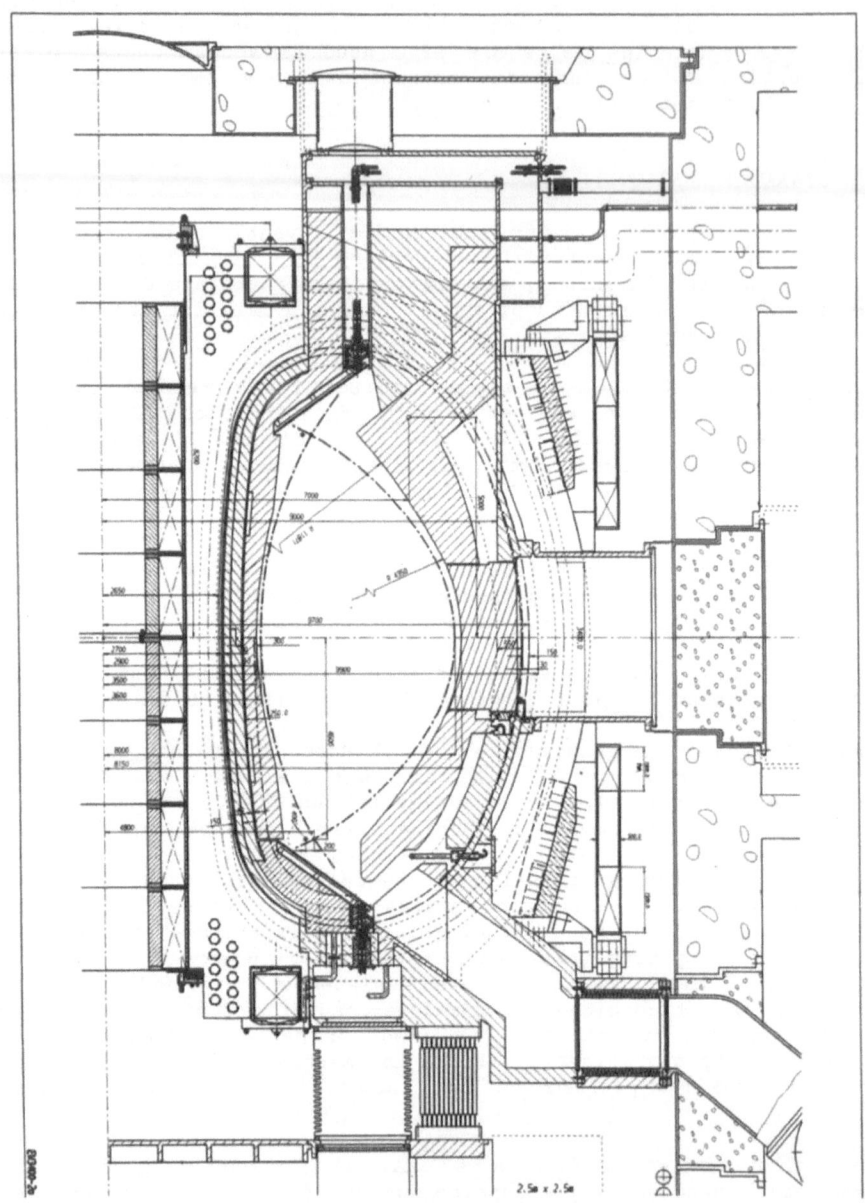

Fig. 4. Access ports

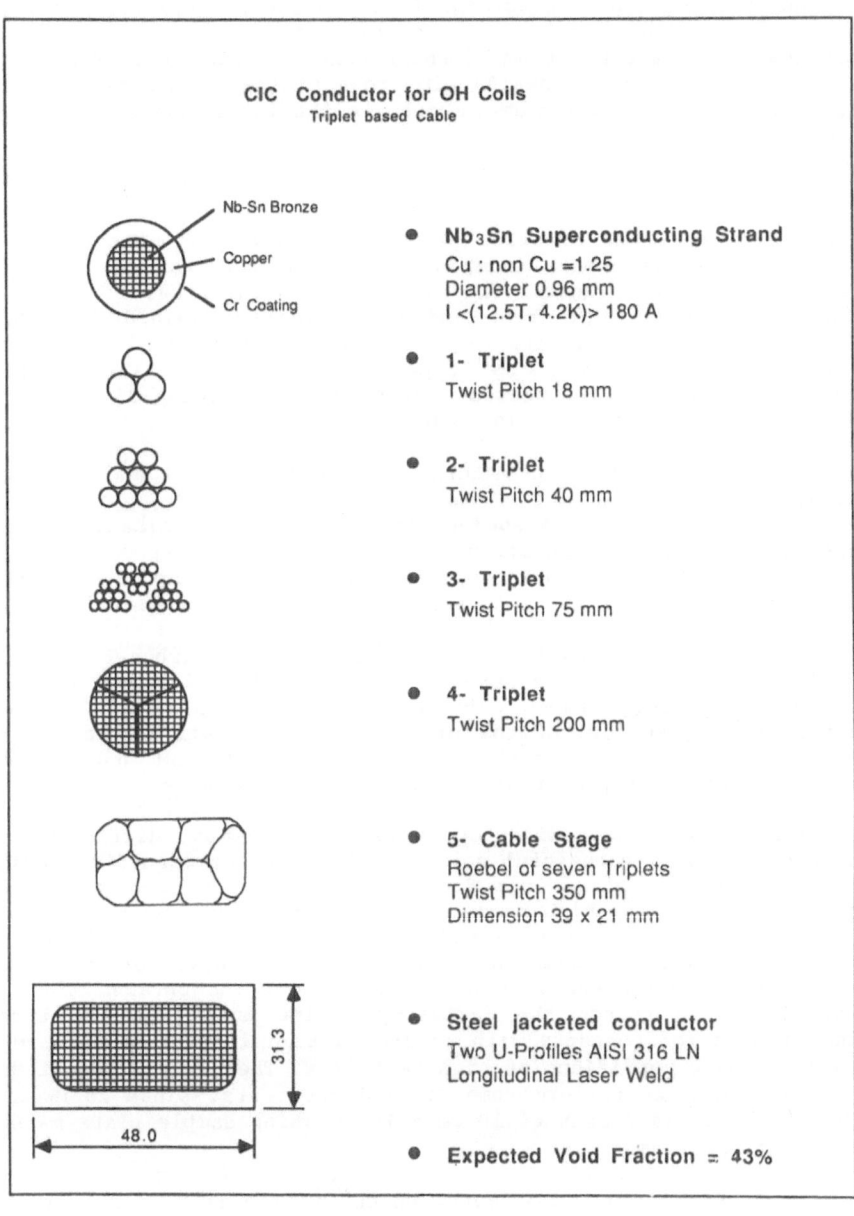

CIC Conductor for OH Coils
Triplet based Cable

Nb-Sn Bronze

Copper

Cr Coating

- **Nb$_3$Sn Superconducting Strand**
  Cu : non Cu =1.25
  Diameter 0.96 mm
  I <(12.5T, 4.2K)> 180 A

- **1- Triplet**
  Twist Pitch 18 mm

- **2- Triplet**
  Twist Pitch 40 mm

- **3- Triplet**
  Twist Pitch 75 mm

- **4- Triplet**
  Twist Pitch 200 mm

- **5- Cable Stage**
  Roebel of seven Triplets
  Twist Pitch 350 mm
  Dimension 39 x 21 mm

- **Steel jacketed conductor**
  Two U-Profiles AISI 316 LN
  Longitudinal Laser Weld

31.3

48.0

- **Expected Void Fraction = 43%**

Fig. 5. Cable in conduit conductor

The coil systems require a support structure to enable the magnetic forces to be cancelled (the total magnetic force on either TF or PF system is zero, but individual coils have a net force). The TF coils are each subjected to a net centering force of about 35000 tonnes. By forming the "noses" of the coils into a circular vault, the centering force is supported by hoop compression. The overturning force on the TF coils from the poloidal field is supported by friction between coils inside the vault, shear keys at the top and bottom and a system of shear boxes around the outside of the machine. The coil structural design aims to minimise the amount of extra support required, to provide maximum access to the internal nuclear components.

HIGH HEAT FLUX COMPONENTS

The first wall and divertor plates experience the most severe loading and are therefore removable independently from the rest of the components internal to the vacuum vessel. The first wall is an integral part of the inner shielding/breeding blanket, and is typically divided into 32 or 48 toroidal units each consisting of a separate part inboard and outboard of the plasma. In addition to experiencing neutron fluxes of about 5×10^{18} $m^{-2}s^{-1}$ and highly peaked thermal loads reaching 15 MW/m^2, the divertor plate design must be able to handle:

- erosion and redeposition of material, through physical and chemical sputtering;
- disruption loading, due to sudden termination of the plasma, giving electromagnetic forces and exceptionally high local heat deposition;
- gas (particularly hydrogen isotope) retention in the surface material;
- contamination of the first wall and plasma by divertor material.

A typical concept consists of a 5-10 mm thick carbon-fibre-reinforced graphite armour brazed onto a water cooled heat sink of copper or molybdenum alloy (Fig. 6). Radial tubes carrying coolant at up to 100°C connect to collectors in the shadows of the plates and first wall/blanket. During remote handling the divertor plate connections must be cut and rewelded from the outside, while the plates are removed from the inside.

Testing of potential divertor designs is the most difficult aspect of the development of a successful solution. The present magnetic configuration concentrates all the power lost from the plasma in a thin scrape-off layer ~ 1 cm, leading to very high power density on the divertor plates. Different approaches have to be attempted in order to ease the working conditions of the divertor. The true operating conditions of the divertor, with regard to power density, erosion and redeposition, can only be obtained in a burning tokamak. For this reason the next-step device must plan for substantial testing of divertor concepts with several design iterations. However, tests can be made under conditions of high heat flux. The neutral beam test bed at JET has been used to perform some preliminary tests, since 15 MW/m^2 can be delivered to a surface area of 10 cm x 10 cm using sample plate materials as a target for the beam.

Other plasma facing components, which, if required, also have very demanding operating conditions, are the lower hybrid launchers (Fig. 7) and ion cyclotron antennae. In addition to the requirements of the first wall, these components have to satisfy electrical requirements. They will very likely be the closest components to the plasma separatrix, and their design must be such as to resist the high heat particle and neutron flux, while maintaining the mechanical precision and electrical characteristics required.

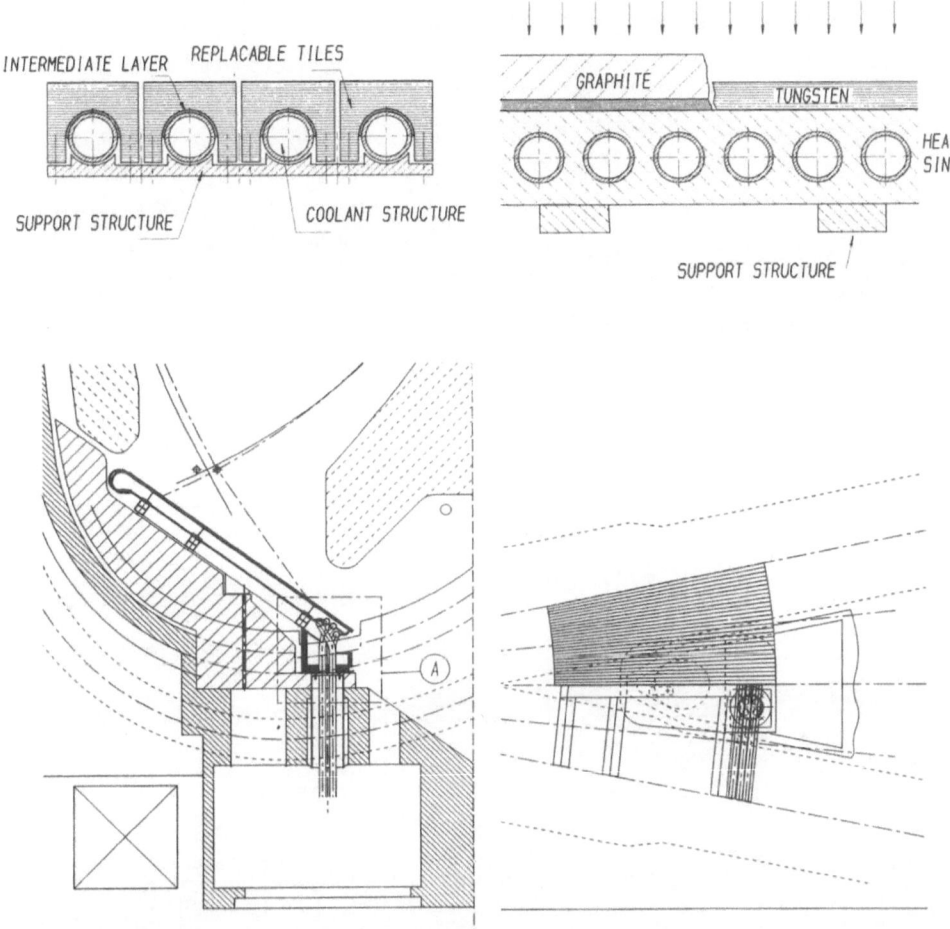

Fig. 6a. Divertor plate design

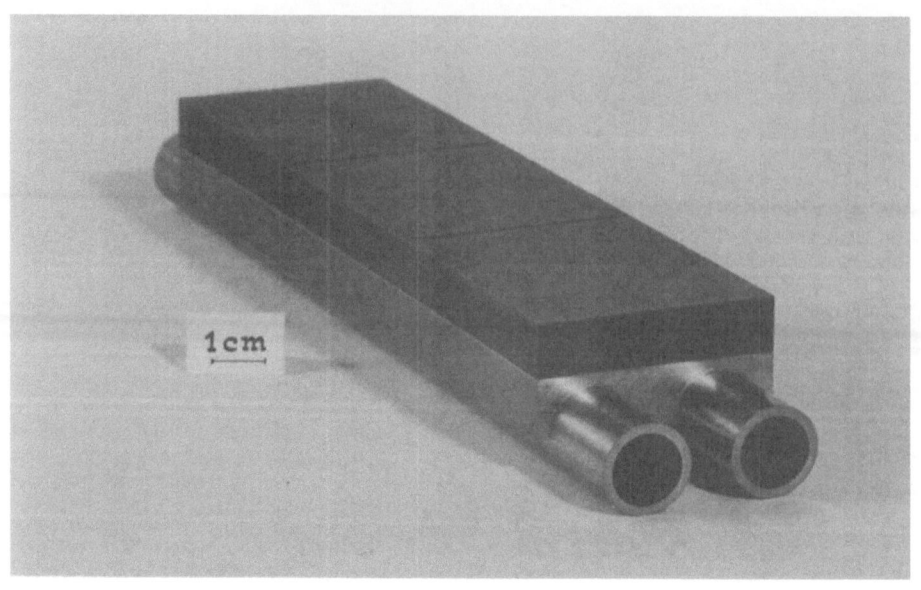

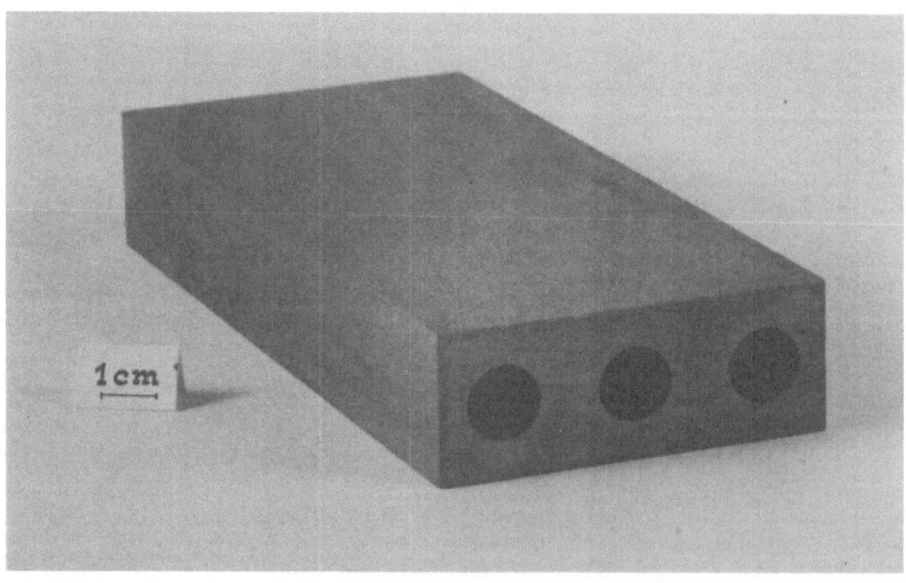

Fig. 6b: Divertor plates specimens

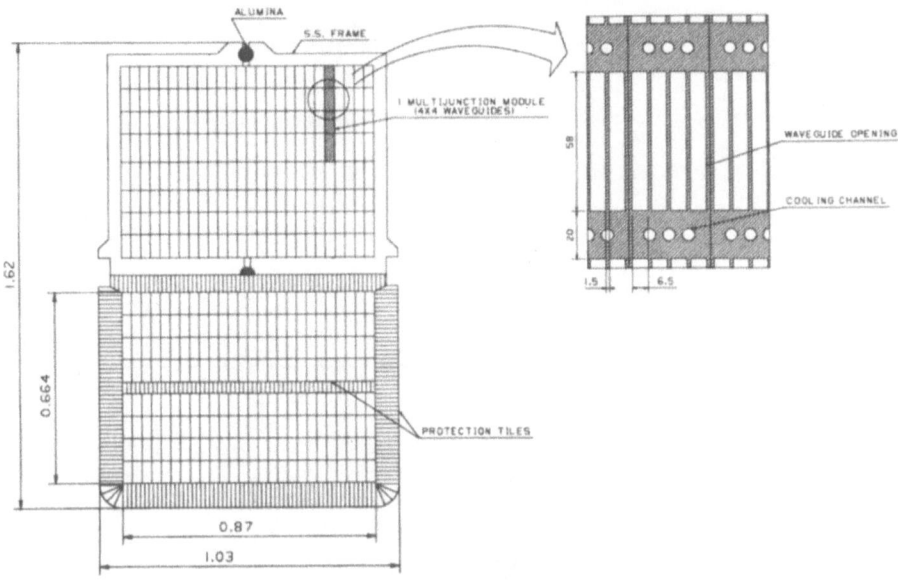

Fig. 7. Lower hybrid launchers

The distance from the striking point on the divertor to the null point of the separatrix has a strong influence on the machine layout and performance. The bigger the distance, the more current and energy is needed for the PF coils. This has a strong influence on the out-of-plane loads of the TF coils, which is one of the parameters driving the design concept, as well as the power density and the temperature of the striking particles (Fig. 8a, 8b).

Plasma disruptions affect the machine via heat deposition and induced eddy currents. Fig. 9 shows the effect of a plasma disruption from the electromagnetic point of view. The induced loads can be relatively easily supported by the semi-permanent components, while for the divertor plates and inboard blanket quite a high degree of electrical segmentation is required (Fig. 10). Relativistic electrons in the range of hundreds of MeV, experienced in JET during disruption[3], could cause severe damage to the plasma facing components. On the other hand, plasma disruption for a reactor must be regarded as an accident, because ample control margins will be provided.

CONCLUSIONS

The development of a fusion reactor poses formidable problems in physics, technology and engineering. The solution can only be found with progress in the three areas and early experiments in a proper environment where synergetic effects can be analysed. At the same time the reward for this achievement would amply justify the effort. The main issues still not quantified are linked to the uncertainties in the plasma edge physics and plasma wall interaction. As far as other design issues are concerned (e.g. superconducting coils), a satisfactory solution is to be expected in the near future.

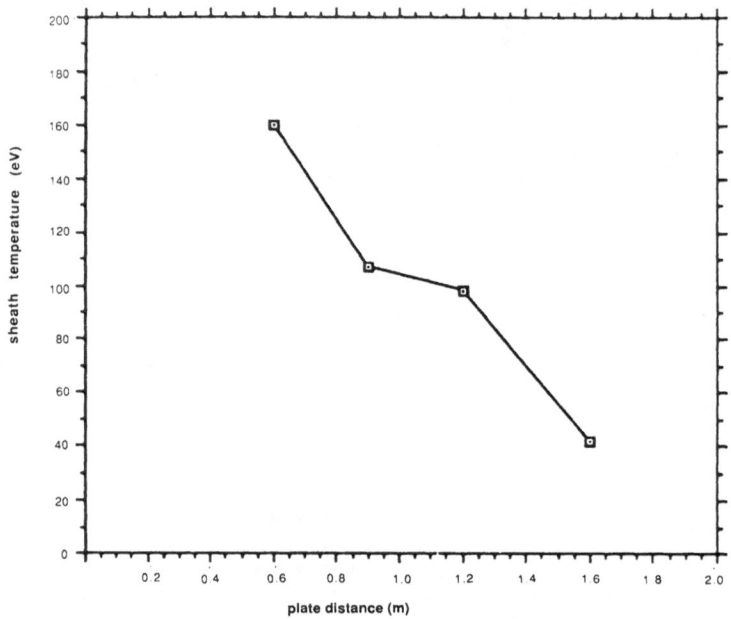

2-D Braams Code (PPPL)
"H-mode", $\chi_\perp^e = 1$, $\chi_\perp^i = D_\perp = 0.33$ m^2/s
$n_{sep} = 3 \times 10^{19}$ m^{-3}

Fig. 8a. Sheath temperature VS plate distance

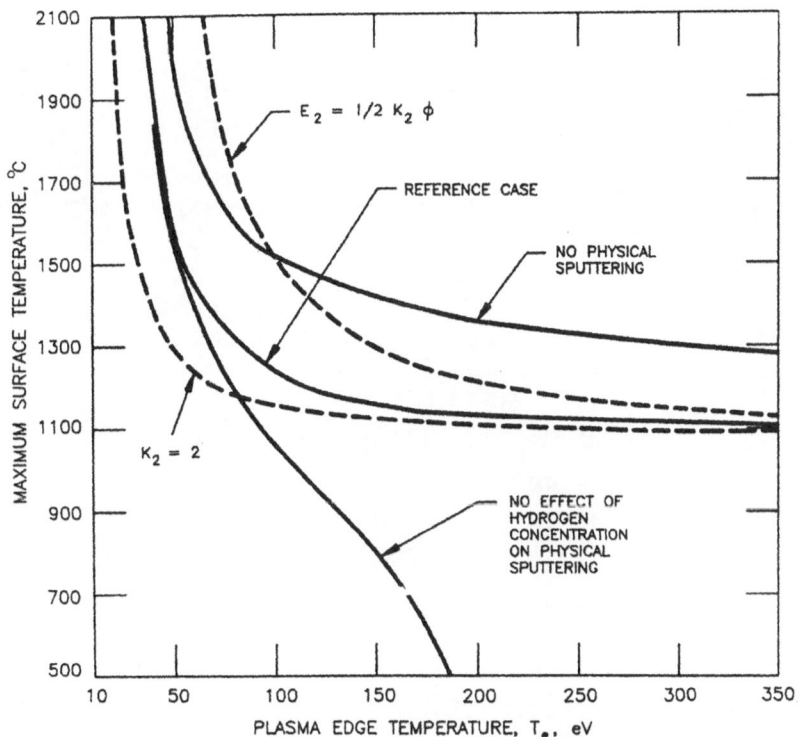

Fig. 8b. Max. surface temperature VS edge temperature

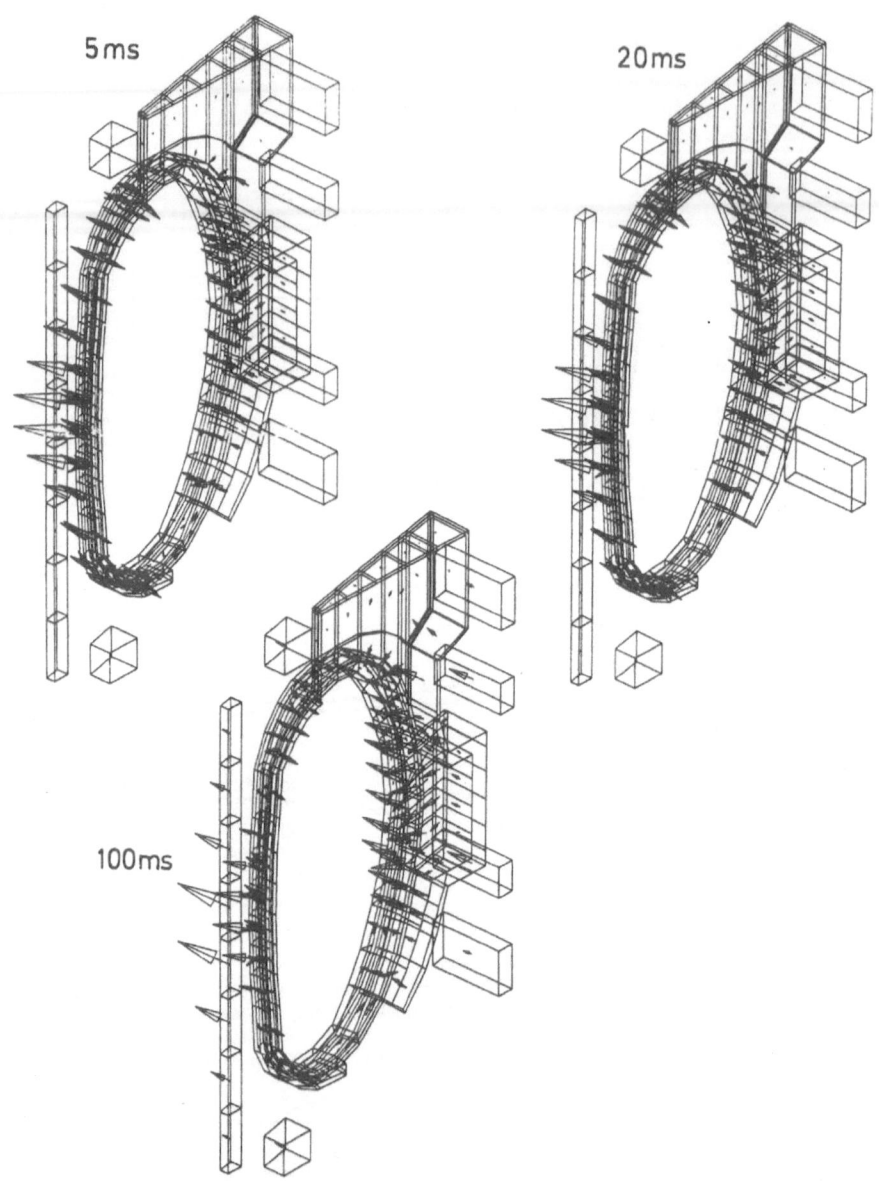

5ms

20ms

100ms

Fig. 9a. Current in structure due to centered disruptions

Plasma current: 20 MA

Disruption Time: 20 ms

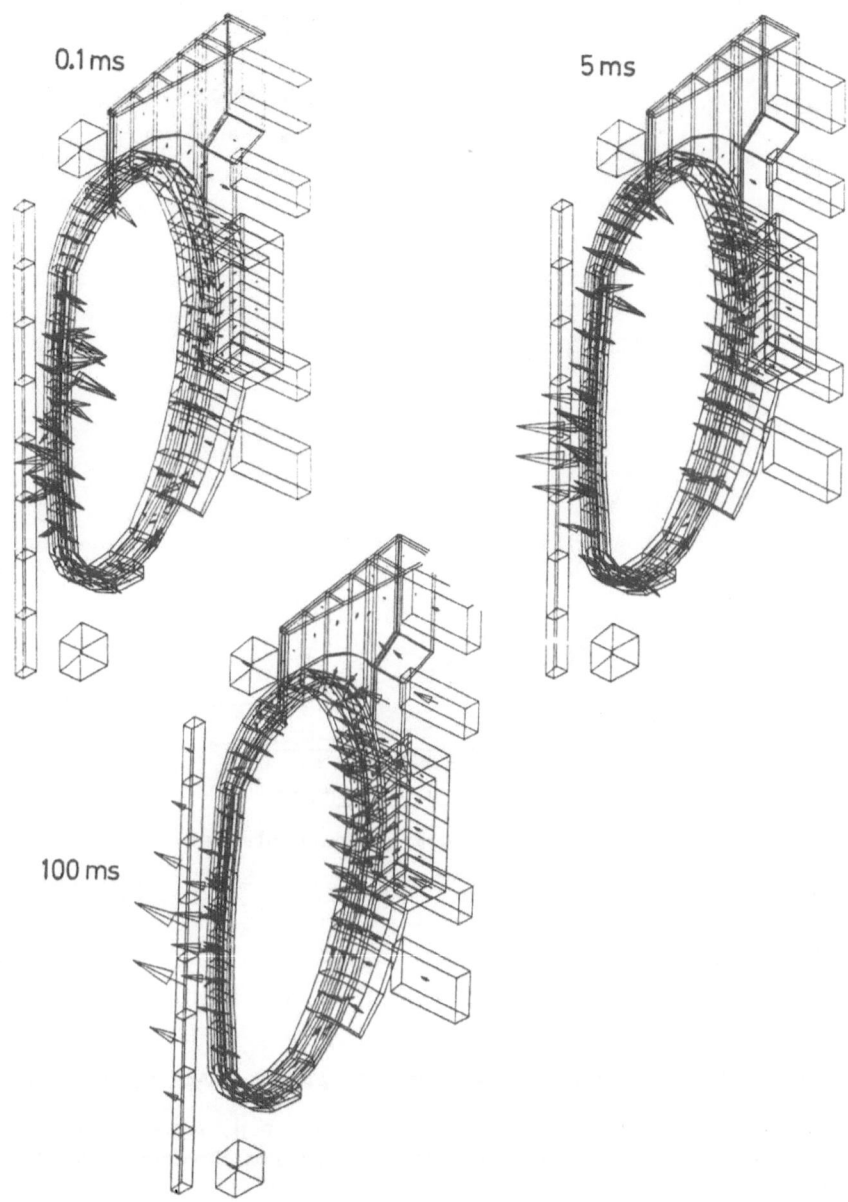

Fig. 9b.   Currents in structures due to vertical disruptions

Plasma current: 20 MA

Vertical displacement: 4.1 m

Disruption time constant: 20 ms

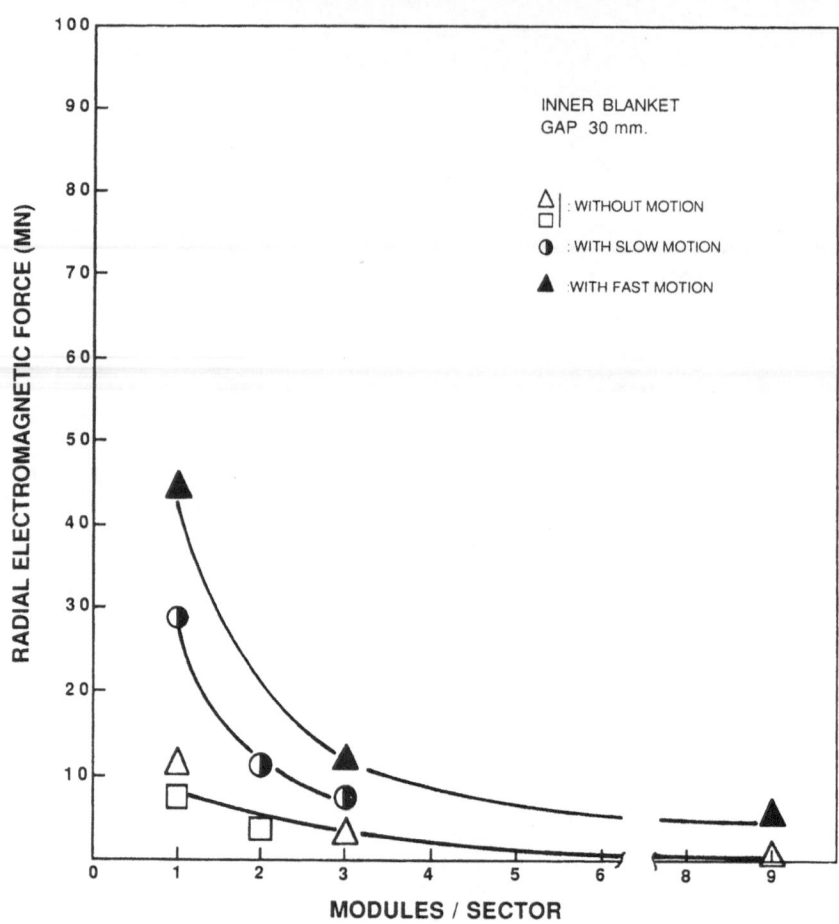

Fig. 10. Forces in inner blanket

REFERENCES

1.   Next European Torus (NET), special issue Fusion Technology, July 1988, Vol. 14, No. 1.

2.   ITER Concept Definition Phase Report, ITER-1, IAEA Vienna, Oct 1988.

3.   JET Results and the Prospects for Fusion, P-H. Rebut, P.P. Lallia, Fusion Technology, Proceedings 15th Symposium Utrecht, The Netherlands 19 - 23 Sept. 1988.

SPECIAL MATERIALS FOR FUSION REACTORS

Carlo Ponti

Commission of the European Communities
Joint Research Centre 21010 ISPRA - ITALY

INTRODUCTION

Fusion technology raises a number of new problems to be solved that require new materials or the adaptation of known materials which allow proper operation in the new conditions. The material requirements will evolve with the design of the plant and may change significantly from NET to DEMO /1/. However there are many common aspects, typical of the Tokamak reactors. Intense magnetic fields, superconducting coils, cryogenic temperatures, high vacuum, intense neutron fluxes, pulsed operation, thermal gradients and stresses, are a few of the specific features of the fusion environment. The materials required to operate in this field must provide the best performance, the maximum safety and their cost must be acceptable.

The physical and chemical parameters that characterize the working conditions of the materials are not always well known, and sometimes are rather uncertain. A striking example is offered by the plasma facing components: there are many uncertainties in the plasma-wall interaction phenomena namely in the charged and neutral particle fluxes, in the number and quality of disruptions, in the peak values of the thermal loads, etc. Other uncertainties exist on the behaviour of materials operating in the new and very stressing conditions (e.g. fast neutron damage, erosion of the first wall, fatigue, compatibility with tritium).

New information is required and new research and experimental facilities that simulate the reactor operating conditions. The investment costs may be considerable, as is the case for intense neutron sources. The materials must provide safe operation in both normal or abnormal situations. The latter must be analyzed on the basis of accident scenarios that depend on the design of the plant.

The safe operation of materials is very much related to the neutron induced radioactivity that affects the operational exposure, the release of radioactive effluents, the possible accidental release of radioactive materials, the maintenance operations, the decommissioning of the plant and the waste management.

Safety, Environmental Impact, and Economic Prospects of Nuclear Fusion
Edited by B. Brunelli and H. Knoepfel
Plenum Press, New York, 1990

In the next sections we will deal with the problems of materials in different components of a Tokamak reactor, starting from those relatively more conventional, such as the radiation shields and the SCM (superconducting magnets), and then considering the breeding blanket and the plasma facing components. This is a short review which is intended to consider the most important items, the feasibility problems and the R&D needs.

The neutron induced radioactivity is often the most relevant issue in safety analysis: the development of fusion materials tends more and more to coincide with the development of LAM (low activation materials). These will be the subject of the last section, which deals with the possibility of developing materials with reduced activation by the readjustment of their elemental composition, and of new materials with low activation properties.

MATERIALS OF THE SUPERCONDUCTING MAGNETS

The technology of SCM is quite independent from the other components and relatively well established. Satisfactory results have been achieved in the international LCT /2/ (Large Coil Task) programme. A Tokamak with SC magnet is now in operation (Tore Supra) in France. Size and performances close to those required by the next generation machine ($B_M \simeq$ 10-12 Tesla) have been achieved.

From the viewpoint of the performance required from the materials of the SCM and of their feasibility, the main issue for the impact on the machine lifetime and cost is related to the radiation damage that can be tolerated by the insulators. Low temperature and ionizing radiation concur to produce embrittlement and damage to the mechanical and dielectric properties of the insulator.

These effects are irreversible and finally limit the lifetime of the component. The best performances are at present reported for organic materials (epoxy-glass composites) that can withstand a dose in the range $5 \cdot 10^6 - 5 \cdot 10^7$ Gy. /3/.

The development of superconducting magnets able to work at higher temperature, and/or of materials (particularly insulators) that tolerate higher doses, would allow a considerable economy in the overall cost of the reactor. (See also the next section).

SHIELDING MATERIALS

There is not much literature on shielding materials for fusion reactors. The problem is not substantially different from that found in the context of fission reactors; namely one must attenuate the radiation below levels compatible with proper operation of components or allowing personnel access, with materials of minimum size (and cost).

The purpose of the inboard shield is the protection of the SCM; more specifically the radiation level in the SCM causes damage to the SC material (reducing the critical current), to the stabilizer (increase of resistivity) and to the insulator; furthermore it will deposit nuclear heat, which must be removed in an expensive way from a cryogenic system.

The most critical parameter is at present the radiation level to the insulator (see section 2). An increase of a factor 10 of the allowable radiation level, would save 15-20 cm in the thickness of the inboard shield. The importance of saving this space is very great: 1 cm saving is worth more than 1% of the cost of the plant. Actually, the plasma power density is proportional to B_0^4 (B_0 is the toroidal field on the plasma axis), the maximum field B_M on the inner face of the SCM is limited by the magnet technology, and the toroidal field B decreases as $1/R$ (R=major radius). This determines the high cost of the space between the magnet and the plasma, which is shared by the inboard blanket and the shield. In all the ships with nuclear power engines built so far, the radiation shield is a structure of alternate iron and water layers. This solution optimizes size and weight. To obtain shields with still lower thickness one must employ sophisticated and expensive materials. Tungsten has been suggested /4/ as shielding material for the inboard shield, to save up to about 10 cm, which would compensate the higher cost of the material and the increased complexity of the design and manufacture. However, tungsten has a serious drawback related to safety: its decay heat is more than a factor ten higher than that of steel; it can be used in regions of relatively low flux, or in reduced quantities.

The shields are very heavy (about 3000 tons for NET) and large radioactive pieces; at the end of the life of the plant they represent a relevant fraction of the radioactive waste. The activation behaviour of the shielding material should be carefully considered. Cast iron performs well either as a shielding material or as a low activation material. It could be recycled after cooling times of 50 to 100 years according to the position of lower or higher flux.

BREEDING BLANKET MATERIALS

It is useful to distinguish between blankets with solid and blankets with liquid breeder materials. Of the solid materials the preferred options are Li_2O, Li_4SiO_4 and $LiAlO_2$.

Lithium oxide has the greatest Li density and has been most extensively studied. An extensive date base exists /5, 6/ on physical properties, on tritium solubility, diffusivity and other items related to its recovery; this is probably the most difficult issue and is not yet completely solved. Tritium recovery is performed by purging He gas at high temperature, to increase diffusivity and reduce the inventory.

The temperature, however, is limited by radiation damage effects, chemical stability and compatibility with the structural material. Below 500 °C and for fluences not exceding 30 dpa (INTOR-NET), austenitic steels may be used. For higher fluences ferritic steels are favoured. A blanket with Li_2O, He cooled, HT9 structural material has the highest ranking among solid breeder blankets in the BCSS study /7/, if tritium effluent control is good.

The use of Be as neutron multiplier is necessary if tritium self sufficiency is required. Usually it is in close association with the breeder material to improve neutronic performances. Most of the issues identified /8/ are constraints and not feasibility issues, since alternative design solutions generally exist. Dominant issues relate to irradiation and chemical performance.

A solid breeder material (Li_4SiO_4) in the form of small pebbles has been proposed /9/ for NET, using Be as multiplier and He as coolant and purge gas. The silicate was chosen because of its "potentially very good tritium release properties". AISI-316 is the structural material; the operating temperature is below 500° C.

Despite the extensive data base developed for solid breeder materials, a better understanding of the mechanical properties is thought to be required /10/ to assess the tendency of these materials to crack during power cycling and cause breeder-structure mechanical interaction.

Liquid metals (Li, 83Pb-17Li) are candidate breeder materials in self-cooled solutions or in association with water coolant. Self cooled Li with Vanadium is the top ranking blanket proposed by the BCSS (Blanket comparison and selection study). The main concerns with this material are the need to avoid any possible interaction with air or water, corrosion of structural materials, and MHD pressure drop.

If chemical reactions of Li with air or water cannot be adequately controlled, then the eutectic 83Pb-17Li could be used instead of Li. The reaction with water is more moderate /11/ and is proposed also with water cooling. In this case its circulation is slow (only for the purpose of tritium recovery). However corrosion is still a concern, and limits the operating temperature severely (400 °C with AISI-316). This is the main reason to replace steel with vanadium (or V-alloys) in the BCSS /7/ proposal. Another concern is tritium penetration into the water coolant which means that a coolant DTS (detritiation system) is necessary.

The simplest tritium breeding material is a solution of a Li salt in water. In this way a neutron shield may become a breeding blanket, when there is a need to breed tritium, by the addition of the salt to the coolant. $LiNO_3$ and LiOH have been studied for the use in the NET breeding blanket /12/. The first is preferred for its lower corrosion but is more susceptible to radiolysis and produces ^{14}C in amounts hardly compatible with safety requirements. A pressure around 1 MPa would reduce bubbles and keep the radiolytic gas in solution. The overall TBR is lower than unity, which makes the concept not reactor relevant. Tritium inventory is also a concern. The level of tritium in the water is around 10 Ci/l: a very tight coolant circuit is needed to keep leakage and routine tritium releases to acceptable levels.

All the breeder materials mentioned above (with the exception of $LiNO_3$) are LAM: their activity decreases rapidly with time. The presence of nitrogen should be avoided because of the biological hazard related to the production of ^{14}C and of the mobility of gas with which it can be associated (e.g. CO_2).

Lead radioactivity also decays rapidly with time, its decay heat is negligible. However the possible presence of bismuth as an impurity or transmutation product should be checked and possibly removed. The presence of Bi generates ^{210}Po, an alpha-decay nuclide of very high biological hazard and mobility.

As for the waste management /13/ the only concern, which is not serious, is that of ^{205}Pb produced by lead. Its halflife is $1.4\ 10^7$ year; it decays by electron capture with emission of low energy X rays.

PLASMA FACING COMPONENTS

The anticipated operating conditions of the plasma facing components of NET /14/ are reported in Table 1. For the first wall AISI-316 is proposed as structural material, with water coolant and a graphite protection armour. The choice of AISI-316 is based on the following advantages:
- extended data base
- acceptable damage with the low fluence expected
- no ferromagnetic effects
- ease of manufacturing, welding, brazing.

The lifetime is limited by the thermal fatigue and the allowable number of cycles depends on the peak heat flux and on the design of the armour (heat transfer by conduction or radiation). The armour is required to protect the FW surface from the energy deposited by the plasma and the runaway electrons during start up and disruptions, from the neutral beam shine, and also to improve impurity control. At least 1 cm refractory low Z material (carbon) is at present proposed.
This solution may not be extended to DEMO and power reactors since:
- the radiation damage (swelling, embrittlement) of AISI-316 is acceptable only below $\sim$ 50 dpa
- the radiation damage also limits the use of carbon based material to less than 3 MWa/m^2 (swelling)
- a substantial amount of dust is produced

Martensitic steel is not suitable for NET FW since the neutron irradiation at the low temperature causes high embrittlement. However it is more reactor relevant than AISI-316 since it has better dimensional stability, resistance to fatigue and embrittlement in reactor conditions (higher temperature and fluence). Breeding blanket modules with martensitic steel as structural material will be tested /15/ during the technological phase of NET.

TABLE 1

Expected operating conditions for plasma facing components of NET (14).

	Physics		Technology	
	FW	Divertor	FW	Divertor
1. Nominal operation				
average neutron wall load MW/m^2	1	.4	1	.4
heat flux – average "	.15	1.5	.15	1.5
– peak "	<1	10	<1	10
average neutron fluence MWa/m^2	.03	.015	.8	.4
average neutron damage – dpa	.3		9	
– He (appm) in AISI 316	4		118	
number of pulses (10^4)	1		8	
pulse duration s		100		300
total burn time h		300		7000
2. Disruptions				
number	100		10	
thermal quench time ms	.1 – 3		.1 – 3	
current " " ms	5 –50		5 –50	
peak energy depos. MJ/M^2	1	5	1	5

152

No solution for the protective armour of a DEMO or power reactor has yet been proposed. Better plasma performance is expected which will allow the effects of runaway electrons and disruptions to be neglected or substantially reduced, the number of cycles to be reduced. It is possible to replace the FW (and the blanket) of NET after having optimized the plasma performance during the physics phase. It is probably not feasible to replace the FW (and/or the blanket) of a power reactor: the price, in terms of down time, cost, operational exposure, volume of wastes, would be very high.

The presence of the graphite armour also raises a safety concern because of the possible fire hazard in the case of water or air ingress into the plasma chamber.

The step from NET to DEMO will become feasible, once the experience gained with NET has provided better control of the plasma burn and confinement.

A double null (DN) divertor is proposed for NET at present, with operating conditions summarized in Table 1 . Water-cooled copper (or TZM) heat sink with a protection armour similar to that of the FW is actually proposed /14/. The lifetime is limited by the net erosion (global erosion minus redeposition). Its design allows replacement with a limited down time.

LOW ACTIVATION MATERIALS

The most significant factor determining the safety and environmental characteristics of a fusion reactor is the intense radioactivity of the materials surrounding the plasma /16/. Unlike the radioactivity associated with the fuel cycle of fission reactors, this is not an intrinsic feature of the thermonuclear reactions themselves but arises from the nuclear reactions induced by plasma neutrons on surrounding materials. LAM (low activation materials) developement is an important means of enhancing the advantages of fusion as a safe and environmentally benign energy source.

As long as the radioactivity remains confined within stable and immobile materials, it does not raise serious concern. There is concern however when the material, or part of it, is moved across the different containment barriers or through circuits and finally is released into the reactor hall and/or into the environment. LAM are those materials that either decay in short times; or have a negligible probability in any normal or abnormal event, during or after the operation of the plant, to be released into the reactor hall and/or into the environment; or the effects of their possible releases remain within "acceptable limits". The first condition depends only on the nuclear properties of the material; the second one depends also on the chemical and physical parameters (melting temperature, reactivity with air, water or other fluids, corrosion, sputtering etc.) and on the possible accident scenario; the last depends also on what is considered to be an acceptable limit. The level of radioactivity may be very different in different components; a material which is not suitable for the blanket could be considered suitable for the shield or for the magnet.

In order to know whether a given material satisfies the above conditions, with reference to a particular component for which it is designed,   one must determine the possible releases into the environment, considering all the possible routes:
- accident scenarios
- waste management
- routine operation, scheduled and unscheduled maintenance
- disassembly and decommissioning
- recycling of materials

Unfortunately, it is not always clear what the "acceptable limits" are, or what are the appropriate radiological criteria to apply are. This is true in particular for waste disposal: there is no regulation applicable to all EC countries /17/, but only acceptance criteria different for each site and different from those applicable in the US.

The EC program for LAM developement is divided into seven chapters:
- criteria for LAM
- nuclear data base
- development of LA martensitic steel
- development of LA austenitic steel
- development of non-ferrous LA alloys

- identification and development of LA advanced materials
- production routes of high purity LAM

The importance of LAM has been widely recognized /18, 19/ but it is difficult to specify which nuclear and non-nuclear properties the material should exhibit, and to establish priorities.

More attention has been devoted to the waste problem and to the long-life radioactivity; recommendations for material composition have been given by several authors /13, 20, 22/. In line with these recommendations studies have been carried out to produce and characterize low activation austenitic (23, 24) and martensitic steels /25/. The main criteria were to achieve shallow land burial conditions according to US norms and to extrapolate them to fusion. Actually the analysis of the European situation /17, 26, 27/ shows that these norms are not applicable to many states on this side of the Atlantic.

In FRG all nuclear wastes must be disposed of in deep geological repositories.

In May 1987 the UK government announced that although it is believed that a safe near surface disposal facility for low level waste (LLW) could be developed, for economic reasons this option would no longer be considered.

In France the Centre de Stockage sur la Manche accepts only LLW and MLW cointaining nuclides with less than 30 years half-life.

Two geological repositories are currently operated or planned for the management of nuclear waste in Sweden.

Only relatively small repositories within nuclear sites are available in other European Countries. Shallow land burial (SLB), when applicable, is limited to solid waste that decays to very low levels while institutional controls are operating.

The final disposal of wastes from the FW and blanket of European fusion reactors will be geological disposal /28/. According to this line, the criteria for LAM may be completely different: decay heat could be more restrictive than activity or dose, the allowable concentration of radioactive nuclides could be quite different from that of US shallow land burial criteria, more relaxed, and site dependent. A large amount of data is now available on the activation behaviour of the irradiated elements /20, 21, 29,30/ mainly focused on long term activation.

The attempt to reduce activation will emphasize other items, related to short-medium term radioactivity, and in particular to accidental release.

Data and experimental information on accident situations, mobilization and transport of radioactive material are being developed at present /31, 32/ but the state of the art is far from being satisfactory and there is still much to learn.

The presence of tramp impurity elements (e.g. Ag, Bi, Ir, Tb, Eu) is relevant only in the framework of the long-life activity and of waste management, if the goal is SLB. It is much less important in all other aspects.

The radioactivity of a material is the sum of the radioactivities of the elements that compose the material. LAMs prefer elements that produce radioactive isotopes that decay in short times, and with low potential environmental impact. They should also be physically and chemically stable in any normal or abnormal situation.

Of the more common elements those with best activation performances are H, He, Li, Be, B, C, O, Mg, Si, V, Cr; elements to be avoided are N, Ni, Co, Cu, Nb, Mo, Ag, and Bi.

The development of LAMs will go through more steps. The first will be the production and characterisation of reduced activity steels by the replacement of troublesome elements with better ones. The second step will be the developement of non-ferritic alloys mainly based on vanadium and chromium /33/. Finally possible new types of LAM could be studied (e.g. SiC).

CONCLUSION

The projected material requirements for NET and DEMO are not the same. For NET radiation damage will play a minor role, while the limited experience of plasma engineering will imply severe problems of protection against plasma-FW interactions and disruptions.

In a DEMO reactor the problems of radiation damage will be emphasized, while plasma control and confinement will be better established.

NET will essentially rely on present day technology, while DEMO will use new materials and low activation materials, which have yet to be developed.

Fusion material R&D should be focused on a given device and planned in agreement with the design schedule of that device. This is the case with the European programme, which is focused on NET.

No major experimental facility is planned for the R&D program for NET. An intense, 14 MeV neutron source will be required for the development of a DT fueled DEMO reactor.

REFERENCES

1. P.Schiller. and J. Nihoul., "Projected materials requirements for the NET and for long term tokamak-based DEMO fusion devices". Int. Conf. on Fusion Reactor Materials, Karlsruhe, October 4-8, 1987.
2. P.N. Haubenreich., S. Shimamoto., P. Komarek., G. Vecsey., "International developement of superconducting magnets for fusion power", Fusion Technology 15, 2, 909 (March 1989).
3. C.D. Henning., J.R. Miller., "Design considerations for ITER magnet systems". Fusion Technology 15, 2, 915 (March 1989).
4. A. Laila. El-Guebaly, M.E. Sawan., "Tungsten versus steel in inboard shield of ITER: impact on magnet damage, reactor size and cost", Fusion Technology 15, 2, 881 (March 1989).
5. C.E. Johnson., K.R. Kummerer and E. Roth., "Ceramic breeder materials", Journal of Nuclear Materials 155-157 188(1988).
6. M. A. Abdou et al., "Modeling, Analysis, and Experiments for Fusion Nuclear Technology: FNT Progress Report, Modeling and FINESSE", UCLA-ENG-86-44, University of California, Los Angeles, January 1987.
7. D.L. Smith et al., "Blanket comparison and selection study", ANL/FPP-84-1.
8. M. Abdou et al., "Summary of the ISFNT workshop on ITER". Fusion Technology. 14, 3, 1399 (November 1988).
9. M. Dalle Donne et al., "Pebble bed canister: the Karlsruhe ceramic breeder blanket design for NET". Fusion Technology 14 357 (November, 1988).
10 M. C. Billone., W. T. Grayhack., "Mechanical performance of fusion solid breeder and multiplier materials". Fusion Technology 15, 2, 1205 (1989).
11. O. Kranert., H.M. Kottowski., C. Savatteri., "Large scale Li17Pb83/water interaction studies". Fusion Technology 15, 2, 973 (1989).
12. W. Daenner et al., "Status of NET shielding blanket development", 15th SOFT, Utrecht 19-23 september, 1988.

13. C. Ponti., "Recycling and shallow land burial as goals for fusion reactor materials development, Fusion Technology 13, 157, (January 1988).
14. G. Vieider ., M. Harrison . and F. Moons., "Net plasma facing components", 15th SOFT, Utrecht 19-23 September, 1988.
15. M. Chazalon., W. Daenner . and B., "Blanket Testing in NET", 15th SOFT, Utrecht 19-23 September, 1988.
16. S. Piet., "Safety and environmental challanges in materials selection", IAEA-Yalta 26.5-6.6/1986.
17. C. Ponti., "Disposal of radioactive waste in Europe: norms and practice", Tech. Note I 8908 (January 1989).
18. "Panel report on low activation materials for fusion applications". UCLA report to DOE, March, 1983.
19. G. Casini., C. Ponti., P. Rocco., "Environmental aspects of fusion reactors", EUR 10728 (1986).
20. L. Giancarli & G.J. Butterworth., "The implications of dose rate limits for the recycling of fusion reactor first wall structural material", XIV SOFT, Avignon September, 1986.
21. C. Ponti., "Fusion reactor materials to minimize long living radioactive waste", ISFNT, Tokyo 10-15 April, 1988. Fusion Engineering & Design.
22. S. Fetter., E.T. Cheng., F.M. Mann., "Long term radioactivity in fusion reactors", Fusion Engineering & Design 6, 123 (1988).
23. P. Fenici et al., "Properties of Cr-Mn austenitic stainless steels for fusion reactor applications". Nuclear Engineering & Design/Fusion 1, 2, 167 (1984).
24. G. Piatti., D. Boerman. & J. Heritier, "Developement of low activation Cr-Mn austenitic steels for fusion reactor applications". 15th SOFT, Utrecht 19-23 September, 1988.
25. G.J, Butterworth, K.W Tupholme, J. Orr, D. Dulieu, "A study of the prospects for development of low activation martensitic steels for first wall and blanket structures in fusion reactors". CLM-R264 (1986).
26. J.P. Davis and G.M. Smith., "Radiological aspects of the management of solid waste from the operation of D-T fusion reactors", NRPB-R210 (1987).
27. K.R. Smith and G.J. Butterworth., "The radiological impact of fusion waste disposal", IAEA Technical meeting on Fusion Reactor Safety, Jackson, Wyoming, 4-7 April, 1989.
28. W. Gulden et al., "Waste management for NET". 15th SOFT, Utrecht 19-23 September 1988.
29. R.A. Forrest & D.A.J. Endacott., "Activation data for some elements relevant to fusion reactors". AERE R-13402 (1989).
30. L. Giancarli., "On the radiological behaviour of first wall fusion structural materials". CLM-R275 (1987).
31. S.J. Piet, R. M. Neilson Jr., G.R. Smolik, G.A. Reimann, "Initial experimental investigation of the elemental volatility from steel alloys for fusion safety application". EGG-ESP-8459 (April, 1989).
32. C. Ponti E. Ruedl., G. Casini., "Release of Mn radioisotopes from fusion reactor steels". 15th SOFT, Utrecht 19-23 September, 1988.

33. P. Hokkeling, W. van Witzenburg, "Developement of low activation vanadium alloys for fusion reactor application". 15th SOFT, Utrecht 19-23 September, 1988.

FEASIBILITY ASPECTS OF THE D-^{3}He FUEL CYCLE IN TOKAMAK POWER REACTOR

PLANTS

G. Casini

Commission of the European Communities
Joint Research Centre - Systems Engineering Institute
21020 Ispra (Va) - Italy

1. INTRODUCTION

The recent suggestion that the surface of the moon may be mined for
^{3}He to be used as a fuel in terrestrial fusion reactors /1/ has renewed
the interest on the D-^{3}He fuel cycle for energy production. The basic
reaction of this cycle is:

$$D + {}^3He \rightarrow {}^4He(3.6) + p(14.7) + 18.3 \text{ MeV}$$

There are some attractive features in this reaction, namely:
- fuel components and reaction products are all stable elements;
- 100% of the energy released is in the form of charged particles;
- the total energy per reaction is large.

The main difficulty, which makes this reaction less attractive as
compared to D-T for the first generation of fusion machines, is its unfavourable behaviour from the point of view of maximum value and
dependence from temperature.

The Maxwellian-average $<\sigma v>_F$ for the various fusion reactions are
shown in Fig.1.1. The $<\sigma v>_F$ for the D-^{3}He is much lower than that for D-T
and peaks at a much higher plasma temperature. Of more interest is the
figure of merit, $<\sigma v>_F/T^2$, which is a measure of the power produced for a
given beta (ratio of the volume-averaged plasma pressure to the magnetic
pressure) and magnetic field strength. This is shown in Fig.1.2. We see
that $<\sigma v>_F/T^2$ for D-^{3}He peaks at 55 keV compared with 13 keV for the D-T
reaction. In addition, the peak for the D-^{3}He reaction is about a factor
of 50 smaller than that for the D-T reaction. Also shown for comparison is
the $<\sigma v>_F/T^2$ for the D-D reaction; this is about a factor of 2 lower than
the D-^{3}He value. Other factors (e.g. confinement scaling, volume averaging
and radiation losses) can shift the optimum temperature from the 55 keV
peak, but the optimum will still be in the 50-80 keV range and not at the
very high values implied by the maximum of $<\sigma v>_F$.

In addition to the fusion reactivity, a concern is the degree of
plasma confinement required for ignition; the usual figure of merit for
this is the product of the density and the energy confinement time re-

Safety, Environmental Impact, and Economic Prospects of Nuclear Fusion
Edited by B. Brunelli and H. Knoepfel
Plenum Press, New York, 1990

159

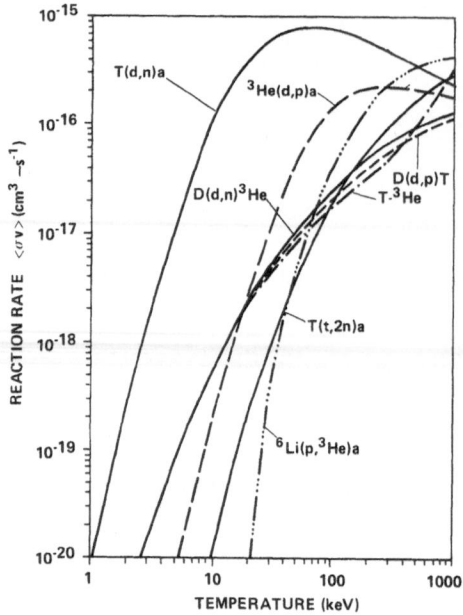

Fig.1.1. The Maxwellian averaged reaction rate $<\sigma v>_F$ as a function of ion temperature for various fusion reactions.

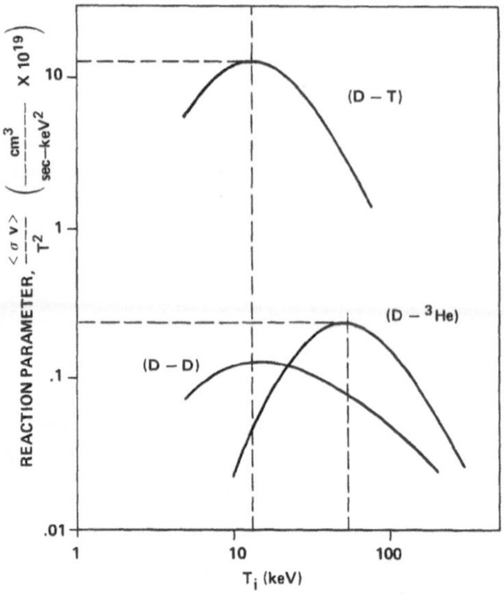

Fig.1.2. The Maxwellian averaged reaction rate parameter $<\sigma v>_F/T^2$ as a function of ion temperature for several fusion fuel cycles.

quired for the plasma to be self-sustaining against power losses (loosely referred to as the "Lawson criterion"). Shown in Fig.1.3 is $n\tau$ required for ignition versus the ion temperature; this energy confinement time measures the loss of energy by diffusion across the confining magnetic field /2/. Bremsstrahlung and synchrotron radiation losses are included separately. Synchrotron radiation loss depends not only on the magnetic field strength, density and temperature, but also on the degree of reabsorption; the latter also introduces an additional magnetic field, density, temperature, size and wall reflectivity dependence. Consequently, one cannot give a generic curve for $n\tau$ versus ion temperature. The density has been eliminated using the definition of beta whereas for the plasma minor radius and the effective reflectivity reactor relevant values have

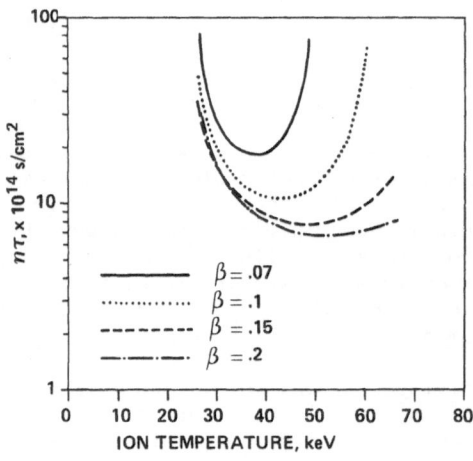

Fig.1.3. Lawson parameter for D–^{3}He versus volume-averaged ion temperature /2/.

been chosen. Also included in this nτ is the effect of the density and temperature profile in the plasma; both the density and temperature in Fig.1.3 are volume averaged. At low temperature bremsstrahlung dominates the losses; at higher temperature synchrotron radiation is more important. This is the reason for the beta dependence of the required nτ being larger at higher ion temperatures. Reducing the reflectivity raises the required nτ: at zero reflectivity the minimum in the beta = 0.1 curve is raised by a factor of about 2.5 and shifts the optimum temperature to a somewhat lower value; for beta = 0.1, the required nτ at the optimum ion temperature of 40 keV is about 1×10^{15} s/cm, which is about five times the required nτ for D-T fusion.

The high operation temperature, coupled with a larger electron density due to the presence of ^{3}He (Z=2) in the fuel, makes bremsstrahlung and synchrotron radiation play a larger role in the overall power balance than they do in the D-T case. In addition, the higher electron density reduces the allowed ion density for a given beta, so further reducing the fusion power density.

In a D-^{3}He plasma, neutrons and tritium are produced directly by the reactions:

$$D + D \rightarrow {}^3He(0.82) + n(2.45) + 3.27 \text{ MeV}$$
$$D + D \rightarrow T(1.01) + p(3.02) + 4.03 \text{ MeV}$$

Neutrons are also produced indirectly by the reaction:

$$D + T \rightarrow {}^4He(3.5) + n(14.1) + 17.6 \text{ MeV}$$

Although the last one is a two-step process, it is not insignificant because of the large D-T cross-section.

Decreasing the deuterium concentration in the fuel, the neutron and tritium production rate are decreased. This is, however, balanced by the fact that the relative electron density and the radiation losses are increased (due to higher ^{3}He fraction) so worsening the fusion power density.

As mentioned the D-^{3}He reaction releases 18.3 MeV of energy per reaction event, while the D-T reaction releases 17.6 MeV. The energy of the alpha particle released in each reaction is almost the same, but the D-^{3}He reaction releases a 14.7 MeV proton instead of a 14.1 MeV neutron. It is important that 100% of the energy released in the D-^{3}He reaction is in the form of charged particles which deposit their energy in the plasma and help maintain the plasma energy balance. For the D-T reaction, only 20% of the energy released is due to alpha particles and is available for plasma heating. Because of the smaller mass of the proton, its gyroradius and banana widths are only twice that of a 3.6 MeV alpha particle. Hence finite orbit effects are not greatly increased over that of the D-T case. Fast ion losses due to trapping in the magnetic ripple is a concern for both D-^{3}He and D-T, but is larger in D-^{3}He because of the higher energy protons. This loss leads to bombardment of the first wall by fast ions and results in an increased local surface heat load. Since all the D-^{3}He fusion yield is in the form of charged particles and is deposited in the plasma, it is transferred to the first wall and divertor as surface heating. In a D-T plasma 80% of the fusion yield is in neutron energy which is deposited in the blanket and shield as volumetric heating. Consequently, the surface heat loads are higher in a D-^{3}He plasma for the same fusion power density in the plasma.

2. POWER REACTOR DESIGN APPROACHES

Several studies have been made in the past of possible power fusion reactors involving the D-^{3}He cycle. Before the announcement of the presence of ^{3}He in the moon, a large fraction of these studies has focused on a solution for the resource problem by pursuing "satellite" D-^{3}He reactors with separate D-D or p-^{6}Li reactors producing the ^{3}He /3,4/. In the next chapter we will discuss, among the feasibility problems, the question of the ^{3}He supply from the moon. The question of the most suitable plasma configuration to exploit the possible advantages of the D-^{3}He cycle was also investigated. In our review we will consider only the case of tokamak configurations even if, in some respect, this does not appear a priori as the most suited plasma configuration for the D-^{3}He cycle /3/.

The assessment of the parameters of a D-^{3}He tokamak power reactor depends on the hypotheses made both on the expected plasma physics and technological constraints. In the following we will show, as a matter of example, two extreme cases: one which assumes the minimum of extrapolation from the present situation, and a second one which exploits some more futuristic assumptions. The first case is that contained in a MIT study made by S.J. Brereton and M.S. Kazimi on "Safety and economics of fusion different fuel cycles" /5,6/, namely D-T, D-D and D-^{3}He. The second case is the APOLLO design by the University of Wisconsin /7/.

2.1 Design Parameters

The set of parameters of the MIT reactor is shown in Table 2.1, compared to the D-T case considered in the same study as reference; in both cases the fusion power was about 3700 MW and a value of beta equal to 0.1 has been assumed. One can note that the maximum toroidal field does not exceed the present limit of 11.5 T taken for NET and ITER. The average ion temperature is 33 keV, the minimum value compatible with the ignition requirements shown in the Introduction. These hypotheses lead to large dimensions of the reactor (10.0 m of major radius and 2.5 m of minor radius, as compared to 6.5 m and 1.6 m, respectively, for the D-T case), as a consequence of the low power density. The shielding requirements are lower in case of D-^{3}He, as a consequence of the lower neutron wall loading

Table 2.1. Reactors Key Parameters

| | MIT study /5/ | | APOLLO*/7/ |
	D-T	D-^{3}He	D-^{3}He
Major radius (m)	6.5	10.0	8.0
Minor radius (m)	1.6	2.5	2.0
Toroidal field on axis (T)	3.7	7.6	12.9
Max coil toroidal field (T)	6.7	11.7	20.0
Plasma current (MA)	12.5	39.3	47.0
Beta	0.1	0.1	0.063
^{3}He/D density ratio		1.0	0.77
Average ion density (10^{15} cm^{-3})	2.2	2.8	1.3
Average ion temperature (keV)	11.3	33.3	69.0
Fusion power (MW)	3645	3763	2872
Total wall loading (MW/m^2)	5.2	2.31	1.1
Neutron wall loading (MW/m^2)	4.16	0.18	0.1
In-board blanket/shield (m)	0.95	0.64	0.85
Electrical power (MW)	1225	1213	1204

*Case LSS: microwave conversion only, stainless steel shield.

(0.2 MW/m^2 compared to 4.2 MW/m^2 for the D-T case). The same applies to the total wall loading, even if the heat flux at the surface of the first wall is higher (2.1 MW/m^2 as compared to 1.0 MW/m^2 in the D-T case). In the design here considered the hypothesis is made that the power from the charged particles and from bremsstrahlung and synchrotron radiation is recovered as heat through a standard thermal cycle, as in the case of the D-T systems.

The main parameters of the APOLLO reactor are shown again in Table 2.1. The fusion power is about 2900 MW. Also in this case a low beta factor is assumed (0.063).

The main differences in the design constraints and solutions are the following:
- the maximum toroidal magnetic field is now 20 T;
- the average ion temperature is 69 keV;
- synchrotron power, which is $\sim$ 60% of the total one, is converted directly to electricity by use of solid-state rectifying antennas (rectennas);
- the ^{3}He/D density ratio is 0.77.

The hypothesis on magnet field leads to a more compact reactor, the major radius and minor radius being 8.0 and 2.0 m, respectively, as well as to a higher plasma and power density. The ion temperature is also much higher (69 keV as compared to 33 keV); this enables to maximize the fraction of the amount of synchrotron reaction which is directly converted into electricity.

The average first wall heat flux is 0.9 MW/m , including all the bremsstrahlung and one third of particle loss. At the high magnetic field strengths considered, the plasma current is large (47 MA) and transport across the magnetic field is small ($\sim$ 10% of the fusion power). The average neutron wall loading is 0.1 MW/m^2, even lower as compared to the previous case. This leads to important advantages on first wall lifetime and safety as compared to D-T reactors, namely:
- The maximum neutron fluence is $\sim$ 3 MWy/m^2 (40 dpa), a value much lower as compared to the case of D-T power stations (10-20 MWy/m^2); this low radiation damage, coupled with the expected relatively low structural temperature, should enable the first wall, even if made of stainless steel, to stand without replacement for the whole life of the plant (30 years).
- The low decay heat makes the reactor inherently safe. Indeed it was shown that in case of the worst possible accident, i.e. the loss of coolant accident (with plasma still active for 10 seconds) the temperature increase of the first wall after one week would not exceed 400°C (adiabatic case), so leading to a maximum value of the order of 600°C.
- The level of radioactive wastes after decommissioning is low. In the APOLLO study a Manganese-Chromium austenitic stainless steel (TENELON) is assumed for the first wall and blanket structures. Ths steel is similar to that under investigation at the Joint Research Centre as a low activation material /8/. The European Mn-Cr steel has been supplied by the French firm Creusot-Loire (now UNIREC) in two batches (AMCR and IFS); the experimental data base implementation at Ispra for such a type of steel is expected to be completed by the end of 1991. In case of the APOLLO design the Waste Disposal Rating for the first wall and shield are so low that these components can be disposed, according to the USA criteria (NRC-10CFR61), $\sim$1 metre below surface with minimal restrictions on the container (Class A). Unfortunately in Europe no unified rules exist for Low Level Waste Disposal, so today the concept of Shallow Land Burial cannot be taken as a reference for such type of analysis /9/.

Tritium inventory as well as the net loss of tritium from D-^{3}He plasma are low (22 g and 24 g/full power day, respectively). Even if we assume that all tritium could be released in an accident to the environment, the exposure to a member of the public at the side boundary would result to be 1 Rem, i.e. below the limit for evacuation (5-15 Rem).

2.2 Economic Assessments

Economic analysis has been carried out for both D-^{3}He reactor designs. Let us start with the MIT study. In Table 2.2 a comparison of the major contributors to the Cost of Electricity (COE) for the D-^{3}He and the reference D-T reactor systems is presented. As expected, the COE of the D-^{3}He power station is much higher than that of D-T. The reasons of this increase are related to the conservative assumptions made and explained previously. In case of APOLLO the situation is reversed (see Table 2.2). The direct capital cost is lower, even when compared to the D-T case. This reflects the assumptions made in the APOLLO design, namely the high magnetic field and the fact that a large fraction of the fusion power is directly converted to electricity. The influence of the ^{3}He fuel cost for the APOLLO case is shown in Table 2.2. The impact on COE is relevant; therefore, the question of fuel supply, which will be discussed in the next chapter, plays, as expected, an important role in the economical evaluations related to D-^{3}He power stations.

As known, the economic evaluation on power generation of future power systems is always very questionable, because it depends largely on assumptions related to components whose technological development is undefined. As a matter of example, in APOLLO an optimistic scaling relationship has been used for the high field toroidal coils which leads to an increase in the cost of only a factor three as compared to the reference cost used in the ESECOM study /10/. Therefore, one can derive only little conclusions from such a type of analysis. In showing these results, we would like simply to stress the fact that some basic assumptions of the type of those made in APOLLO are mandatory in case of D-^{3}He systems in order to get values for the cost of electricity at least comparable with those of the D-T systms. This applies, in particular, to the need of high field superconducting coils and to the direct conversion of energy.

On the other hand, in case of D-^{3}He systems the advantages related to the reliability of the nuclear components and to the safety and environmental aspects are without any doubt relevant as compared to the D-T

Table 2.2. Costs Breakdown

| | MIT study | | APOLLO° |
	D-T	D-^{3}He	D-^{3}He
Direct costs (M$)	2165	3667	1444
Indirect costs (M$)	1495	1504	433
Contingency (M$)	461	776	282
Total Capital Cost (M$)	3922	6954	1259
Operation and maintenance (M$/y)	58	47*	+
Fuel cycle (M$/y)	21	107*	+
Cost of Electricity (mills/kWh)	57	90	46*(40)

+not available
*based on ^{3}He fuel cost of 700 $/g (in brackets: value for 200 $/g)
°nuclear grade construction

reactors. This confirms the fact that it is along this direction that one should find the interest in pursuing the effort on D-^{3}He reactor concepts.

3. FEASIBILITY ASPECTS

In the Introduction we have already mentioned the fact that the physics of D-^{3}He cycle is harder as compared to the case of D-T cycle. Furthermore, in the chapter on the design concepts, we have seen that, in order to be economically attractive, D-^{3}He power reactors must rely in several areas on advanced technology. Here we will try to detail the aspects which look as the most critical from the point of view of a possible exploitation of this type of fusion reaction for commercial power production. Therefore, we will not analyse in detail the feasibility of the problems related to ignition in D-^{3}He systems. We will limit ourselves to some general remarks.

3.1 Ignition Problems

Several studies on this subject have been presented in the last few years. Of particular relevance are those related to the high magnetic field experimental tokamak proposed by B. Coppi to test ignition of D-^{3}He fuel, named CANDOR /11,12/. In these studies the proposal is made to raise the plasma temperature up to the values required for D-^{3}He ignition by burning first a D-T plasma to ignition. Thus, tritium is used as a "match" to increase the plasma temperature and, as this temperature increases, is gradually replaced by ^{3}He. Radiofrequency heating at a frequency corresponding to the first harmonic of the cyclotron frequency of ^{3}He, was proposed. The basic parameters proposed for CANDOR are shown in Table 3.1.

A numerical study of ignition scenarios has been developed in /12/ showing that central D-^{3}He ignition may be possible if the thermal transport is not too large. Burning of D-T provides an important boost to the beginning of the ignition process, but may not be sufficient in itself to ignite D-^{3}He, because the allowable concentration of tritium is severely limited by the large neutron yields that result at temperatures near D-^{3}He burning levels. The neutrons must be taken into account in the machine design, and means should be used to minimize their production. On the other hand, relatively large amounts of auxiliary heating are required to ignite a pure deuterium plasma into which ^{3}He is fuelled. In general, ^{3}He can only be added after the temperature has reached the point where some heating from D+^{3}He reactions is possible (e.g. T $\sim$ 20 keV), to avoid prohibitive external heating requirements. Other problems not yet investigated are those related to the ideal MHD stability of specific equilibrium configurations and those related to fuelling and ash removal. Effective

Table 3.1. CANDOR-18 Parameters

Plasma current	$I_p \simeq 18$ MA
Mean poloidal field	$B = 2I/(5/a+b)) \simeq 4$ T
Minor radii	$a \simeq 0.65$ m, $b \simeq 1.15$ m
Major radius	$R_o \simeq 1.7$ m
Toroidal magnet current	$I_M \simeq 115$ MA-turn
Engineering safety factor	$Q = (ab/R_o^2)(I_M/I_p) \simeq 1.65$
Toroidal vacuum field	$B \simeq 13.5$ T
Plasma current density	$(J_{||}) \simeq 766$ A/cm^2
Plasma volume	$V_o \simeq 25$ m^3
Plasma surface area	$S_o \simeq 67.6$ m^3

central fuelling without ash removal should be sufficient to achieve self-substaining D-^{3}He burning, although the central region must be supported by other sources of heating while the cold ^{3}He is being added, if the energy confinement is sufficiently degraded over ohmic. In case of D-T approach, this heat is supplied by the D-T burning.

The possibilities for breakeven and ignition when using D-^{3}He fuel in an experimental tokamak reactor such as NET/INTOR has also been recently investigated /2/. It was found that, depending on the energy confinement scaling law, energy breakeven may be achieved without significant modification to the NET-I design /13/. The best results are for the more optimistic ASDEX H-mode scaling law. Kaye-Goldston scaling with a modest improvement due to the H-mode is more pessimistic and makes achieving breakeven more difficult. Significant improvement in Q (ratio of the fusion power to the injected power), or the ignition margin, can be achieved by taking advantage of the much reduced neutron production of the D-^{3}He fuel cycle. Removal of the tritium producing blanket and replacing the inboard neutron shield by a thinner shield optimized for the neutron spectrum in D-^{3}He allows the plasma major radius and aspect ratio to be reduced and the magnetic field at the axis to be increased without changing the magnetic field at the toroidal field magnet. This allows the plasma to achieve higher beta and Q values up to about 3. On the experimental side, it is interesting to mention that some aspects of the helium-particles physics in deuterium plasmas have been recently simulated by radiofrequency (ICRH) heating of ^{3}He minority ions in JET /14/. The behaviour was broadly found as expected.

3.2 Fuel Supply

The ^{3}He consumption of a power reactor of 3000 MW-th will be about 110 kg/year. Analyses of the terrestrial natural and man-made ^{3}He resources have been made recently /1,15/. Several thousand tons of ^{3}He have been identified in various terrestrial reservoirs. The ^{3}He exists, however, as a dilute component 10^{-9} to 10^{-12} volume fraction, of host gases such as the atmosphere or natural gas (methane). Furthermore, it needs to be separated from ^{4}He, the isotopic ratio ranging from 10 ppm to 0.1 ppm. If the ^{3}He were recovered from the gases currently used, 6 kg/yr would be obtained. In case of a large expansion in the use of natural gases containing significant quantities of ^{3}He, the production rate could be increased up to $\sim$25 kg/yr by the year 2000. No evaluation, however, has been made up to now of the cost associated to the helium separation from the host gases and from ^{4}He.

^{3}He is also formed as decay product from tritium (half life: 12.3 yr). Table 3.2 shows the reserves of ^{3}He that could be available in the year 2000; apart from the existing inventories, the main non military supply could come from the decay of tritium produced in the Canadian reactors, estimated to 2 kg/yr by the year 2000. The question of ^{3}He from the decay of tritium in thermonuclear weapons should be also raised. In

Table 3.2. Reserves of ^{3}He that could be available in the year 2000 from T-decay (kg/y)

USA (Dep. of Energy)	1.3
Canada reactors	2.0
US-weapons	15.0
Total	18.3

Table 3.1 a guess was made of a possible production rate, at the year 2000, of up to 15 kg/yr. More recently /14/, a value of 5 kg/yr was given. In conclusion, even in the most optimistic case (and disregarding for the moment the financial charges), the maximum terrestrial production rate of ^{3}He by the year 2000 could not exceed 40–45 kg/y. This means that, whereas for an experimental reactor such as NET/ITER ^{3}He supply could be insured by terrestrial reserves, no possibility of this kind appears for the case of commercial power stations.

As mentioned in the Introduction, an investigation of the extraterrestrial resources of ^{3}He has been made in the last few years /1/. This was derived from the analysis of samples of lunar soil returned by the American Apollo astronauts and by Russian probes confirming that the lunar soil contains helium with an isotopic ratio near to that of the solar wind ($\sim$ 500 ppm). Starting from this observation, a calculation was made of the potential quantity of lunar ^{3}He /1/. The characteristics of lunar soil (regolith) that makes it an effective helium collector is its extremely fine grain size which facilitates its dispersion in the soil. The potential ^{3}He associated with the lunar surface soil was conservatively estimated as 10^9 kg. The average concentration is likely to be 30 to 40 ppm of helium in the lunar mare and an isotopic ratio of about 300 ppm. In the same report, an analysis was also made of the energy required to extract ^{3}He from the lunar regolith and transport it to the earth. A value of 2400 GJ/kg was found which has to be compared to an energy of $6x10^5$ GJ produced by D–^{3}He reactions, which corresponds to an energy payback ratio of 250. No attempt was made to estimate the final cost of the fuel.

In a recent workshop organized by the NASA on this subject /16/ it was concluded that mining, beneficiation, separation and return to the earth of ^{3}He from the moon is possible. This would require a large-scale infrastructure and improvement in technology. Lunar oxygen production plants could provide an early technology demonstration around 2010 to 2020 for ^{3}He production by demonstrating lunar soil mining and processing techniques and by providing an opportunity to produce some ^{3}He as a by-product of the lunar oxygen production process. No evaluation of the cost of fuel mining and transport to the earth was considered possible at this stage of the studies, the target being that of 1000 $/g of ^{3}He.

3.3 Plasma Refuelling

Refuelling of a D–^{3}He power reactor will be a very difficult task. The techniques used today or envisaged for the D–T power reactors become more difficult, if not impossible, for various reasons, in particular due to the higher densities and temperatures.

Pellet fuelling looks hard to be realized both from the point of view of fabrication and ablation in the plasma. Fabrication of a helium pellet is extremely difficult because the critical temperature (3.2 K) is the lowest of any substance. As the temperature is decreased further, it remains a liquid, requiring a pressure of $\sim$ 3 MPa to cause solidification. Because a fuel pellet could not be delivered into the plasma at such a high pressure, the fuel pellet must contain liquid He. An exposed liquid droplet could neither be accelerated nor remain intact during its permeation into the plasma; therefore, the liquid must be encapsulated. Any encapsulating material will contaminate the plasma. Therefore, the Wisconsin people suggest that the liquid He be encased in a thin-walled polymer sphere which is coated with D_2-ice /2/.

Refuelling by neutral beam injection could be also considered as a possibility. Since helium does not charge exchange well with hydrogen isotopes one might think the required injection energy may be considerably less for a helium beam than it is for a deuterium beam. It was calculated /2/ that the ^{3}He fuelling at the centre of the core can be provided by a 10 MW, 300 keV Neutral Beam Injector. However, the technical feasibility of such an injector was not assessed.

Fuelling by plasma injection would have the advantage that it is not isotope-dependent and appears suited for extrapolation to power reactors. Initial modelling of the slowing down of an injected plasma in a tokamak environment indicates that this method would be feasible. However, the experimental data base is almost non existant.

In conclusion, the question of plasma refuelling, which is already very serious for D-T systems, looks as one of the main feasibility problems of the D-^{3}He power reactors.

3.4 Power Density

As mentioned in the Introduction, at constant plasma beta and magnetic field, the fusion power density of a D-^{3}He reactor would be about 50 times lower than that of a D-T reactor. Therefore, it is out of the question that any increase in beta and in the magnetic field is helping for D-^{3}He systems. However, in judging about the feasibility and economics of the system, a better parameter is the total reactor electric power divided by the reactor core mass. This ratio in a D-^{3}He reactor is increased, as compared to the D-T case, by the reduced shielding requirements and by the possibility of a direct conversion of fusion power to electricity. As seen in the previous chapter, the results from system studies made up-to-date indicate that one can device designs where the reactor mass electric power densities is comparable to D-T reactors. In conclusion, the low fusion power density cannot be considered a priori as a critical aspect of the D-^{3}He systems.

3.5 Heat Fluxes

As mentioned in the Introduction, in the D-^{3}He reaction 100% of the produced energy is associated to charged particles as compared to 20% in the D-T case. Therefore, one would expect that the heat flux to the first wall could represent a critical design problem. In reality this does not obligatory happen because the fusion power density is lower as compared to the D-T case. Furthermore, an important fraction of the particle energy is released to the first wall uniformly as electromagnetic radiation. The corresonding heat flux is $\sim$ 1 MW/m^2, then manageable without major problems. The importance of the particle and heat flux to the divertor plates is related to the plasma parameters chosen. As seen in the previous chapter, in case of the APOLLO design the power associated to the charged paticles leaving the plasma into the divertor area is about 200 MW on a total fusion power of about 3000 MW, then comparable (or even lower) to that expected in D-T power reactors. On the contrary, the erosion rate due to the high energy particles could be serious. This point, however, has not yet been quantified in the existing designs. As a matter of example, in Fig.3.1 a summary of the average heat fluxes to the first wall and divertor zones of recent D-T tokamak reactors compared to APOLLO is shown /7/.

3.6 Energy Conversion into Electricity

As mentioned, one of the attractive features of the D-^{3}He systems is the possibility of directly converting the fusion energy into electricity. When this possibility is exploited, one can expect a cost of the energy

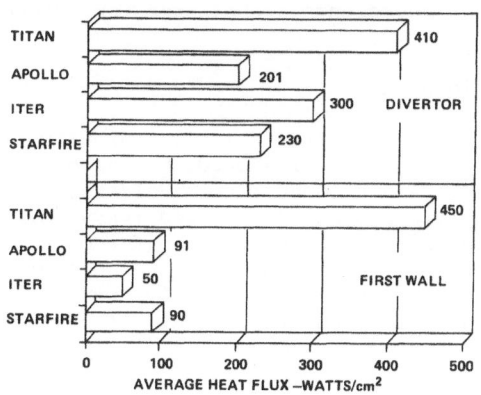

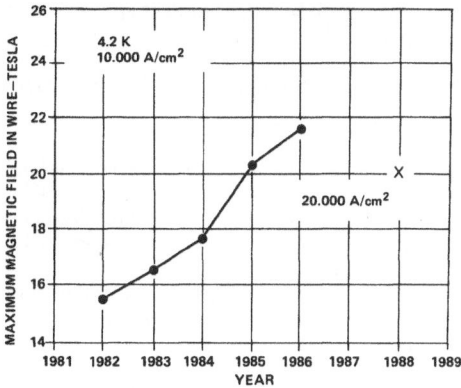

Fig.3.1. Summary of average heat
fluxes to the first wall and divertor zones of recent toroidal
reactor design /7/.

Fig.3.2. Progress in high field Nb_3Sn
super conductors at MIT /7/.

produced by the D-^{3}He power reactors comparable to that of the D-T
systems. Possible approaches of direct conversion include direct
electromagnetic or electrostatic conversion. The second approach is not
applicable to toroidal geometries. An example of direct electromagnetic
conversion was given in the previous chapter, for the APOLLO design. In
this case the synchrotron radiation, which escapes from the plasma at high
frequencies (typically at over 2500 GHz) would be carried by overmoded
wave guides to chambers with rectennas tuned to a selected harmonics.
Fabrication of the rectennas would require the technology of very large-
scale integrated circuits; the specific techniques needed for large-scale
production have not yet been demonstrated.

Another way proposed is that of in-situ MHD conversion /17/. In this
case the synchrotron radiation can either by used to superheat the working
fluid of a MHD generator at its entrance, or can be applied along the
length of the generator to enhance the performance of the MHD energy conversion process. A view of the generator arrangement in tokamaks is shown
in Fig.3.2. Also in this case a number of design and technology problems
have to be better analyzed before getting firm ideas about the feasibility
of the approach. No attempt to evaluate the associated costs has been made
up to now.

In conclusion, the possibility of exploiting the direct conversion of
electromagnetic radiation into electricity does not seem unrealistic but
requires an important design and technological effort.

3.7 Toroidal Magnetic Field

Even if the D-^{3}He reactor concepts are at a preliminary stage, the
need of operating with high toroidal magnetic field appears mandatory. In
the APOLLO design a magnetic field at the coil of 20 Tesla is taken. The
question of the feasibility of such an assumption has to be addressed. The
progress in the last few years on high field Nb_3Sn superconductor wires
has been impressive, as shown in Fig.3.3 /7/. Since a D-^{3}He power reactor
will not be built before at least 30-40 years, the assumption of toroidal
magnetical fields around 20 Tesla does not appear irrealistic. The possibility of coil fabrication is also related to the expected plasma current
and to the question of plasma disruptions. The capability to withstand the

forces induced in such type of accident represents a challenge for future designs.

3.8 Plasma Current and Plasma Disruptions

In designs assuming a low beta (first stability regime), typically of 1%, plasma currents are inevitably high. In case of APOLLO, values of the plasma current ranging from 50 MA to 80 MA are found. These values are of big concern because of the problems with plasma disruption (very high electromagnetic forces in the structures) and power required for current drive if no boostrap or synchrotron current drive mechanisms are possible. In APOLLO an attempt was made to reduce the plasma current by increasing the aspect ratio from 2.5 to 4.0 with a penalty on the cost of electricity of 10%.

If the second stability regime of the plasma will be proved (values of beta 20% or more), plasma currents comparable to those of D-T reactors, of the order of 25 MA, could be envisaged. However, as known, up to now the physics bases for such an assumption are still too weak.

As a conclusion, the question of the high plasma current appears today as a critical one in the D-^{3}He designs.

4.   CONCLUSIONS AND FUTURE DEVELOPMENTS

Recent studies on conceptual designs of commerical fusion power stations based on D-^{3}He fuel cycle have enabled us to better clarify potential advantages and disadvantages of this approach to fusion energy production, as well as to identify the related feasibility aspects.

First of all, it is out of question that the D-^{3}He fusion has to be seen as a further step after D-T and by no means as an alternative.

As compared to D-T power stations, relevant design and operation simplifications of important components such as first wall and blanket, due to the low neutron and tritium production, have been quantified. The same applies to the production standards related to these components and to the safety and environmental aspects of the fuel cycle. It is along this line that D-^{3}He systems could be really attractive.

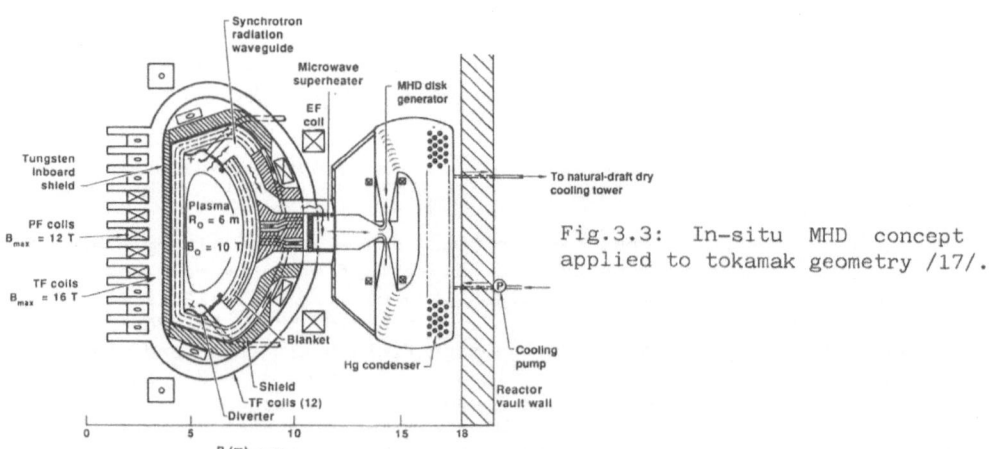

Fig.3.3: In-situ MHD concept applied to tokamak geometry /17/.

On the other hand, it has been shown that, in order to be economically competitive with the D-T stations, the D-^{3}He power reactors will have to rely on advanced design solutions and technology of some of their basic components. This applies in particular to the toroidal field magnets, which should be able to operate at high magnetic field and to systems directly converting into electricity a relevant part of the fusion energy (electromagnetic radiation).

The questions related to the fuel cycle represent the major concern for the D-^{3}He systems. Fuel supply from the lunar soil depends on a number of boundary conditions which are difficult to be defined now. According to the last analyses and discussions, also involving NASA, lunar mining and transport to the earth looks technically feasible. However, all other aspects, in particular the economical ones, are far from being clarified.

Plasma refuelling appears already now to be a very serious concern in D-^{3}He systems because it will require new and difficult approaches as compared to what is foreseen in the D-T systems. Little is known about the question of ash removal. The solution of these last problems will be strongly related to the development of the knowledge of the physics of D-^{3}He plasmas. Ignition represents the first critical point for the whole development in this delicate area.

The route towards getting such an ambitious target, even if very long, seems rather well defined. The exploitation of the capabilities of the existing machines, in a similar way to what is already found in JET, to get first experimental insights in the physics of ^{3}He, is the first step. The CANDOR-type of machine could be the way to investigate the ignition of the D-^{3}He systems; in a later stage machines along the line of NET/ITER could represent the test bed for the plasma engineering aspects, in conditions not far from those of a reactor. The time scale for such a development strictly depends, among others, on the fuel supply capability. Indeed, according to the present evaluations, fuel supply from the moon, which is mandatory for the power stations, could hardly be insured before the second quarter of the next century. On the contrary, ^{3}He supply for experimental machines operation seems not impossible from terrestrial sources after the year 2000.

In any case the present interest on D-^{3}He power reactor analyses will help in defining the future strategy. In this frame one has to mention the scoping activity started recently in the ARIES programme on D-^{3}He fuel cycles /18/. Safety and Environmental Studies on D-^{3}He systems are also foreseen in Europe as a part of the 1989-1991 Fusion Technology Programme. These studies should have to be supported by further evaluations on the feasibility and cost of the lunar ^{3}He supply, in a period where extraterrestrial exploration seems to become again appealing.

ACKNOWLEDGEMENTS

The author would like to thank S.J. Brereton, B. Coppi, G.L. Kulcinski, F. Najmabadj and K.R. Schultz for supplying information of relevance for the present review.

REFERENCES

1. L.J. Wittenberg, J.F. Santarius and G.L. Kulcinski, "Lunar source of He for commercial fusion power", Fusion Techn., Vol.10, No.2 (1986).
2. G.A. Emmert, L. El-Guebaly, R. Klingelhoefer, G.L. Kulcinski, J.F.

Santarius, J.E. Sharer, I.N. Sviatoslavsky, P.L. Walstrorn and L.J. Wittenberg, "Possibilities for breakeven and ignition of D- He fusion fuel in a near term tokamak", KFK-4433 (1988).

3. G.H. Miley, "Advanced fuel concepts and applications", Fusion Reactor Design and Techn., Vol., IAEA, Vienna (1983).

4. G.W. Shuy, A.E. Dabiri and H. Gurol, "Conceptual design of a deuterium- He fuelled tandem minor satellite/breeder system", Fuion Techn., Vol.10, No.6 (1986).

5. S.G. Brereton and M.S. Kazimi, "A comparative study of the safety and economics of fusion fuel cycles", Fusion Engineering and Design, Vol.6, No.4 (1988).

6. S.J. Brereton, J.E. Manida and M.S Kazimi, "safety comparison of fusion fuel cycles", Fusion Techn., Vol.15, No.2 (1989).

7. G.L. Kulcinski, G.A. Emmert, J.P. Blanchard, L. El-Guebaly, H.Y. Khater, J.F. Santarius, M.E. Sawan, I.N. Sviotolaski, L.J. Wittenberg and R.J. Witt, "APOLLO - An advanced fuel fusion power reactor for the 21st century", UWFD-780 (1988).

8. G. Casini and P. Rocco (Eds.), "Joint Research Centre Progress Report 1987 on Fusion Technology and Safety", EUR-11697 (1988).

9. C. Ponti, "Special materials for fusion reactors", this Seminar (1989).

10. J.P. Holdren, Chair, "Summary of the Report of the Senior Committee on Environmental, Safety and Economic Aspects of Magnetic Fusion Energy", UCRL-537/66 (1987).

11. B. Coppi, "Near-term feasibility of 'CANDID' fusion reactors", MIT Report PR-80/24 (1980).

12. B. Coppi and L.E. Sugiyama, "Questions in advanced fuel fusion", MIT Report PTP-88/6 (1988).

13. R. Toschi, Chair, "Next European Torus (NET)", Fusion Techn., Vol.14, No.1 (1988).

14. R.J. Bickerton and the JET-Team: "Latest Results from JET", Europhysics News, Vol.20, No.1 (1989).

15. J.L. Wittemberg, "Terrestrial sources of helium-3 fusion fuel. A trip to the center of the Earth", Fusion Techn., Vol.15, No.3 (1989).

16. G.L. Epstein, J.B. Plescia and E.A. Gabris, "Summary of the NASA Lunar Helium-3/Fusion Power Workshop", April 1988, Fusion Techn., Vol.15, No.1 (1989).

17. M.A. Hoffman, R.B. Campbell and B.G. Logan, "Advanced fusion MHD power conversion using the CFAR cycle concept", Fusion Techn., Vol.15, No.3 (1989).

18. F. Najmabadi and K.R. Schultz, personal communication (1989).

FEASIBILITY, SAFETY AND ENVIRONMENTAL ASPECTS
OF D-He³ FUSION

Manfred Heindler

Alternate Energy Physics Program
Institute for Theoretical Physics
Graz University of Technology
A-8010 Graz, Austria

INTRODUCTION

Out of the long list of potential fusion fuels that have been studied in the past, D-He³ emerged as clear favorite among the various competitors to D-T as the fuel of future fusion reactors. D-He³ seems to represent the only attractive compromise between aspects related to power balance, radioactivity load, neutron production and confinement requirements. Assuming the feasibility of both fuel cycles in their respective optimum confinement concept (maybe steady-state Tokamaks for D-T, and Field-Reversed Configurations for D-He³), the respective engineering, safety and environmental features and the aspect of dependence on a lunar resource have to be carefully evaluated.

The outcome of such an assessment and the conclusions derived therefrom will not be unique. The various countries are likely to have a different perspective on anticipated public acceptance patterns in a few decades from now, on the speed of technological progress to be expected, on the desirability of the exploitation of lunar resources and on the consequences of unequal access to space shuttles. In addition, the interweaving between energy and defense interest extends from the application of defense developments to civilian fusion energy – e.g. large scale laser isotope separation to produce isotopically taylored materials, high gain inertial confinement pellets – to the defense application of civilian tritium, and this may be seen differently by the various nations.

In any case, the data base which is presently available for the assessment of future fusion reactors is too poor even for the most advanced concept, the D-T based Tokamak, to allow a sound comparison between the appeal of fusion vs. that of fission power. The data base for alternatives to D-T fueled Tokamaks is even poorer, in particular if one does not only consider the unfavourable combination of D-He³ with low-beta Tokamaks but also confinement concepts which are more adapted to the intrinsic properties of D-He³.

Safety, Environmental Impact, and Economic Prospects of Nuclear Fusion
Edited by B. Brunelli and H. Knoepfel
Plenum Press, New York, 1990

173

INTRINSIC FEATURES OF D-T AND D-He³ AS FUSION REACTOR FUELS

In addition to the well known basic differences between D-T and D-He³ that are related to their respective nuclear properties, it is interesting to explore the behaviour of these fuel mixes in a reactor environment. One aspect is the flexibility and the sensitivity to variations in the fuel mix ratio. In the case of D-T, deviations from the optimum 1:1 mix ratio to higher or to lower tritium concentrations both result in a reduction of the power density, with only marginal effect on the shape of the fusion power vs. plasma temperature curve or on the position of its maximum. Since the only viable source of tritium seems to be in situ blanket breeding, it is of interest to note that the sensitivity of the fuel mix ratio and, in consequence, of the power density to the blanket breeding performance is disproportionally high. If the breeding ratio falls short of the value required to support a 1:1 mix by ten percent, the pertinent power density in the plasma drops by almost one order of magnitude.[1,2,3,4,5]

In contrast, D-He³ is found to be a more flexible and tolerant fuel. The power density varies roughly by a factor of three for He³:D mix ratios between He³-poor fuels (such as He³-catalized deuterium where only deuterium is externally provided) and He³-rich fuels (e.g. He³:D of the order of 2:1), and has a very flat maximum as a function of the operating temperature. Therefore, D-He³ offers the possibility to optimize the fuel mix with respect to considerations other than maximum power density. As a result, the possibility to achieve high power density, low neutron load and low tritium inventory simultaneously is found to be one of the most appealing features of D-He³. The sensitivity of the width of the temperature window for ignited operation to the fuel mix ratio is small for mix ratios between those for He³-catalized deuterium and ratios around 2:1. Thus an increase of He³-availability goes along with decreasing tritium inventories and decreasing neutron yield. Above the 2:1 He³:D ratio, the low reactivity parameter of He³-He³ reactions closes the ignition domain in the fuel mix ratio vs. temperature parameter space and makes tritium inventories below 1 Ci per GW and neutron power fractions below a few percent of the total fusion power unaccessible. The flat extrema of fusion power, tritium inventory and neutron yield as a function of the operation temperature give room for the optimization of further aspects such as the charged particle to radiation power ratio in the power exhaust.[5,6]

This clearly shows that the higher operating temperature needed in the case of D-He³ may be outweighed, in a broader analysis, by the larger flexibility in the choice of the operating point in the parameter subspace for ignition, without losing any of the potential advantages of D-He³ over D-T.

PROLIFERATION ISSUES

Nuclear fusion has always been considered an important candidate for powerful neutron sources, favoured over its competitors by its low power-per-neutron figure. A world of neutron abundance has been envisaged [7] in which the combination of fusion as neutron source and fission as power source would play a major role.[8] These systems would be characterized by highly coupled neutron flows, fissile and fusile fuel flows and energy flows, and seem to have the important potential to bridge the time span and the developmental gap between experimental reactors and economical, reliable fusion power reactors. However, this fusion-fission symbiosis and the potential use of fusion neutrons to breed defense-related materials has also given rise to proliferation concerns.

In addition, an assessment of the potential near-term defense application of fusion seems to show the appeal even of magnetically confined "pure" fusion research in that respect. [9]

"Aneutronic" fusion has consequently been praised as the "peaceful" alternative, where neutrons as the main energy carrier are replaced by charged particles – with their direct conversion potential – and by electromagnetic radiation. Even aneutronic fuel is not free of worry, however. For confinement times which are of the right order to permit ignition – too short implies unsufficient energy confinement, too long implies excessive ash accumulation – He^3 was found to reach burn-up fractions of a few to more than ten percent, while the tritium burn-up is around fifty percent. This motivates an investigation of the impact of various tritium recycling strategies (full recycling, no recycling, recycling as He^3 after radioactive decay). It is found that tritium recycling does little to improve the power balance or the ignition margin in the parameter range of interest, but strongly increases the adverse effects of neutron load and tritium inventory.[6,10] Therefore the optimum solution would be to store the exhaust tritium and to recycle it as He^3. This however implies huge storage requirements of tritium and the pertinent risk for the environment and of diversion for use as weapons material. According to our calculation, between 5 and 50 kg tritium can be recovered per year and per GW fusion power, depending on the tritium burn-up fraction and the D-He^3 mix ratio. The maximum is achieved for He^3-catalized deuterium (all He^3 and no tritium recycled). This represents tritium quantities of the order of, or largely in excess of the estimated annual tritium requirement of the US defense program.[6,11]

D-He^3 vs. D-T IN TOKAMAKS

Most recent studies have investigated, or are investigating the feasibility of D-He^3 as fuel in Tokamak reactors. D-He^3 is considered either as precursor of D-T operation in a low-radioacitivity physics phase or as potential substitute to D-T in the next generation of experimental Tokamak reactors – examples are studies for NET [12] and announcements for ITER [13] – or as fuel for specially designed commercial power reactors.[14,15] Clearly, the low beta value of Tokamaks makes this concept not very well suited for any fuel other than D-T: For D-T the material related limitation to neutron wall loading and the stability related limitation to plasma density seem to merge into the same constraints for the fusion power density. In the case of alternate fuels such as D-He^3, the low nuclear power-density-parameter $<\sigma v>/T^2$ requires a compensating increase of the confinement power-density-parameter βB^2, either by an increase of beta or of the magnetic field strength B or of both. The Tokamak in the second stability regime (SSR) and the high field Tokamak (HFT) – both under investigation in the Advanced Reactor Innovation and Evaluation Study (ARIES) – represent the two approaches which, however, were not motivated by the requirements imposed by the use of advanced fuels, but by the quest for an attractive commercial D-T fueled Tokamak reactor. In the case of D-T fueled HFT concepts, the synchrotron power density is found to be well above that of bremsstrahlung and radiation becomes an important power flow mechanism. This is equally true for D-T fueled standard low-beta low-field confinement approaches if the first wall is made out of low-activation ceramic materials which have high electric resistivity and consequently low reflectivity for synchrotron radiation.

Thus synchrotron radiation, for some time considered to be an essential drawback of advanced fuels with their high operating temperatures, has entered the arena of high field, low beta devices regardless of the fuel that is envisaged. Thus it is no longer a "privilege" of advanced fuels but also affects D-T fueled devices, in particular if low reflectivity materials are used for first walls and if high central temperatures result from the choice of strongly peaked temperature profiles. As a consequence, solutions have to be found, and are being developed, even in the case of advanced D-T concepts. Not surprisingly, this recognition has revived some old ideas and produced some novel ones about how to use synchrotron radiation in a beneficial way, e.g. as passive mechanism for current drive and as a power extraction mechanism with high efficiency potential.

Within this new scenario of D-T and D-He^3 reactor concepts, synchrotron radiation has become associated with certain types of confinement concepts rather than with certain types of fusion fuels. As a matter of fact, the synchrotron radiation to bremsstrahlung density ratio is lower for D-He^3 in a second stability Tokamak than for D-T in a high field Tokamak. The reflectivity coefficient plays a major role in this comparative assessment and works in favour of advanced fuels: In the case of D-T, safety considerations urge the use of low activation materials; some candidate materials have wall reflectivities well below 0.8 which results in an increase of the synchrotron power density by roughly an order of magnitude.[17] This results in a combination of both high neutron fluences and high radiation loads on the first wall in the case of D-T. In case of aneutronic fuels such as D-He^3, neutron fluences and activation problems are of minor concern and thus D-He^3 offers more freedom in the design of the first wall and of the blanket.[19,17,5,18,20]

D-He^3 IN FIELD-REVERSED CONFIGURATIONS (FRCs)

With respect to the aspect ratio, Field-Reversed Configurations are on the lower end among the various members of the family of toroidal confinement devices. FRCs offer several important advantages over Tokamaks and other concepts with higher aspect ratio.[16] Among these advantages are the open field lines around the toroidal closed field line region – which are expected to act as natural divertor and to focus charged particles into direct convertors – and the high beta which is expected to be between 0.5 and 1.0. Thus, FRCs are considered to ideally match the intrinsic properties of advanced fuels, and of D-He^3 in particular. However, the data base of this confinement concept is so weak that only exploratory assessments can be undertaken. The uncertainty of the extrapolation from existing empirical data to those needed for reactor relevant dimensions, and the uncertainties associated with theoretical scaling expressions are tremendous. Rather than evaluating the expected behaviour of D-He^3 in FRC reactors, one is thus lead to calculate the requirements which an FRC would have to meet in order to qualify as a fusion reactor.[17,18]

If a D-He^3 fueled FRC is calculated for such a parameter set, the neutron power fraction is found to be as low as expected, i.e. of the order of a few percent, a minimum reactor size and fusion power is identified (about 1 m separatrix radius and several hundred MW), the confinement time requirement is of the order of a second (bounded with respect to low and to high values), the operating temperature can be chosen quite freely , the optimum reversal factor would be around one, and magnetic fields resulting in admissible wall loadings would be of the order of 8 Tesla. The major finding is that among the various theoretical and empirical confinement scalings for FRCs, only the optimistic ones would be good enough, and that a bad confinement quality could not be sufficiently compensated by going to higher magnetic field strengths due to the pertinent wall loading figures.[18]

PASSIVE SAFETY AND OTHER ADVANTAGES OF D-He³

With respect to passive safety, D-He³ as a reactor fuel offers several important advantages. For one, the neutron load is reduced by one to almost two orders of magnitude. The remaining neutrons need not be transported and multiplied, and no special materials such as tritium breeding ones need to be exposed to them. This strongly increases the potential for low radioactivity inventories, for more economic use of the space within the magnets, and for first wall structures to last the full operational life-time of the reactor. Secondly, the tritium inventory and thus the tritium handling requirement is reduced by one to two orders of magnitude, and components required for tritium breeding are not required.

It is obvious, however, that these findings can also be seen as disappointment for a fuel which is generally designated as "clean" alternative to D-T: There are still enough neutrons to impose remote handling of the reactor componenents and to require heavy shielding of the magnets, and there is still enough tritium envolved to require the pertinent installations for handling, purification etc.. And since neutron minimization suggests to strive for a tritium burn-up as low as possible, large amounts of tritium would become available and would have to be safeguarded. This picture is further clouded if tritium has eventually to be used to ignite the D-He³ fuel and to bring it to the higher operating temperatures.

The potential for efficient energy conversion is another asset of D-He³, in particular if charged particle power and synchrotron power can be converted directly, eventually making fusion power more than just another way of boiling water. New safety problems may arise in this context, e.g. if mercury had to be used as the working fluid in the energy converters as is being suggested in some current proposals.

CONCLUSIONS

In the competitive world of commercial power production, the relative advantages of D-He³ over D-T may turn out not to be large enough to ensure fusion a warm welcome on this market. But they may as well turn out to be the crucial elements in the endeavour towards public and utility acceptance.

Generally, the scientific, engineering, operational and capital requirements of a fusion based power economy seem to hamper the exploitation of fusion by countries with an unsufficient infrastructure and with limited economic strength, probably to a much larger extent than in the case of fission. The large-scale use of He³ as fuel would certainly introduce a new dimension in so far as it further reduces the accessibility of the various nations to fusion power, which is very much in contrast to the advantage originally claimed for fusion, namely to rely on a fuel which can be extracted from every water pool. D-He³ fusion would leave countries without space programs to the mercy of a market for He³, with all consequences of such a dependence on few countries with moon mining and space transportion capabilities.

In addition there appear new trends in the fission power economy, i.e. in the domain of that competitor to fusion which is very similar to it in many respects, certainly more similar than a solar based power economy would be. Therefore with the announcement of fission power to have its second generation rely extensively on passive safety features, the challenge for fusion to prove its future superiority and to justify its very development did not become easier.

This seems to be a difficult moment for fusion policiy makers: Should the potential use of D-He^3 be further disregarded in the mainline research programs in order to avoid the repeated reminding of D-T specific problems which might be avoided with D-He^3? Or should He^3 be seriously considered as a fuel alternative to D-T for which, on top of all scientific and engineering progress that would have to be achieved, the proof has yet to be made that it can be brought from the moon both efficiently and economically and that it would be made as universally available as the other fuel component, deuterium?

ACKNOWLEDGEMENT

This paper relies heavily on work performed by my coworker Winfried Kernbichler and by G. H. Miley and his coworkers. Their input is highly appreciated. Thanks are due to R. Feldbacher and A. Schönfelder for valuable comments. Financial support was granted by Fonds zur Förderung der Wissenschaftlichen Forschung, Friedrich Schiedel-Foundation, International Atomic Agency, Department of Science and Research (all Austria).

REFERENCES

[1] W. Kernbichler and M. Heindler, Compound light ion fuel cycles: an approach to optimization, in: "Alternative energy sources", T. N. Veziroglu, ed., Hemisphere Publication Corp., Vol.3 (1987) 393-404.

[2] W. Kernbichler, M. Heindler and R. Feldbacher, The promise of tritium lean fuels, in: Proc.14th Symp. Fusion Technology, Sept. 8-12, 1986, Avignon. Pergamon Press (1986) 1869-1874.

[3] W. Kernbichler and M. Heindler, D-based fueled cycles in MCF-Reactors: Trade-offs between neutron-, radioactive- and power-production, in: Proc. 4th ICENES, June 30 - July 4, 1986, G. Velarde, ed., World Scienific Publishing Co. (1987) 447-450.

[4] W. Kernbichler, R. Feldbacher, M. Heindler, A. Nassri and K. Schöpf, D-based fueled cycles in MCF-Reactors: Trade-offs between neutron-, radioactive- and power-production, in: IAEA-CN-47/H-III-9, IAEA STI/PUB/723 Vol.3 (1987) 373-385.

[5] M. Heindler and W. Kernbichler, Advanced fuel fusion, in: Proc. 5th Int. Conf. on Emerging Nuclear Energy Systems, Karlsruhe, FRG (1989).

[6] W. Kernbichler, G. H. Miley and M. Heindler, D-He^3 fuel cycles for neutron lean reactors, Fusion Technology 15 (1989) 1142-1147.

[7] Proc. IIASA Workshop on a perspective on adaptive nuclear energy evolutions: Towards a world of neutron abundance, May 25-27, 1981, IIASA, Laxenburg, Austria. Springer 1983.

[8] A. A. Harms and M. Heindler, "Nuclear energy synergetics", Plenum Press, New York (1982), Tokyo (1986).

[9] D. L. Jassby, Tokamak fusion generators for nuclear radiation effects testing, IEEE Transaction on Nuclear Science Vol.NS-29, No.5, Dec. 1982, 1519-1524.

[10] W. Kernbichler, "Fuel alternatives in nuclear fusion: a physical assessment of the possibilities"(in german), PhD Thesis, Graz University of Technology, Austria (1987).

[11] W. Kernbichler and M. Heindler, Neutron-lean fusion reactor studies for thermal plasmas, Nucl. Instrum. & Methods in Physics Research A217 (1988) 65-70.

[12] G. A. Emmert, L. El-Guebaly, R. Klingelhöfer, G. L. Kulcinski, J. F. Santarius, J. E. Scharer, I. N. Sviatoslavsky, P. L. Walstrom, J. L. Wittenberg, Possibilities for breakeven and ignition of D-He^3 fusion fuel in a near term Tokamak, report KfK 4433, FPA-88-2, Kernforschungszentrum Karlsruhe (1988).

[13] ITER Management Committee: ITER concept definition, report ITER-1, International Atomic Energy Agency, Vienna (1988).

[14] G. L. Kulcinski, G. A. Emmert, J. P. Blanchard, L. El-Guebaly, H: Y. Khater, J. F. Santarius, M. E. Sawan, I. N. Sviatoslavsky, L. J. Wittenberg, R. J. Witt, Apollo – an advanced fuel fusion power reactor for the 21st century, Fusion Technology 15 (1989) 1233-1244.

[15] The ARIES Tokamak reactor study: The collection of papers presented at the 10th Int. Conf. on structural mechanism in reactor technology, Anaheim, CA, Aug. 14-18, 1989, University of California report UCLA-PPG-1244 (1989).

[16] M. Tuszewski, Status of the Field-Reversed Configuration, Fusion Technology 11 (1987) 436-450.

[17] W. Kernbichler, G. H. Miley and M. Heindler, Comparison of the physics performance of D-He^3 fusion in high- and low-beta torodial devices, AEP report 89.028 (1989), Alternate Energy Physics Program, Instiute for Theoretical Physics, Graz University of Technology, Austria.

[18] W. Kernbichler, G. H. Miley and M. Heindler, Comparison of the physics performance of D-He^3 fusion in high- and low-beta torodial devices, in: Proc. 5th Int. Conf. on Emerging Nuclear Energy Systems, Karlsruhe, FRG (1989) 192-196.

[19] M. Heindler and W. Kernbichler, Advanced fuel fusion, AEP report 89.026, Alternate Energy Physics Programm, Institute for Theoretical Physics, Graz University of Technology, Austria.

[20] W. Kernbichler, G. H: Miley and M. Heindler, The role of synchrotron radiation in. future fusion devices, in: Proc. 2nd Int. Symp. on aneutronic power, April 28-29, 1989, Washington, D.C., USA, to be published.

IV. COMPONENT RELATED SAFETY AND ENVIRONMENTAL PROBLEMS

FIRST WALL AND BLANKET SAFETY

Mujid S. Kazimi

Department of Nuclear Engineering
Massachusetts Institute of Technology
Cambridge, MA 02139

INTRODUCTION

For fusion to become a desirable source of energy in the next century, it must meet the increasingly demanding societal standards for environmental and safety acceptability. A case can be made for a potential advantage for fusion in the safety area because of the current understanding that there is no serious threat of power runaway transients. On the other hand, the impact of undercooling transients will depend on the design of the fusion reactor, particularly on the first wall load and blanket materials.

In this paper an attempt is made to explore the thermal design conditions that will ensure passive decay heat dissipation following both loss–of–flow and loss–of–coolant events for a variety of experimental and power reactor blankets. The effects of neutron flux modifiers, both absorbers and multipliers, will also be explored. The details of the work summarized here can be found in the reports by Massidda[1] and Parlatan[2].

It should be clearly recognized that safety concerns are only part of the design optimization matrix. The final choice of a design will also include economic and waste–management factors. These consideration do not necessarily lead to the same set of choices, and it is important to establish some criteria for the minimum desirable features. For passive safety considertions, both protection of the public and protection of the plant can be highly desirable targets for future fusion reactors. The limits implied by both of these criteria also will be explored in this work.

REFERENCE DESIGNS

The reference designs for this study are summarized in Table 1. The design parameters, such as wall load and material lifetime, are somewhat optimistic in view of today's technology base, but do not appear unreasonable when compared to targets for the research and development efforts. The tritium breeding ratios and blanket multiplication factors were checked to ascertain that they are in the acceptable range. The essential features of the blankets are displayed in Table 1.

Blanket 1 is a representative self–cooled, liquid–metal Tokamak blanket. It consists of liquid lithium breeder/coolant, with vanadium alloy $V-15Cr-5Ti$ structure. Blanket 2 was chosen as a representative gas–cooled, solid breeder

Safety, Environmental Impact, and Economic Prospects of Nuclear Fusion
Edited by B. Brunelli and H. Knoepfel
Plenum Press, New York, 1990

183

Table 1 General Features of Reference Blankets

Blanket	Material Combinations (Breeder/ Coolant/Structure*)	Reactor Type	Total Thickness (Blanket + Shield, m)	Neutron Wall Load (MW/m^2)	Lifetime (Full Power Years)
1	$Li/Li/VCrTi$	D–T Tokamak	1.280	5.0	3.0
2	$Li_2O/He/$HT–9	D–T Tokamak	1.068	5.0	3.0
3	$Li/Li/VCrTi$	D–T RFP**	1.055	15.0	1.0
4	$Li_{17}Pb_{83}/Li_{17}Pb_{83}/VCrTi$	D–T Tokamak	1.283	5.0	3.0
5	$Li_2O/He/$HT–9/Be†	D–T Tokamak	1.068	5.0	3.0
6	$He/$HT–9††	D–D Tokamak	0.910	1.17	12.5
7	$He/$HT–9‡	D–T Tokamak	0.910	1.0	1.0

*   Structural Materials: $VCrTi$ refers to V–15% Cr–5% Ti alloy; HT–9 is ferritic steel.

** Reversed–Field–Pinch; This design uses copper magnets, and thus does not have a magnet shield. The coils are included in the blanket thickness.

† Beryllium neutron multiplier is added to solid breeder Blanket 2 to improve neutronic performance.

†† D–D fuel cycle does not require tritium breeding.

‡ Experimental reactor conditions with no tritium breeding requirements.

Tokamak blanket. The breeder material is Li_2O, the coolant is helium, and the structure is ferritic steel HT–9. Blanket 3 is representative of a Reversed–Field–Pinch (RFP) reactor. It has the same material composition as Blanket 1, but since the RFP uses copper magnets, it does not require a large shield behind the blanket. Blanket 3 is designed for a very high neutron wall load to achieve the more compact RFP geometry. Both of these differences have an interesting impact on thermal safety, and thus comparisons between Blankets 1 and 3 are useful. Blanket 4 was chosen to contrast with Blanket 1. Blanket 4 is different from Blanket 1 only because it has lithium–lead $Li_{17}Pb_{83}$ as the breeder/coolant material. Thus, comparing Blankets 1 and 4 essentially amounts to comparing the thermal safety characteristics of liquid lithium with those of lithium–lead. Also to examine a materials impact, Blankets 2 and 5 are identical except that Blanket 5 has beryllium in the breeder region, to improve the neutronic performance. Blanket 6 was chosen to examine some of the thermal safety characteristics of the D–D fuel cycle, but without real optimization of the D–D reactor design. Blanket 7 has the same characteristics as those of Blanket 6 except that it is for an experimental D–T reactor that does not require tritium breeding.

THERMAL RESPONSE DEPENDENCE ON BLANKET DESIGN

For the blankets defined in Table 1, two types of undercooling transients were analyzed, the Loss-of-Flow Accident (LOFA) and the Loss-of-Coolant (LOCA). For each blanket the decay heat distribution at the end of life was obtained from the result of the neutron flux distribution of ONEDANT[3] and an activation analysis by REAC[4], as described by Massidda[1]. The temperature response of the inboard blanket to undercooling transients was obtained using the one-dimensional heat transfer code THIOD[1]. The shield surface was assumed to radiate to the magnet coil which remains at 20° C. The results for LOFA and LOCA are shown in Figures 1 and 2, respectively. The results indicate that LOCA is a more severe event than LOFA. It is also clear that the high wall loading of the RFP blanket leads to the highest temperature excursion. In these analyses, radiation surfaces within the blanket were assumed to have an emissivity of 0.1 while the emissivity between the blanket and the shield and from the shield to the surrounding was 0.5. The impact of the assumed emissivity within the blanket can be significant particularly in the case of a LOCA (see Figures 3 and 4 for peak temperatures at various wall loadings).

The above analysis of the *Li/Li/VCrTi* blanket assumes that the manifold region is of HT–9 type steel. If, for economic reasons, the manifold region was made of the much less expensive *Fe1422* steel, a much higher temperature excursion would result. The decay heat of *Fe1422* is much higher than that of HT–9 because of the 15% *Mn* content instead of 1% in HT–9.

The availability of a heat sink behind the shield to which heat can be radiated does not have any impact on the first wall temperature in the first 24 hours. However, the heat sink availability will mean that, in the longer term, it will act to limit the rise in the temperature on the first wall, as can be seen in Figure 5.

EFFECTS OF NEUTRON FLUX MODIFIER

Effect of Neutron Absorbers in Shield Blankets

In early experimental reactors the goal will be a better understanding of the physics and engineering aspects of the plasma rather than on power and tritium production. Thus, a shield–blanket may surround the plasma, one that does not contain lithium. Because of the need to protect the magnets and personnel, the shield blanket may essentially be a massive steel structure cooled by water or

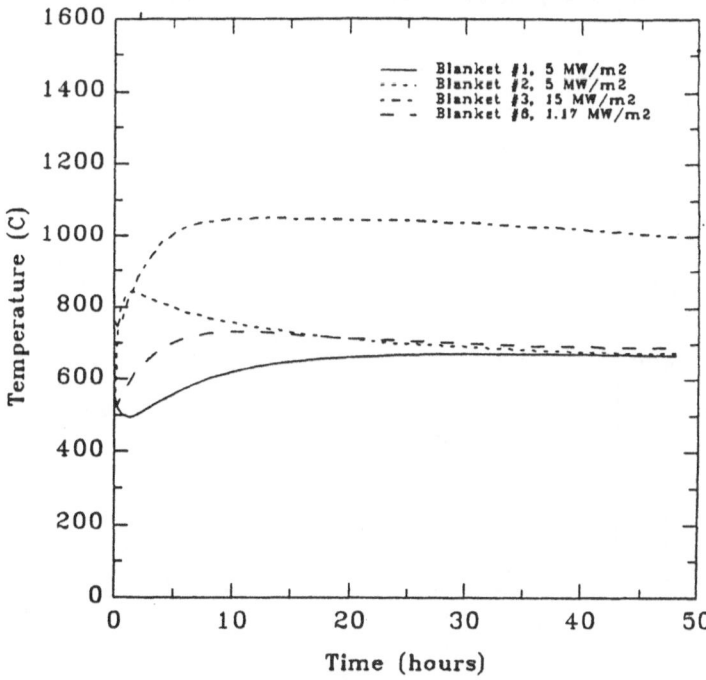

Fig. 1.   First Wall Temperature History Following a LOFA for Four Blankets.

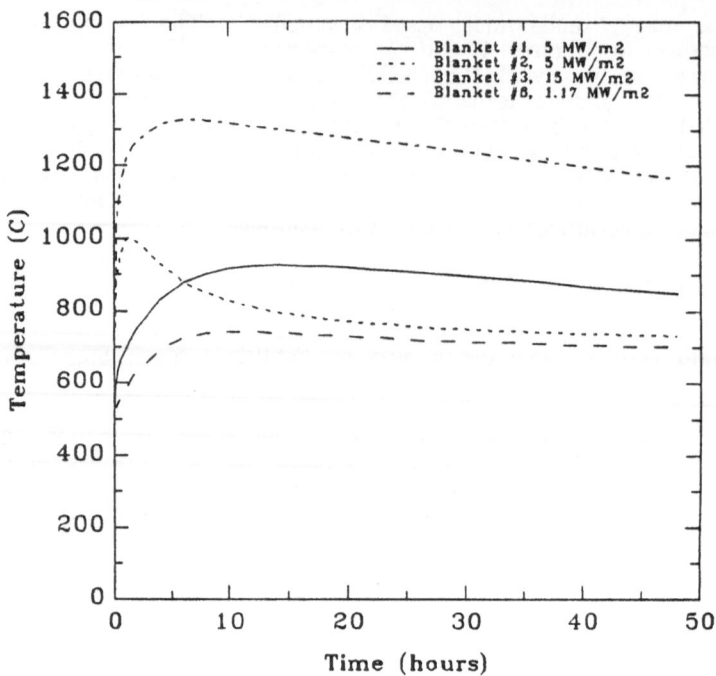

Fig. 2.   First Wall Temperature History Following a LOCA for Four Blankets.

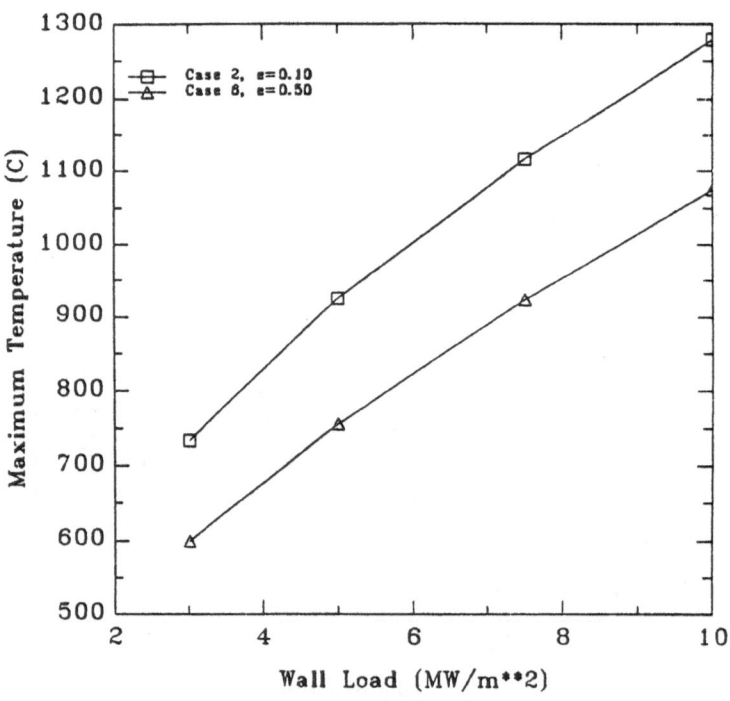

Fig. 3.   Effect of Emissivity of In–Blanket Stuctures on Maximum First Wall
Temperature After LOCA in the *Li/VCrTi* Blanket (Blanket 1).

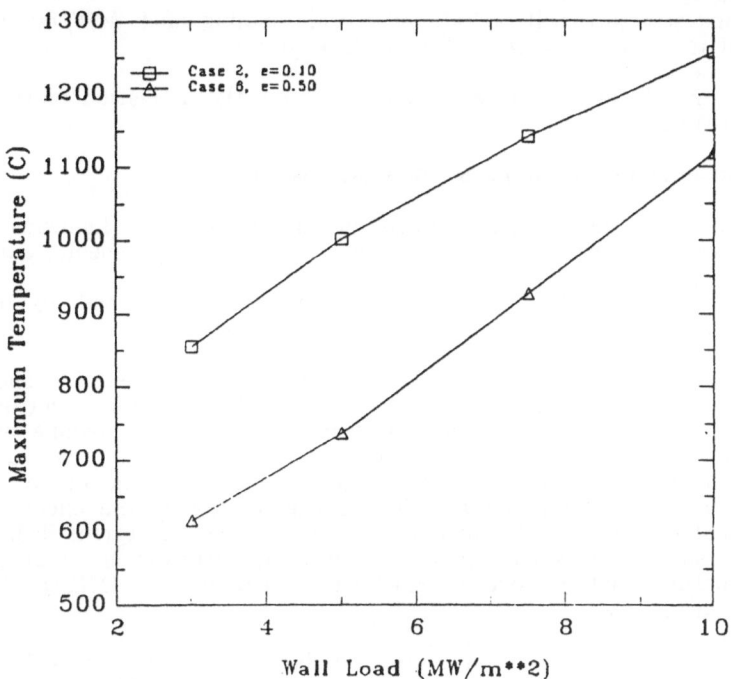

Fig. 4. Effect of Emissivity of In–Blanket Structures on Maximum First Wall Temperature After LOCA in the HT–9/Li_2O/He Blanket (Blanket 2).

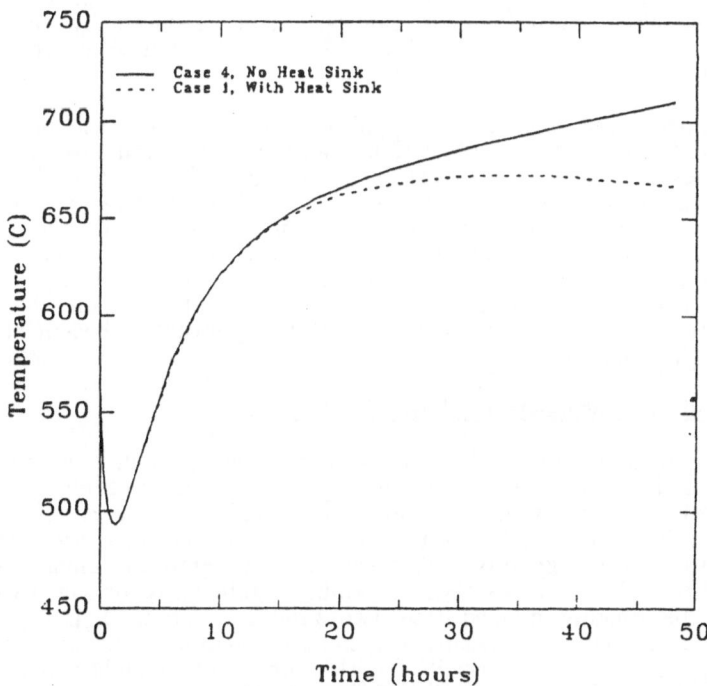

Fig. 5. Effect of Heat Sink on First Wall Temperature History Follwoing LOFA in the $Li/Li/VCrTi$ Blanket (Blanket 1).

187

helium. This structure will operate at a wall loading of 1 MW/m^2 or less, and be subjected to a total fluence of 1 to 3 $MW-yr/m^2$.

It is desirable in this shield blanket to limit the activation of the structures for several reasons:

1.   To diminish the problems of waste handling,

2.   To limit the potential escalation in temperature following shutdown, because of the heat generated by the radioactive elements, and

3.   To minimize the dose rate around the reactor and to maintenance workers.

Thus it would be useful to consider a low activation ferritic steel. The limitation on structural activation maybe accomplished by including strong neutron absorbers in the structure. The effect of such inclusion is particularly striking in a blanket that does not originally contain a neutron moderator near the first wall. In that case, both the total volume of the activated structure and the neutron absorption density of the metal are reduced. This is exemplified by the effect of including B_4C in the shield–blanket (Blanket 7 of Table 1), shown schematically in Figure 6. Five versions of this blanket were examined, with varying amounts of B_4C, using the thermal analysis code THIOD[1], the neutronic code ONEDANT[3] and the activation code REAC[4].

The inclusion of 12% natural boron carbide in version 2 was found to have an order of magnitude effect on the neutron flux level. The additional B_4C in versions 3 and 4 was found to have only a modest further effect. The effect of LOCA was investigated using the one–dimensional THIOD Code[2], and assuming the decay heat was dissipated only by conduction and radiation. The temperature history after a LOCA event is shown in Figures 7 and 8, near the first wall and the back of the blanket, respectively. It is seen that the temperature of the first wall decreases after shutdown, but can be further reduced by the inclusion of B_4C. Again natural boron at a level of 12% by volume of the blanket can achieve most of the desirable reduction effect.

The impact of a moderating graphite liner, separating the plasma from the first wall, on the effectiveness of B_4C has also been assessed. A graphite liner of 2 cm thickness was assumed to exist before the B_4C–free blanket (Version 6) and before the 12.5% B_4C blanket (Version 7). The temperature history at the position 1 cm from the first wall is shown in Figure 9 with the graphite liner (Versions 6 and 7) and without the graphite liner (Versions 1 and 2). It is seen that in both cases the graphite liner has a moderating influence on the temperature rise due to decay heat. The effect of B_4C on the temperature history in the presence of a liner was still significantly beneficial (comparing Version 2 to Version 6).

Effect of Neutron Multipliers in Power Blankets

The use of a neutron multiplier has been considered in a number of blanket studies, primarily as a means of increasing the tritium breeding ratio. Because the $^9Be(n,2n)2$ ^4He reaction is endothermic, and the energy of the incoming neutron must be divided between the two outgoing neutrons, the effect of Be is to increase the low energy flux. This can have a significant impact on the decay heat production. The isotopes that contribute significantly to the decay heat in a blanket can be roughly divided into two groups[5]: group 1 isotopes which are produced by slow neutrons (usually (n,gamma) reactions), and group 2 which are produced by fast neutrons (usually (n,p), (n,d) and (n,alpha) reactions). The group 1 isotopes tend to have shorter half–lives than the group 2 isotopes. Thus, group 1 isotopes generally have a higher initial decay rate (and thus decay power) at shutdown, but they decay more quickly. That is, group 1 isotopes "burn

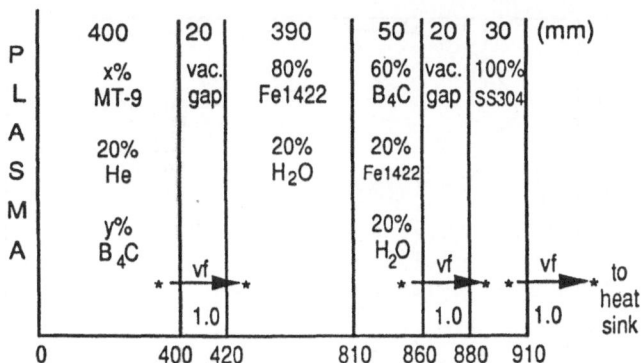

Arrows connecting asterisks indicate radiation paths. The vf parameters indicate the view factor for that particular radiation path.

Version #	x% MT-9	y% B_4C	B^{10} enrich.
1	80.0	0.0	----
2	67.5	12.5	natural
3	67.5	12.5	50%
4	40.0	40.0	natrual
5	40.0	40.0	50%

Fig. 6.   One–Dimensinoal Schematic of HT–9/B_4C/He Inboard Blanket for Neutronics and 1–D Thermal Transport Modeling.

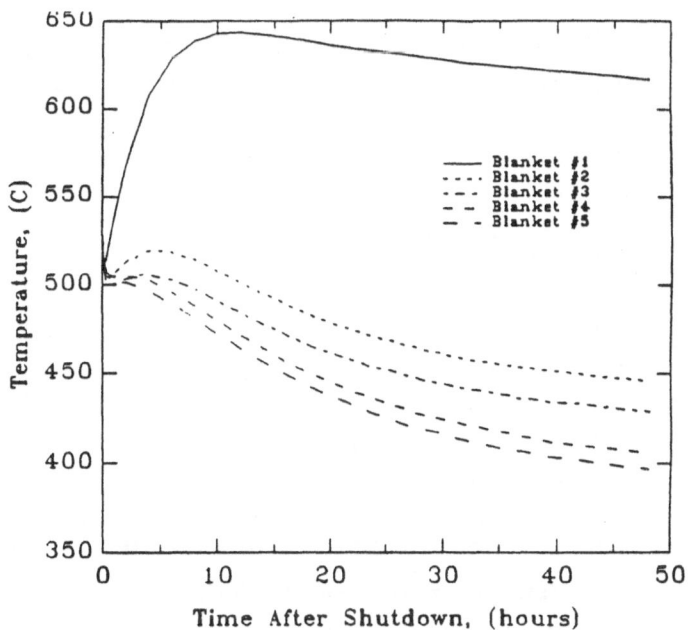

Fig. 7.   Temperature History of Shield–Blanket, Blanket 7, After LOCA at $1 cm$ from First Wall.

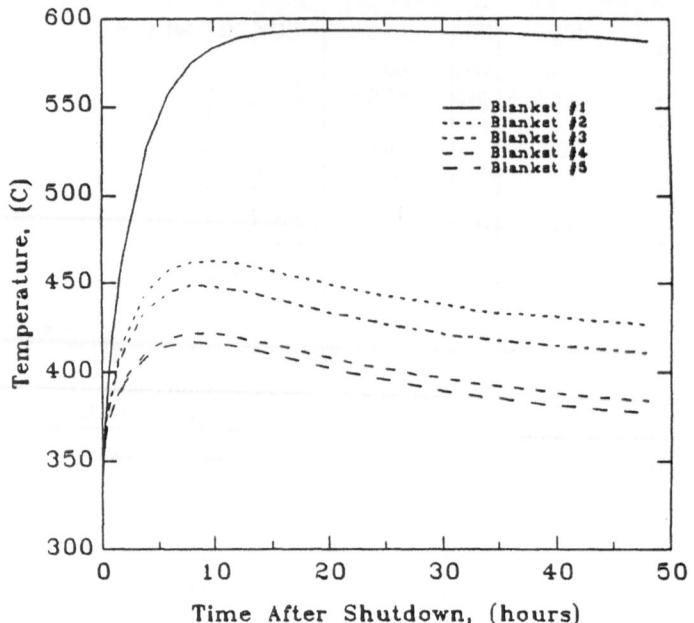

Fig. 8. Temperature History of Shield–Blanlket, Blanket 7, After LOCA at 40*cm* from First Wall.

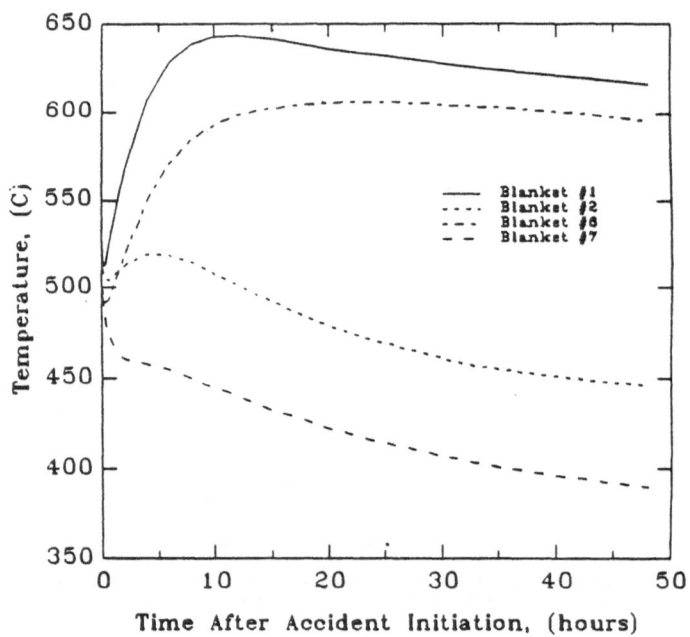

Fig. 9. Temperature History of the Shield Blankets with a 2*cm* Graphite Liner (Versions 6 and 7) and the Identical Blankets without Graphite (Version 1 and 2), Respectively.

hotter for a shorter time." This effect was analyzed by comparing the decay heat and corresponding post-LOCA first wall temperature rise from the Li_2O blanket without Be, Blanket 2, and the Li_2O blanket with Be, Blanket 5.

Figure 10 plots the ratio of the decay heat in the Li_2O blanket with beryllium to the decay heat without beryllium. Note that in the breeder region, where the beryllium is located, the ratio is just about 1.2, and decreases slightly with time. Table 2 displays the significant structural isotopes in this breeder region, along with the decay heat that each isotope produces in two days. The ratio shown is the ratio of the total decay heat produced by the isotope in a two day period in the blanket with Be to that without Be. Note that the group 2 isotopes show the least difference (lowest ratios), while the group 1 isotopes show much greater differences (higher ratios). This is because the group 1 isotopes are produced by the thermal flux, which is much higher in the beryllium system (Blanket 5).

Note also in Figure 10 that the decay heat ratio at 0.44 m from the first wall, which lies in the front of the $Fe1422$ shield, is 50% higher in the Be blanket at shutdown. This again is due to the increase in the thermal flux, which produces a larger amount of ^{56}Mn, a group 1 isotope for $Fe1422$. At longer times, such as one day, after the ^{56}Mn has decayed away, the decay heat in the front of the shield of the beryllium blanket is much lower than that without beryllium. Because thermal neutrons are attenuated more quickly (over a shorter distance) than fast neutrons, the mostly thermal flux in the shield of the blanket without beryllium falls off sharply. The harder flux spectrum in the shield of the beryllium blanket (5) falls off less rapidly, and thus at the back of the shield, the decay heat is higher.

Overall, the total decay heat produced in the beryllium blanket over a two day period is about 15% higher than that of the non-beryllium blanket. At shorter times, the difference is larger, because of the faster decay rate of the group 1 isotopes. After one hour, the decay heat difference is about 25%. This difference will lead to a post-LOCA first wall temperature that is about 25° C higher when beryllium is present. While this difference is not alarming, it would be higher in higher power density machines. Also, other blanket systems might be even more adversely affected by the presence of beryllium. From the $Fe1422$ shield region of Figure 10, it is seen that the difference can indeed be very large for some materials.

Table 2   Comparison of Decay Heat Produced in Two Days by Isotopes in Beryllium Blanket 5 and Non-Beryllium Blanket 2

Isotope	Group	Blanket 5 (J/cm^3)	Blanket 2 (J/cm^3)	Ratio
^{54}Mn	2	2.227 x 10^3	1.966 x 10^3	1.133
^{56}Mn	1	1.750 x 10^3	1.491 x 10^3	1.174
^{58}Co	2	1.548 x 10^2	1.365 x 10^2	1.134
^{99}Mo	1	1.573 x 10^2	1.190 x 10^2	1.322
^{51}Cr	2	9.240 x 10^1	8.497 x 10^1	1.087
^{59}Fe	1	5,420 x 10^1	1.108 x 10^0	48.92
^{55}Fe	2	2.789 x 10^1	2.594 x 10^1	1.075
TOTAL		4.684 x 10^3	4.027 x 10^3	1.163

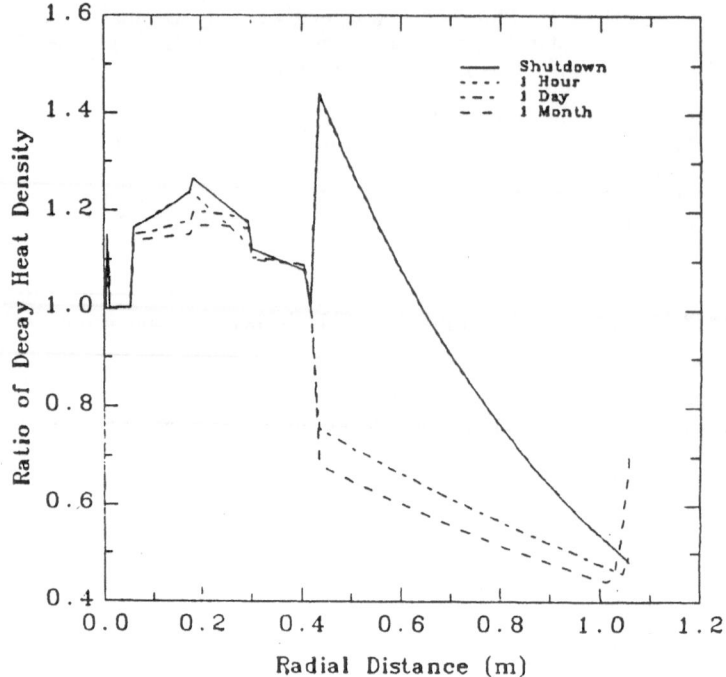

Fig. 10. Ratio of Decay Heat Density in Beryllium Blanket (Blanket 5) Over Non-Beryllium Blanket (Blanket 2).

SAFETY LIMITS

Passive safety is used here to imply that no intervention is required to ensure that an accident does not lead to a large radiological release or substantial plant damage[6]. This definition encompasses minimization of the risks from accidents to both the public and the investors. Both types of risks are serious enough to impact the desirability of energy sources. As evidenced by the TMI accident which ruined the reactor but had negligible public releases of radioactivity, the lost investment does not always have to be associated with harm to the public. On the other hand, it may be true that a major threat to public safety is always associated with substantial, if not ruinous, plant damage.

Limits for Public Safety

The threat to public safety is dependent on the amount of radioactivity that may be released to the environment following an accident. There are two sources of radiation activity in the fusion plant: tritium and neutron activation products. The total inventory of these sources depends on the reactor design (i.e., compactness, structural materials, and fuel cycle). The public consequences of accidents can be minimized by reducing the ability to mobilize the activation products (concentrated in the first wall). Mobilization of the radioactivity occurs primarily through volatilization of the activation products when the structural materials are exposed at high temperatures to an oxidizing environment such as air or steam. Hence, if the temperature history following an accident is such that it allows for limited volatilization, the possible release to the public would be limited. In the ESECOM study[7] a public safety threshold was defined as a release that, under the most adverse weather conditions, would lead to a prompt whole body dose of less than 200 *rem* at a distance of 1 *km* downwind from the release. (This prompt dose is the whole body dose in the first seven days due to external exposure and inhalation during the passage of a plume over a 10 hour period which is different from the two-hour exposure dose usually used in emergency planning for fission reactor accidents.) This threshold will ensure that no acute public fatalities would result in the most severe accidents. If the volatilized material was less than what would result in this threshold dose, mitigating factors such as plateout and precipitation of aerosols along with the plume rise due to its heat content above ground will further limit the public impact.

Limits for Plant Protection

Gross failure of the first wall due to excessive tensile stresses or creep rupture will result in substantial loss of both capital investment and operating time, and may even lead to a more severe propagation of the accident. Hence, protection of the first wall integrity can be taken as an appropriate criterion for plant protection.

Creep Rupture: To simplify the analysis, heuristic estimates of the first wall stress were made, based on the operational conditions and the available data. In the LOFA scenario, the coolant remains in the coolant channel. In the liquid-metal cooled blankets, the non-flowing coolant pressure will be low, since most of the operating pressure is designed to overcome the MHD pressure drop. In this instance, the total post accident stress (due to pressure and temperature) was assumed to remain constant at 10 MPa. In the helium cooled blankets, the non-flowing coolant pressure will be almost the same as the operating coolant pressure (5 MPa), and thus the post-accident first wall stress is assumed to be 120 MPa. In the LOCA scenario there is only a thermal stress, since the coolant is no longer present. In the $VCrTi$ blankets, since no data exists, this thermal stress was assumed to be constant at 1 MPa. The improved data base for HT-9 allowed for a more detailed calculation of the post-LOCA thermal stress based on thermal creep correlations[8]. (It was assumed some annealing occurs during operation, but a reversed stress occurs when the plasma is shut.)

Acute Failure: The high temperature data base for the candidate structural materials $VCrTi$ and HT-9 is quite limited, requiring extrapolation of the lower temperature data. The ultimate tensile stress (UTS) of the $VCrTi$ alloy decreases dramatically above 1200C [1] [9]. For this reason, it is assumed that $VCrTi$ components which reach above 1200C will suffer acute structural failure. Extrapolation of the HT-9 data[1] [10] indicate that the UTS will be very close to zero at 900C, and thus it is assumed that acute structural failure occurs when the wall reaches the 900C level. It must be pointed out that when the stress level is very low, structural failure does not actually occur. However, at temperatures between 1100C to 1200C for $VCrTi$, and 900C to 1000C for HT-9, re-crystalization of the material will occur, and thus the material will lose its ductility, and thus the component will not be reusable. Further, the deformation and/or distortion that occurs may prohibit easy removal of the damaged module. In either case, this would represent an economic penalty, since the affected modules would have to be repaired. Thus the acute failure temperature limits of 1200C for $VCrTi$ and 900C for HT-9 are appropriate. Also, exclusion of oxidizing media is important at this temperature or else $VCrTi$ alloys will form oxides that are molten at these temperatures.

SAFETY LIMITS ON WALL LOADS

The limits on the allowable neutron wall load for four reference cases are summarized in Table 3. The plant protection load limits indicated here are over a local area (since no 2-D effect is considered). Thus the average wall loading will have to be limited to lower values, to allow for the expected peak-to-average ratio in each reactor. Also shown in Table 3 are the wall load limits to protect the public based on the 200 rem limit on the prompt dose criterion (but ignoring in-plant decontamination effects of plate-out and deposition). These loads

Table 3   Maximum Allowable Wall Loads to Ensure Passive Safety

| Blanket | Accident | Neutron Wall Load (MW/m^2) | | |
		Thermal Creep Rupture*	Acute Structural Failure*	Radioactive Release**
1	LOFA	>10†	>10†	≫10‡
	LOCA	~8	8.8	>10
2	LOFA	~3	6.1	>10
	LOCA	~6	3.6	>10
3	LOFA	~17	20	>20
	LOCA	~7	11	~20
4	LOFA	~1	1.8	>2.5
	LOCA	>2.5	1.8	>2.5

* These wall loads correspond to peak values in the reactor, which can be 1.5 times the average load.

** Excluding effects of possible lithium fire (Blankets 1 and 3 only).   These wall loads correspond to average values.

† Wall Load Required is 2 – 5 MW/m^2 greater than maximum analyzed.

‡ Wall Load Required is at least 5 MW/m^2 greater than maximum analyzed.

represent the average wall conditions.   It is clear that plant protection requirements are likely to set more stringent limits on the operating conditions than the public protection requirements as defined here.

There are some important simplifications in the present analysis.   For example, the effect of coolant natural convection from the first wall/blanket on reduction of the temperature rise has been neglected.   If such natural convection would be assured (only at appropriately low magnetic fields), the allowable limits for plant protection may be higher than indicated in Table 2.   Similarly, if the effects of settlement and plate–out of volatilized materials were considered, they may lead to higher allowable limits for public protection.   However, the radioactive release limits are already high with respect to the loads desired for best economic performance of the fusion plants[7].   Finally, for reactors with liquid metal coolants, it is important to realize that a combination of lithium fire and decay heat would impose lower limits for plant and public protection.

EFFECTS OF LITHIUM CHEMICAL REACTIONS

The chemical reactivity of lithium (whether in an elemental form or alloyed with lead) has been of major concern in safety analysis of fusion reactors.   Two major experimental programs have been conducted to date.   The first addressing the lithium and lithium–lead reactions with atmospheric gases (air, steam and CO_2) and was performed in the U.S.[11].   The second concerning the consequences of pressurized water leakage effects inside a lithium–lead blanket and is now being conducted in Europe[12].   The observations at the Hanford Engineering Laboratory indicates that, based on experiments involving 10–100 kg of Li, air and steam reactions with liquid lithium would lead to a maximum temperature of 1300° C. Nitrogen reactions alone lead to a lower maximum temperature (about 1000° C), while carbon dioxide (in the atmosphere or while evolving from concrete) leads to higher temperatures (over 1500° C).   On the other hand, the lithium–lead pool reactions seem to be self–limited due to depletion of lithium on the pool surface.

The potential hazard is very small as the fire is extinguished in a relatively short time.

Calculations performed with the LITFIRE code were performed to determine the consequences of lithium fires[13]. For a large scale fire on the floor of a reactor building, such as schematically depicted in Figure 11, the gas temperature in the building would rise by 200–400° C, depending on the humidity (or steam) in the air (see Figure 12). The pressure rise maybe on the order of 2–3 atmospheres. If the lithium fire is to take place inside the plasma torus which is surrounded by the $Li/Li/V$ blanket and shield, the available heat capacity would limit the maximum temperature to less than 1000° C or 1200° C for simultaneous occurrence of a LOFA or LOCA, respectively (see Figure 13). However, sufficient hydrogen may accumulate in the torus to lead to a few percent concentration in the torus atmosphere (See Figure 14).

The results from the experiments at the Joint Research Center of Ispra, which involve 300 kg of $Li_{17}Pb_{83}$, lead to the following main conclusions[12]:

1. The thermal effects of water injected under pressure in the lithium–lead breeder are inconsequential.

2. The pressure in the breeder zone rises very rapidly to a value below the water injection pressure, but subsides somewhat as expulsion of the breeder occurs.

3. Hydrogen generation in the process may require special measures of collection.

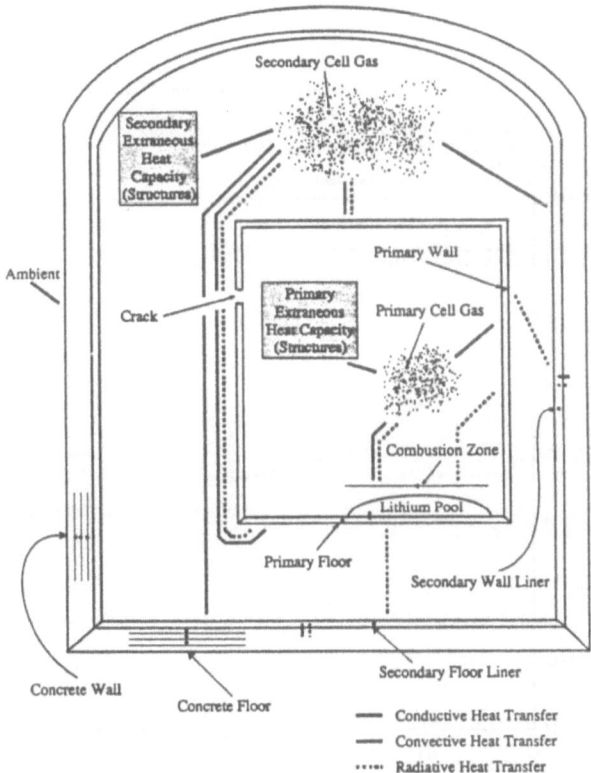

Fig. 11. Energy Flow in Two–Cell LITFIRE.

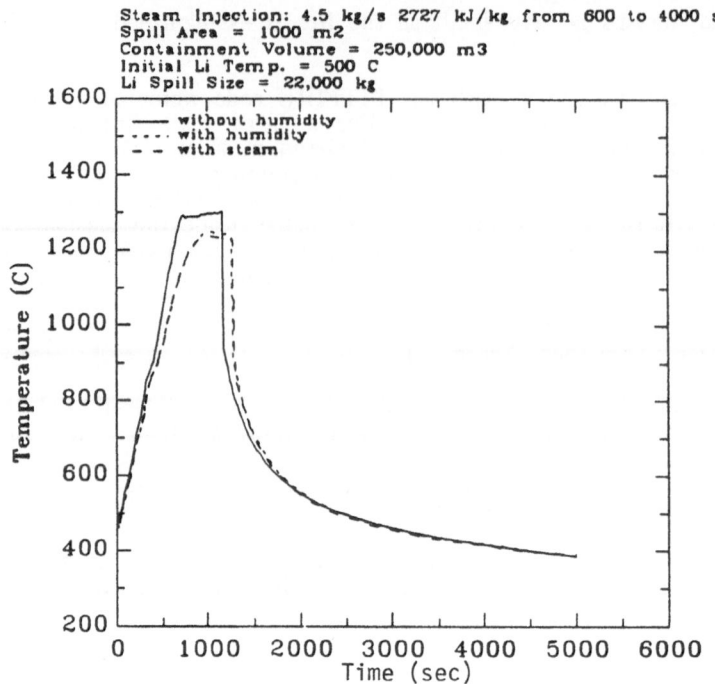

Fig. 12. Combustion Zone Temperature Following a Major Lithium Spill on Containment Flow.

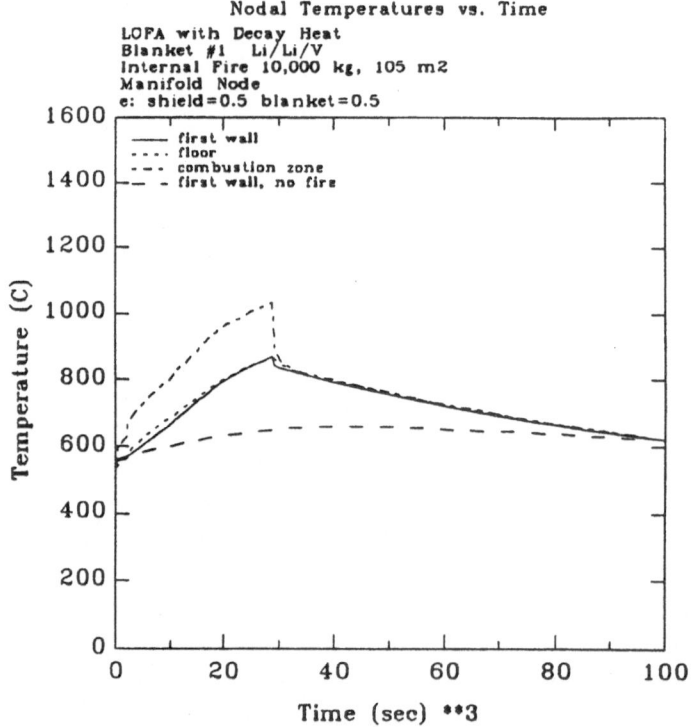

Fig. 13. Temperature History for Simultaneous Loss of Flow and Lithium Fire in Plasma Torus.

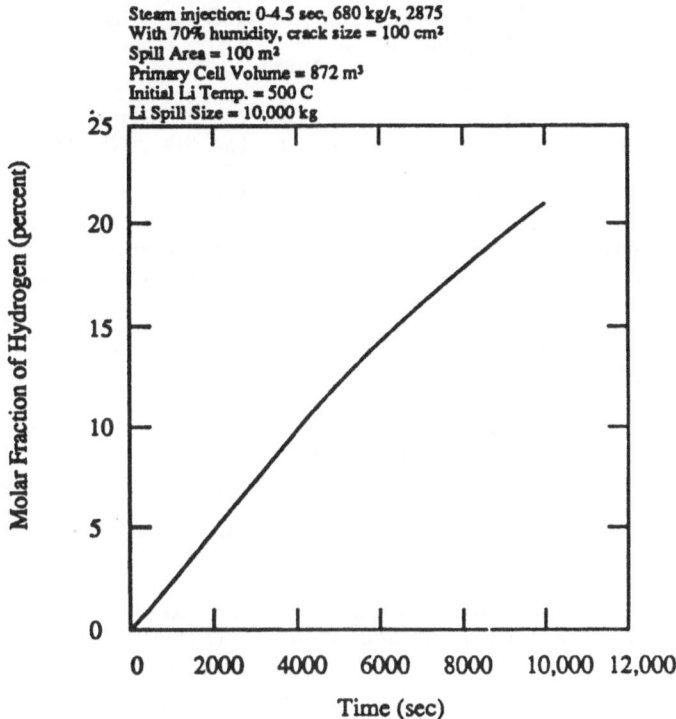

Fig. 14. Hydrogen Concentration in the Plasma Chamber Following Lithium Fire

CONCLUSIONS

The design guidelines to ensure adequate passive heat dissipation seem to limit the peak operating neutron wall load in D–T reactors to values between 3 MW/m^2 and 8 MW/m^2. The plant protection criteria appear to set lower limits than those imposed by radioactivity release criteria. The proper margin to account for the peak–to–average load values is yet to be determined as applicable to both nominal power and overpower conditions. The role of neutron absorbers and multipliers modifying the thermal response as well as the radioactive waste levels in future experimental reactors can be significant, and deserves special attention.

REFERENCES

1.   J.E. Massidda and M.S. Kazimi, "Thermal Design Considerations for Passive Safety of Fusion Reactors," PFC/RR–87–18, Plasma Fusion Center, MIT (October 1987).

2.   Y. Parlatan, "The Impact of Blanket Designs on Activation and Thermal Safety," M.S. Thesis, MIT Nuclear Engineering Dept. (June 1989).

3.   R.D. O'Dell, et al., "User's Manual for ONEDANT: A Code Package for ONE–Dimensional, Diffusion–Accelerated, Neutral–Particle Transport, Los Alamos National Laboratory, LA–9184–M (February 1982).

4.   F.M. Mann, "Transmutation of Alloys in MFE Facilities as Calculated by REAC (A Computer Code System for Activation and Transmutation)," Hanford Engineering Development Laboratory, HEDL–TME–81–37 (August 1982).

5.   J.E. Massidda and M.S. Kazimi, "Aspects of Decay Heat Behavior in Fusion Reactor Blankets," presented at the 8th ANS Topical Meeting on Fusion Technology, Salt Lake City (October 1988).

6.   M.S. Kazimi, J.E. Massidda and M. Oshima, "Thermal Limits for Passive Safety of Fusion Reactors," presented at the 8th ANS Topical Meeting on Fusion Technology, Salt Lake City (October 1988).

7.   J.P. Holdren, et al., "Exploring the Competitive Potential of Magnetic Fusion Energy: The Interaction of Economics with Safety and Environmental Characteristics," *Fusion Technology*, 13 (January 1988).

8.   R.J. Amodeo and N.M. Ghoniem, "Development of Design Equations for Ferritic Alloys in Fusion Reactors," *Nuclear Engineering and Design/Fusion*, 2, p. 97 (1985).

9.   R.E. Gold and R. Bajaj, "Mechanical Property Evaluation of Path C Vanadium Scoping Alloys," Alloy Development for Irradiation Performance – Semiannual Progress Report, DOE/ER–0045/10, pp. 122–141 (October 1983).

10. J.E. Chafey and J.B. Wattier, "Estimation of Allowable Design Stress Values for $12Cr-1Mo-0.3V$ Steel," GA–A14610 (1978).

11. S.J. Piet, et al, "Liquid Metal Chemical Reaction Safety in Fusion Facilities," *Fusion Engineering and Design*, 5 (1987).

12. O. Kranert, H. Kottowski and C. Saratteri, "Large Scale Lithium–Lead/ Water Interaction Studies," presented at the American Nuclear Society 8th Top. Mtg. on Fusion Energy, Salt Lake City, UT (October 1988).

13. D.S. Barnett and M.S. Kazimi, "The Chemical Kinetics of the Reactions of Lithium with Steam–Air Mixtures," PFC/RR–87–3 Plasma Fusion Center, MIT (April 1989).

TRITIUM ENVIRONMENTAL RISK IN FUTURE FUSION REACTORS

Yves BELOT, Pierre ZETTWOOG

Commissariat à l'Energie Atomique
Institut de Protection et de Sûreté Nucléaire
BP 6, 92265 FONTENAY AUX ROSES CEDEX, France

SUMMARY

The rationales for the environmental tritium program carried out in the frame of the technology fusion and safety program of the EC are exposed in part one of the report.

Results obtained in the laboratories and in the field experiments are presented in part two, with the major findings from the risk management point of view.

PART ONE - GENERAL CONSIDERATIONS

The first generation of fusion reactors will be based on the deuterium tritium reaction. The tritium inventory in a future commercial reactor will be in the range of a few kilograms. A 1000 MW(e) reactor will burn 600 g per day, which is a yearly consumption, for 300 days of operation, of 1.8×10^5 grams.

To put this quantity in perspective, let us recall that the total quantity of tritium of natural origin accumulated in the earth, taking in account the annual production of 200 grams of tritium by cosmics rays in the upper atmosphere, and its radioactive decay with a half life of 12.3 years, is 3.5 kilograms.

The French nuclear reactors, with a total production of 200 TWe x h, released 1.7 g of tritium in 1985.

Thermonuclear tests from 1952 up to 1962 introduced 660 kg of tritium into the atmosphere.

Tritium is a slightly toxic form of the hydrogen atom, due to its beta radioactivity, if incorporated into a living organism. It is accepted in the frame of an accepted risk management type hypothesis that the probability of a carcinogenic effect appearing in the long term is proportional to the total incorporated quantity of tritium. The model adopted by the expert group of the International Commission on Radiation Protection (ICRP) shows that the incorporation of 0.8 micrograms of tritium in the form of tritiated water by a man weighing 70 kilograms results in an increased probability of cancer of

Safety, Environmental Impact, and Economic Prospects of Nuclear Fusion
Edited by B. Brunelli and H. Knoepfel
Plenum Press, New York, 1990

199

10^{-4}. An increased probability of cancer occurrence of 10^{-4} per year of exposure is considered for the members of the public as the boundary between unacceptable risks and risks which could be tolerated if justified. This gives a limit of incorporation of 0.8 micrograms per year for the reference man.

Comparing a yearly consumption of 1.5×10^5 grams in the plant and the 0.8 microgram annual limit of incorporation indicates that future fusion reactors do present potentially an environmental tritium risk. The level and therefore the significance of the actual risk will depend on the possibility to safely confine the tritium inside the plant and the efficiency of the transfer to the public through the environment and the food chain.

The possibility for hydrogen molecules to diffuse or to permeate even through supposed tight seals and walls is unfortunate in this respect. This tritium emanation results in the contamination of the atmosphere inside the reactor building and peripherical laboratories, including the tritium handling facility. Another unavoidable source of tritium is the outgassing of tritiated pieces of equipment during repair, maintenance, and dismantling operations, as well as permanently from the intermediate solid waste storage unit. Leakages of tritiated water from the cooling water circuitry are also a contribution to the tritium pollution of the atmosphere.The use of intermediate confinements around the sources of tritium, the detritiation of the water in the cooling circuit and of the atmosphere of the building help in reducing the tritium concentration in the gaseous releases to the environment through the ventilation systems and the stack.

Nevertheless, the tritium concentration in the routinely released air through the stack will not be zero. A reason is that the performances of the detritiators will be less and less effective as the tritium concentration decreases ; anyhow there is no longer a sanitary justification for further expenses when the alleged residual becomes less and less significant, if existing at all. More of concern is that the occurrence of an accidental release of a sensible part of the total tritium inventory has, from the point of view of the risk management theory, a non-negligible probability, with detriment levels which could exceed the acceptability levels for the members of the public. Safety analyses of various failure scenarios have demonstrated this situation.

What is needed is to provide the designers of future reactors with exact figures (probability and released quantity) for permissible releases in the event of an accident, and for routine releases (yearly admissible released quantity), on the basis of which they will be able to design facilities that will be appropriate both technically and economically and in terms of health for the public and for the workers of the plant.

To assess the health impact on the public of a release of toxic material in the environment, one has first to use transfer models and numerical codes for each of the pathways to man identified as critical, through the different compartments of the local ecosystems and through the food chain. Secondly one has to evaluate trough dosimetric models the level of risk of the exposed individuals, taking in account the incorporation modes of the considered toxic, in our case, tritium, according to its chemical speciations.

The main limiting factor used in determining the permissible release is the exposure of the individuals of the most exposed group in the considered site, which has to be less than the regulatory limit. In the absence of other artificial sources of radiation, this exposure should lead to an excess of the probability of carcinogenic effect less than 10^{-4} per year. A further requirement is to keep the collective risk of the total exposed population to a level as low as reasonably achievable, technical and economic considerations being taken into account, and assuming a linear relationship without threshold between the probability of cancer occurrence and the individual doses.

Attention has also to be given to the extent and level of the contaminated area due to chronic and accidental releases, and to the level of the local contamination of the water and of vegetation in the long term.

In the case of an individual being exposed to air contaminated with tritiated hydrogen, irradiation occurs mainly from the beta particles in the lungs. The possibility and level of oxidation of tritiated hydrogen in the lungs are under investigation.

Tritium in the form of tritiated vapor is much more harmful than in the form of tritiated hydrogen, due to the possibility of incorporation of the vapor into the body fluids through the skin and the lungs. The dose of irradiation for an individual exposed to a tritiated atmosphere is, according to different authors, 10000 to 25000 time higher if the tritium is in the form of tritiated vapor than if it is in the form of tritiated hydrogen.

Organically bound tritium, which may form inside the plant, and later in the environment, is even more harmful than tritiated vapor. This is another possible concern which still has to be quantitatively investigated.

In a number of scenarios, the tritium released from the stack, particularly in case of an accident, will be in the form of tritiated hydrogen that is the less harmful form. Unfortunately, the possibility of oxidation of the tritiated hydrogen form to the tritiated water vapor form does exist in the environment due to various chemical and biological phenomena. This oxidation increases the level of exposure and the level of risk of the individuals, according to the higher toxicity.

It was realised in the early eighties that the literature on the behaviour of tritiated hydrogen in the environment was scarce and, in many cases, the conclusions drawn were not consistent.

The half-life for the conversion rate of the hydrogen form deduced from differents studies on the behaviour of tritium in the environment ranged from 1350 ± 150 seconds (obtained from 5 controlled release experiments at ground level) to abouts 6 years (considering the conversion rate of the tritium of the global atmosphere).

If the first number is taken into account, one has to assume a complete oxidation of the hydrogen form even in the immediate vicinity of the plant. With the last number, no oxidation occurs before complete atmospheric dilution to radiologicaly non-significant levels. This leads to a variation of 10000 to 25000 for the level of permissible authorised tritium release.

A first proposal for an investigation program on the behaviour of tritium in the environment, including laboratory and field experimentation, was made by the CEA-team in 1984. The EC decided to carry out such a program in 1987-88 with its fusion technology and safety program.

The results of this program are reported below, as well as results obtained in a cooperative program on the same line between Canadian and European research laboratories. Readers who would like to have more understanding of the radioactive risk management philosophy and principles may be interested in the appendix. Arguments are developed for a sound policy of the risk limitation, focusing on the exclusion of costly requirements not related to any sanitary concern.

Concerning the health effects of tritium incorporation (for what is known with certainty, and what is not known), the reader may refer to L. Jeanmaire and J. Piéchowski (1986), and to J. Lafuma (1986).

PART TWO - A REVIEW OF RECENT STUDIES ON TRITIUM BEHAVIOUR IN THE ENVIRONMENT

Background

The purpose of this review is to present the major findings of recent experimental field studies of the behaviour of elemental hydrogen HT, which were carried out in the period 1986-1987. The first study involved a puff release of 260 TBq of HT from the stack of a tritium facility of the French Commissariat à l'Energie Atomique at 20 km to the south of Paris, the second and third studies a 30 min release of 3.5 TBq of HT from a source situated at ground level in the vicinity of the Chalk River National Laboratories in Ontario, Canada. 1 TBq represents 0.0027 g of tritium. The main purpose of these experiments was to study the conversion of HT to HTO in the environment, by tracking the release over several km in the first case, 400 m in the second, and measuring the distribution of elemental hydrogen, tritiated water and organically bound tritium in different compartments of the environment, including atmosphere, soil and vegetation. This series of field experiments was motivated by the necessity to assess the radiological consequences of an accidental release of tritium from a hypothetical fusion reactor, and consequently to collect all the information required to determine the radiation doses on a scientifically sound basis.

Before the present field experiments, it had been observed that the chemical form HT presents a relatively small radiological hazard unless it is converted to HTO which is much more readily absorbed by the body through inhalation, moisture exchange through the skin or ingestion. In early experiments, the attention was consequently focused on the conversion of HT to HTO, in the body and in the environment. It was found that, in spite of a slight oxidation of HT in the body (Peterman et al. 1985), the dose arising from exposure to a given concentration of HTO in air is approximately 10,000 times that for exposure to an equivalent concentration of HT. The impact of a release of HT into the environment was then recognized as dependent on the extent to which HT is converted to the oxidized form HTO. Two processes were expected to play a role in the oxidation of tritium in the environment: a gas phase reaction in the atmosphere and a biochemical process at the atmosphere-land interface. The gas phase conversion was shown by theory and laboratory experiments to be extremely slow and without previsible incidence on a local scale (Ogram 1982a,b). On the contrary, many studies demonstrated that soils, and in particular, the bacteria in the top few cm of soils, were able to catalyse the oxidation process. Tritium oxidizing activity was found in a wide variety of soils, showing that the oxidation can be attributed to the action of ubiquitous soil microorganisms (McFarlane et al. 1978, Garland and Cox 1980, Sweet and Murphy 1981). The HTO formed at soil surface was shown to be gradually evaporated to the atmosphere and scavenged into underground waters through bulk water mass flux and diffusional processes. Studies on plants exposed to HT in laboratory chambers demonstrated that the rate of oxidation of HT in plant tissues and the rate of incorporation of tritium into the organic fraction of the plants were extremely low, and that the contamination of plants was mainly due to the absorption of HTO formed at soil surface (Garland and Cox 1980, Sweet and Murphy 1984, Dunstall et al. 1985, Spencer and Dunstall 1986).

In recent years, it was envisaged to concentrate efforts on the identification of all important processes, the characterization of process rates and, finally, the validation and improvement of predictive models. The first idea was to obtain information from actual releases that had occurred in the past, but it appeared that the data base available from actual releases was inadequate for identifying processes and validating models. These releases had not been carried out with the objective of determining the rate of oxidation of HT in the environment and the results might have been influenced by uncontrolled factors, such as the composition of the effluent, the HTO background, the difference between the plume duration and the sample averaging time. The only way to get valid information was then to carry out controlled release experiments under closely monitored conditions and parallel small-scale experiments to gain a better knowledge of the basic processes and, particularly, of the tritium reemission to the atmosphere through evaporation and diffusion.

A large-scale release experiment and small-scale parallel experiments were then carried out in France in September 1986. These studies were jointly sponsored by the Commission of the European Communities (CEC) through the Next European Torus (NET) Team and by the French Commissariat à l'Energie Atomique (CEA). They included the participation of scientists from France, Germany, Sweden and Canada and from the Joint Research Center of Ispra. The Canadian studies then occurred in two phases, a pilot study in August 1986 to test experimental procedures and to obtain preliminar data, and a subsequent intensive experiment in June 1987, involving participation by investigators from Canada, France, Germany, Japan, Sweden, United States. Funding for the experiments was provided by the Canadian Fuels Technology Project (CFFTP), the Atomic Energy of Canada Limited (AECL), the Japan Atomic Research Institute (JAERI) and Los Alamos National Laboratory (LANL). Each laboratory participating in the experiments in France or Canada covered the travel, labour, sampling and analytical costs of its own investigators (Djerassi and Gulden 1988, Burnham et al. 1988).

The presence of scientists from different countries and laboratories allowed more extensive studies to be performed, different sampling techniques to be tested, and a firm international consensus to be established on the conversion of HT to HTO in the environment. The conditions of safety of the large-scale release experiment were checked by pre-operational calculations and agreed upon by the responsible authorities. The local population was informed in advance on the nature and purpose of the experiment.

The broad objectives of the experiments were: (i) to verify that the conversion of HT at soil surface was the dominant process in the conversion of HT to HTO in the environment; (ii) to quantify the deposition and reemission processes under the real conditions of a field release; (iii) to determine the transfer processes between different compartments of the environment, including atmosphere, soil and vegetation; (iv) to validate new dispersion codes based on the deposition of HT at soil surface and the reemission of tritium as HTO; (v) to test and compare sampling equipments and analytical techniques.

Dispersion in the atmosphere

Experimental conditions : The reported field release experiments were designed to simulate accidental releases from a fusion power plant, determine the conversion of HT to HTO, and observe the movement of tritium in different components of the environment. The first experiment carried out in France was intended to simulate a puff release at a significant elevation above the surroundings, which led to envisage the release of a relatively large amount of tritium and the arrangement of a sampling grid extending over several km from the source. This was obtained by releasing 260 TBq of DT from a stack of 40 m height and sampling air and soil to 4 km from the stack (Paillard et al. 1988a). The Canadian experiments, which were organized later, put the emphasis on the contamination of soil and vegetation and on the subsequent reemission phenomena and were designed to simulate a 30 min continuous ground level release. Consequently, the released amount was reduced to 3.5 TBq only, and the sampling range limited to 400 m from the source (Brown et al. 1988, 1989). As one of the primary objectives was to study the oxidation of HT, it was necessary in both cases to minimize the ratio of HTO to HT at the source.

This was achieved by outgasing tritium from a uranium bed, avoiding any prolonged contact of the tritium with air and, for the Canadian release, further purifying the tritium by discharging it through water bubblers. The ratio HTO/HT measured in the immediate vicinity of the sources was 2×10^{-4} for the French source and much less than 10^{-4} for the Canadian source. These values were probably overestimated due to an unperfect discrimination of HT versus HTO in the air samplers.

HT and HTO during and after release : in all the experiments, time averaged air concentrations were measured during a period of 30 min after start of the release. The sampling period coincided with the release period for the 30 min release of the Canadian experiments, but was obviously much larger than the release period of about 8 min for the puff release of the French experiment. In all cases, the HTO concentrations measured during this first sampling period was much lower than the HT concentrations. In the French experiment, the ratios HTO / HT were comprised between 1.4×10^{-4} and 1.1×10^{-2}, without any obvious relationship to the distance from the source (Paillard et al. 1988b). All the Canadian results showed HTO / HT ratios increasing with distance from 1.5×10^{-5} at 5 m from the source, up to about 5×10^{-4} at 400 m. The increase of this ratio with distance would correspond to an apparent oxidation rate of HT in air of 1.5 % per hour at the maximum. But, as we will see later, oxidation truly in the atmosphere must be much slower than this apparent rate as most of the HTO observed in the atmosphere originates from the soil (Brown et al. 1989).

The two tritium species were also measured in the air during a period of 30 min beginning 30 min after the start of the release, that is after the plume had passed over the field. It was then observed that HTO persisted at levels not much lower than those during the first sampling period including the release, while HT had diminished to a very small fraction of the original level. This indicated that most of the HTO in air was not produced from gas phase oxidation, but rather from oxidation at the soil surface.

The concentrations of HTO in the air, observed after the release in the Canadian experiment, remained elevated and showed a strong diurnal variation with maximum values occurring in the middle of the day and minimum values during the night. This suggested that evaporation and transpiration could be responsible for emission of HTO from soil. During the first night after the release, HTO concentrations in air decreased by about an order of magnitude from the peak values observed just after the release, but increased again the following day to values similar to those immediately after the release. Rainfall that occurred the second night after the release was observed to reduce air HTO concentrations greatly, however a diurnal cycle was reestablished a day later (Ogram 1988, Brown et al. 1989).

Interaction with soil

Deposition of HT in the soil : On both the sites, measurements showed that tritium was deposited from the airborne HT plume as HTO in soil water. Segmented cores taken from different types of soil just after the release indicated that most of the HTO was in the 0-5 cm top layer, with a maximum of activity generally situated at a depth of 2-3 cm as shown on Fig. 1. The presence of HTO in the upper centimetres of soil showed that HT did not diffuse very far into the soil before being oxidized (Täschner et al. 1988, Noguchi et al. 1988).

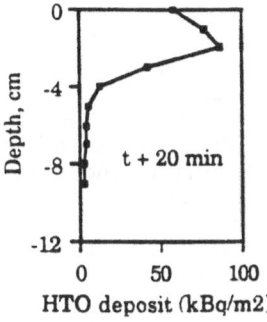 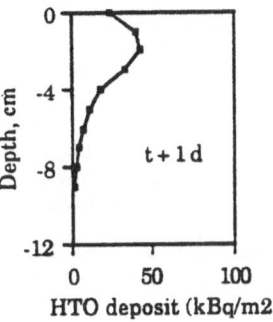

Fig.1. French experiment - Variation of HTO profile with time after release for a homogeneized soil plot at 800 m from the source. After Täschner et al. (1988).

The distribution of the HTO deposit over the field closely followed the distribution of HT in the plume, as would be expected if the deposition velocity did not vary too greatly over the field. The conversion and deposition of HT was parameterized by the deposition velocity defined as the ratio of the amount of HT deposited as HTO per unit area of soil surface to the time-integrated concentration of HT in the atmosphere above the soil surface. All values of the deposition velocity measured in France and Canada for different types of soil were comprised between 3.3×10^{-4} and 1.2×10^{-3} m/s (Paillard et al. 1988b, Täschner et al. 1988, Ogram et al. 1988). The deposition velocity did not seem to be related to the soil humidity, although this parameter waried between 2 percent for a coarse sand in Canada, up to about 30 percent for sandy or silty loams sampled in France and Canada. This could be explained by the fact that the biomass of microorganisms and the air space available for gas diffusion were probably similar in the soils that were encountered in the experiments.

Soil cores had been also taken from the field before the release and sent to the laboratory to perform chamber measurements of deposition velocity for comparison with values determined in the field (Förstel 1986, 1988; Förstel et al. 1988a,b). the values obtained in the laboratory were consistent with those obtained in the field for the same types of soil, and also with measurements made in various countries by means of laboratory and field chambers under non-winter conditions and indicating deposition velocities in the range 10^{-4} - 10^{-3} m/s.

Soil HTO loss rates : From cores taken at different times post exposure (see Fig.1), it can be seen that a part of the deposited tritium diffuses down into deeper layers, and that another part is retained by vegetation or escapes to the atmosphere. As the fraction retained by vegetation is negligible, the rate of loss of tritium can be compared to its reemission rate to the atmosphere. The time history of the tritium contained in the soil was determined by sampling cores of generally 15 cm depth at prescribed times after the release, plotting the tritium content versus time, fitting an exponential function to the experimental points and estimating the average loss rate from the slope of the exponential line, as shown in Fig.2. The average loss rate so determined was of about 2 percent per hour over a 20 hour period on both sites (Paillard et al. 1988b, Brown et al. 1989). It was of about 0.7 percent per hour over a period of 2 weeks on the Canadian site. The general decline in the soil HTO loss rate could be explained by the downward displacement of the HTO concentration peak in the soil, leading to an inhibition of the exchange at the soil-atmosphere interface. The constancy of the average loss rate after the initial more rapid decline over the very first days could be a consequence of HTO removal from the root zone to the atmosphere by transpiration of vegetation.

The use of an average exponential loss rate is only a convenient empirical approximation which is valid only over a sufficiently long period of time. Determinations of the reemission rate over short periods of time, as low as half-an-hour, were made by using a new method derived from the "energy balance method" commonly employed in agricultural micrometeorology (Ogram 1988). In the hours following the Canadian release the reemission rate peaked at about 7 percent per hour in the mid afternoon, then progressively decreased to 1.6 percent per hour in the evening. The same tendency was observed in the following days. Although restricted to a few data, these results seem to indicate that the tritium reemission rates follow a diurnal cycle similar to that of ordinary water evaporation. This is further substantiated by the observation presented above that the concentration of HTO in air followed a diurnal cycle with maximum values near midday and minimum values at night.

Attempts were made to obtain the reemission rates from small-scale experiments carried out in the field. The experiments consisted in covering a small area of undisturbed soil by a field chamber, exposing the enclosed soil to HT, then determining the changes in the HTO vapour content of a measured air-stream passing through the chamber. The initial reemission rates obtained from this method were similar to those obtained in the open environment, but further progress is to be made to simulate more closely the diurnal variation of the reemision rate, and its long-term evolution (Belot et al. 1988).

Labelling of vegetation

Tissue free water tritium : The field release experiments in France and Canada gave the opportunity to expose natural or potted vegetation to releases of HT. Natural vegetation was composed of grasses and trees; potted vegetation comprised corn, soybean, red maple and white spruce saplings in Canada, wheat, bean and cabbage in France (Diabaté and Honing 1988, Spencer et al. 1988).

In both countries, the vegetation was first sampled one hour after start of the release. The mean HTO concentration in tissue water was much smaller than that of the upper layers of soil and ranged from 15 percent to 80.5 percent of the HTO concentration in air moisture. At this time, the HTO deposited into the top layer of the soil had not yet migrated to the root zone and thus was not available for root absorption. However, some HTO that had been emitted from the contaminated upwind soil surfaces was present in the atmosphere. The relatively low level of HTO in leaf water was typical of contamination by HTO from the atmosphere, and indicated that there was little, if any, conversion of HT in the leaves. This was in agreement with previous chamber exposure experiments where conversion of HT in the leaves, if any, was shown to be extremely slow.

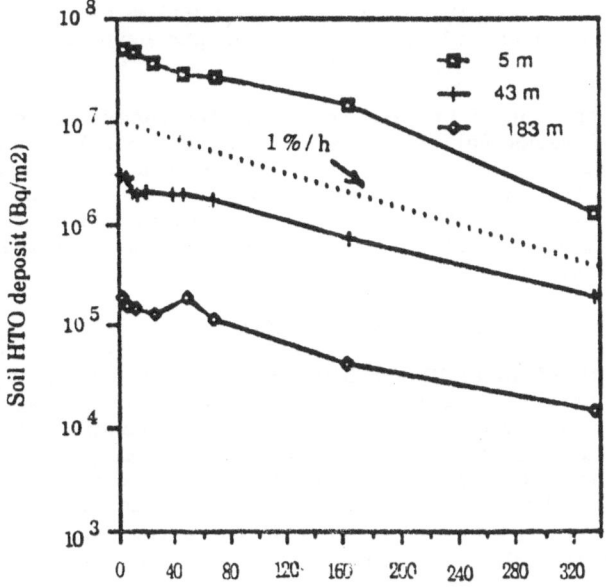

Fig.2. Canadian experiment - Variation of field centreline deposit with time after release, at different distances from the source. After Ogram, Spencer and Brown (1988).

In Canada, plant samples were collected at greater times post exposure, i.e., from 1 h to 336 h from start of release. The concentration of HTO in the water of the leaves was initially near background and was observed to reach a maximum at 4 to 24 h after the HT release and thereafter to decrease in parallel to the decline in the soil water tritium content. During this phase of decline, the half-life of HTO in the vegetation was closely related to the half-life of HTO in the soil, which meant that most of the tritium measured in vegetation was then being derived from uptake of soil water. Within this period, since tritium has migrated into deeper layers of the soil, the transfer of HTO to the atmosphere by transpiration is likely to predominate over its transfer by evaporation (Spencer et al. 1988).

Organically bound tritium : In the Canadian release experiment, the plants which had been exposed to HT and sampled at different times after exposure were assessed for non-exchangeable organically bound tritium (Spencer et al. 1988). This form of tritium is closely attached to the carbon atoms of organic compounds and cannot be removed by exchange with the hydrogen atoms of water molecules. It was measured by removing the exchangeable tritium by dry steam extraction, combusting the sample in a flow of oxygen and measuring the activity of the combustion water. One hour after start of the release, the activity of the combustion water was less than a few percent of the activity of the tissue free water. The percentages of incorporation observed in the field were similar to those which had been observed in the laboratory for plants exposed to HTO alone. This suggested that direct uptake of HT into organic matter was negligible and that the tritium measured in the organic matter of plant samples originated from the photosynthetic assimilation of tritium from HTO.

The concentration of tritium in combustion water from samples collected in the hours and weeks following the release was observed to decrease with time, probably due to the incorporation of lower activity specific hydrogen during plant growth. However the decrease was not so rapid as that of tritium in tissue free water, with the consequence that, after a few weeks, the specific activity of the tritium of the exposed plants was greater in organic matter than in free water.

Evaluation of doses from field measurements

The concentrations of HT and HTO measured in the field during the French experiment were used to determine the committed effective equivalent dose to the most exposed individual as a function of exposure duration. The most exposed individual (MEI) is defined as an individual placed at the point of maximal air concentration, in this case located on the plume axis at about 800 m from the source. Several sampling points were situated at this distance from the source, but were not placed exactly on the plume axis. The concentrations of tritium measured at these sampling points were therefore corrected to obtain the values that would have been measured on the axis. These values were multiplied by the conversion factors of the International Commission of Radiation Protection (ICRP 1979) and divided by the amount of tritium emitted from the stack (0.7 g) to obtain the doses per g of tritium emitted as HT.

Figure 3 represents the dose so calculated as a function of the time spent on the field by the MEI. The dose that was delivered during plume passage was 8.7×10^{-9} Sv/g and increased to 4×10^{-7} Sv/g after a one-day exposure, and to 6.8×10^{-7} Sv/g after a 4-day exposure. This last figure was consistent with the value obtained from the urine analyses of a NIR-scientist who remained continuously in the field at the 800m distance for five days after the release (Wiener et al. 1988. Täschner et al. 1989).

Figure 3 shows that the dose delivered during the passage of the plume is negligible in comparison to the dose delivered after plume passage, which means that most of the dose is due to the secondary emission of HTO occurring after release. The maximal dose obtained for a release of 1 g of tritium and for a man standing in the field for 4 days is smaller than the annual limit of 5×10^{-3} Sv by four orders of magnitude. The long term component which results from the contamination of locally produced food and of local water supplies was in the present case quite negligible in comparison to the short and medium term components arising from the inhalation of HTO, but this could not be generalized to other sites.

The doses to the MEI obtained from the experimental data of the French experiment correspond to the specific conditions of the release (release height 40 m, weather stability class B). The doses to the MEI under worst-case conditions, and hence the maximum permissible amounts for accidental releases, can only be derived from code calculations (Täschner et al. 1989).

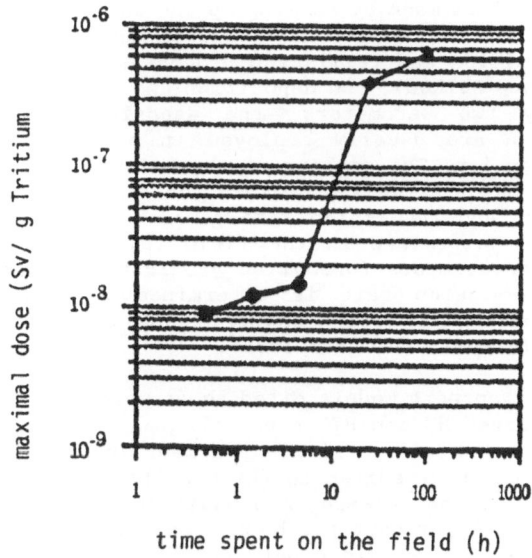

time spent on the field (h)

Fig.3. French experiment - Committed effective dose (per gram of tritium released to the atmosphere) to the most exposed individual as a function of the time spent on the field after release.
The tritium had been discharged from a 40 m-high stack for 2 min. The ratio of HTO and HT activity concentrations at the source was 2×10^{-4}. The calculations have been made by Laylavoix (1988) and by Täschner et al. (1989).

Modelling codes

Until recently the modelling of atmospheric dispersion and transfer of HT in the biosphere had not been treated correctly. Either tritium was assumed be entirely in the form of HTO and the dose obtained was greatly overestimated, or tritium was assumed to be in the form of HT without any conversion to HTO, thus leading to an underestimation of the dose. This was one objective of the recent experiments carried out in France and Canada to develop new codes describing the time-dependent behaviour of tritium deposition, conversion and reemission in the environment, and to compare the predictions obtained from these codes to the experimental data acquired in the field experiments. The experimental data obtained during the French experiment were compared to the predictions of codes from Canada (OHTDC), France (DISTRAIR), Sweden (TRILOCOMO) and from the Joint Research Center of Ispra (SCIROC). The data obtained in the Canadian experiments were compared to the predictions of the OHTDC code.

In most of these codes, the local environment downwind from the release point is divided into rectangular elements whose size has been chosen to optimize the accuracy of calculation and computing time. Each rectangular element is characterized by parameters related to the disposition and reemission of tritium and represents a small area source from which HTO escapes to the atmosphere. The calculation of the concentration of HTO in air at any point of the field is made by summing up the contribution of the ground level area sources so defined. Differences between models occur mainly in the treatment of the deposition and reemission processes. In the simplest codes (DISTRAIR, TRILOCOMO and SCIROC), the deposition and the reemission processes are characterized by two parameters, the deposition velocity and the reemission rate, which are usually employed in problems of radionuclide deposition and resuspension. The decline of the emission of HTO from soil is taken into account by using a reemission rate changing with time. The values of deposition velocity and reemission rate given as input to the code are those which have been determined in the field release experiments and in parallel small-scale experiments (Crabol et al. 1988). In the most advanced code (OHTDC), the reemission rate is determined in a subprogram which describes the processes of conversion and transport of tritium in the soil-atmosphere continuum (Russel and Ogram 1988).

All the tritium transport models cited above were tentatively evaluated by comparing the predicted HT and HTO concentrations with those obtained in the field from the French HT dispersion experiment in October 1986. Class C or D stabilities were the best suited to the conditions of weather that were experienced on the day of the release. A deposition velocity of 2×10^{-4} m/s and a reemission rate of 4 per cent per hour were selected as representative of values obtained from small-scale experiments and were taken as input to the codes. The calculation of the plume centerline ratio of HTO to HT air concentrations at 2500 from the source, averaged over the first 30 min, post exposure, was theoretically about 2×10^{-4} compared to an experimental value of 8×10^{-4}, obtained after deduction of the ratio of HTO to HT at the source. The prediction appears to give only "order-of-magnitude" estimates. Such an uncertainty is within the limits usually accepted for the dispersion codes.

An example of calculation with DISTRAIR is given in Figure 4.

The OHTDC transport model was used for prediction purposes in the Canadian experiments of August 1986 and June 1987. The weather category was estimated to be Class A or B. The HTO / HT concentration ratio for the first 30 min was calculated to vary from 1.4×10^{-6} at 5 m downwind to 3.4×10^{-4} at 400 m. As can be seen on Fig. 5, the theoretical values are similar to the observed ones within a factor of two. The HTO emission rate was predicted by the model to be about 6 percent per hour at one hour after the release, 2 percent per hour at 12 hours and 1 percent per hour at 24 hours. These

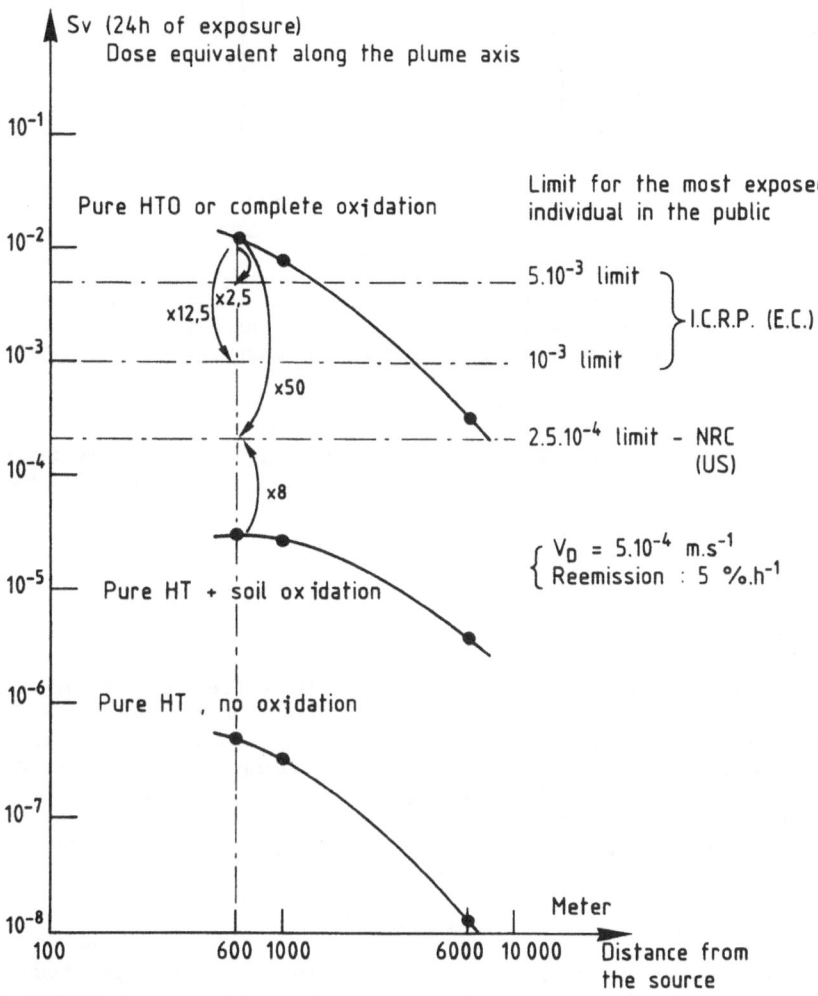

Dose conversion factor for immersion in HTO :
Breathing rate : $1.2 \ m^3 .h^{-1}$
+skin absorption : $3.1.10^{-11} \ Sv.h^{-1}$ for one $Bq.m^{-3}$
Dose conversion factor for immersion in HT :
$1.25.10^{-15} \ Sv.h^{-1}$ for one $Bq.m^{-3}$

DOSE CALCULATION FOR A PUFF RELEASE OF TRITIUM
Released quantity : 100 grams
Duration   : 30 mn
Height of stack : 40 meters
Wind speed   : $3 \ m.s^{-1}$
Pasquill stability class D

Fig. 4. Dose equivalent for a 24 exposure as a function of distance downwind of the source. A puff release of 100 grams of tritiated hydrogen HT
is considered, with three assumptions:

1. no oxidation of HT
2. HT oxidation in soils and reemission of HTO
3. complete oxidation of HT immediately after the release

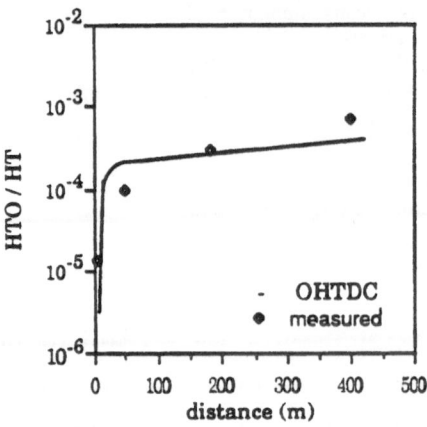

Fig.5. Canadian experiment - HTO / HT plume centreline concentration ratio as a function of distance downwind from the source during the experiment of June 1987. Comparison between model predictions and experimental data. After Russel and Ogram (1988).

reemission rates agree well with the observed HTO loss rates from soil. As time-averaged parameters were given as input to the model, the diurnal variation in HTO reemission rates and HTO concentrations in air could not be simulated (Russel and Ogram 1988).

Implications of the experiments

Conceptual model : The results obtained from the field experiments described above can be explained by the following model which was agreed upon by all the participating experimenters. In the atmosphere, the molecular hydrogen (referenced as HT) is transported along the mean wind trajectory, while diffusing laterally and vertically into the contiguous air masses. At the soil-atmosphere interface, the plume is gradually depleted by penetration of HT into the air space of soil pores and its subsequent conversion to HTO. The sorption of HT by vegetation appears to be quite negligible. The flux of tritium deposited to the soil can be parameterized by the deposition velocity usually employed in the description of the transfer of radionuclides to the soil. Once in the soil, HT is rapidly oxidized to HTO. The conversion is made in the top few cm of soil through the action of micro organisms which are generally present in a sufficient amount to ensure the conversion without being a limiting factor. In fact, the transfer of HT to the soil and its subsequent conversion are most often limited by the diffusion of HT through the air space of soil pores. When the soil moisture is very high, the soil pores are clogged and no longer available for HT penetration, which results in low deposition velocities. In the present field experiments on well aerated soils, and in parallel experiments on similar soils, the measured deposition velocity was comprised between 2×10^{-4} and 4×10^{-4} m/s.

The tritium deposited into the soil and converted to HTO is gradually reemitted to the atmosphere by evapo-transpiration from soil and vegetation. The amount of tritium that returns to the atmosphere per unit time, termed as the reemission rate, can vary between 2 and 8 percent per hour immediately after exposure, depending probably on the evaporation rate which occurs at that time. At longer times post exposure, the HTO emission rate tends to decrease, but may remain significant over several days. The emission rate can be reduced drastically by the occurrence of rainfall which dilutes the HTO present at the soil surface and moves it to a greater depth in the soil.

The results obtained here are more certain than those inferred from actual releases which occurred in the past. The conditions of some early observations were not well determined and the corresponding data were interpreted by using a model which now appears to be inadequate. In the early model, the conversion of HT to HTO was expected to occur only as a gas phase reaction in the atmosphere. We are now aware that the dominant mechanism is diffusion of HT into the soil, oxidation in the soil mediated by soil microbes, and finally reemission of tritium from soil and vegetation in the form of HTO.

Consequences for risk evaluation : The major finding of the French and Canadian experiments is that the fraction of HT oxidized to HTO in the atmosphere during plume travel at distances of a few kilometers is small, typically less than 10^{-3}. This was in agreement with predictions derived from codes based on the premises that the gas phase oxidation of HT is negligible and that the conversion of HT to HTO occurs in the top layer of soil. The experimental results were obtained at two different heights of release and under conditions of weather and soil which were favorable to the transport and oxidation of HT in the air space of soil pores. The experiments were therefore likely to be conservative in regard to the formation of HTO (Ogram 1988). Apart from release height and weather conditions, factors such as time of day, season and soil conditions could be further investigated as will be seen later.

Due to the small amount of HT converted to atmospheric HTO, the inhalation dose during plume passage at distances typical of site boundaries is much less for a release of pure HT than for an equivalent release of HTO. The dose obtained by considering the real conversion of HT to HTO is much more realistic and much lower than that obtained by using the conservative assumption that the release occurs in the form of HTO. As HTO in soil and in air persists for weeks after the passage of the plume, the dose which can be delivered to a man standing continuously on the field is much greater after than during the passage of the plume.

Needs for further research

The recent experiments reported above and many early studies have been centered on the processes which chemically transform tritium into water that could be easily taken up by vegetation and man. The key processes controlling the conversion of tritium in the environment and the short range behaviour of HT releases are now well established. Nevertheless, it appears that the large-scale field experiments that have taken place up to the present have been carried out under rather similar environmental conditions and that it could be useful to perform new field experiments under significantly different conditions, such as winter conditions (Ogram 1988). Many scientists agree that the large scale experiments will be necessarily limited in number, due to the difficulties encountered in obtaining acceptance for field release experiments, and that it could be of value to conduct, instead, small scale experiments in the laboratory or in the field. These experiments would permit an extension of the range of explored environmental conditions and, associated with a modelling effort, would establish methods of predicting the rate of deposition and reemission from a knowledge of the conditions in the soil and the atmosphere. It would be of particular interest to study the reemission of tritium from the soil as a function of environmental conditions and of their fluctuations over diurnal or seasonal cycles. This could be achieved in the laboratory by using soil columns placed in a wind-tunnel arrangment, or in the field by means of field chambers placed on the soil, taking care that the environmental conditions inside the chambers not be too different from those outside.

While it is necessary to understand the processes controlling the conversion of HT to HTO at the soil surface to assure that we can predict the consequences of tritium inhalation, it is also important to determine the rate of passage of tritium into the food chain to determine the dose from ingested food. The exchange between HTO and plant water is now well documented (Belot et al. 1979, Garland 1980), but the incorporation of tritium into the organic fraction of vegetation is not entirely understood, in spite of the importance of this pathway in the dermination of dose to man. We know that some tritium can be incorporated through the photosynthesis of carbon dioxide and water (Guenot and Belot 1984), but in many circumstances and particularly in the case of industrial releases, an excess of organically bound tritium was observed that could not be explained by this process only. The excess was attributed by several investigators to the direct incorporation of HT into vegetation (Sweet and Murphy 1984), but this assumption has not been confirmed by any of the short term experiments done to date (Murphy 1989). Further studies are necessary to clarify this point and to be able to predict the contamination of foodstuffs and the resulting dose to man.

CONCLUSIONS

1) A current conceptual model which describes the conversion of HT to HTO in the environment has been verified for the particular conditions of the field experiments presented above. Different codes based on this model have been validated within the uncertainties usually accepted for such types of codes.

2) The present results can be extended to other situations by changing the values of the deposition velocity and reemission rate which have been chosen as input parameters to the codes. It was shown that these parameters can be correctly determined by means of chamber experiments carried out on undisturbed sections of soil in the field, and on small soil samples in the laboratory. The prediction of HT oxidation on various sites does not require the multiplication of large-scale field release experiments.

3) The codes which are derived from the agreed model are able to predict the concentrations and doses associated with the "molecular hydrogen pathway" and with the "tritiated water pathway" in the event of an accidental release of tritium by any future fusion reactor. Values of the committed doses thus obtained are much more realistic and much lower than those obtained by using the conservative assumption that all releases occur in the form of HTO.
4) It will therefore be possible to provide the designers of future reactors with exact figures for permissible releases in the event of an accident, on the basis of which they will be able to design appropriate facilities.

REFERENCES

Belot Y., Gauthier D., Camus H. and Caput C., 1979
Prediction of the flux of tritiated water from air to plant leaves
Health Phys., 37: 575-583

Belot Y., Guenot J. and Caput C., 1988
Emission to atmosphere of tritiated water formed at soil surface by oxidation
of HT
Fusion Technology, 14: 1231-1234

Brown R.M., Ogram G.L and Spencer F.S., 1987
Field studies of HT oxidation and dispersion in the environment
I. The 1986 August experiment at Chalk River, Canadian Fusion Fuels Technology
Project
Report No. CFFTP-G-87004

Brown R.M., Ogram G.L and Spencer F.S., 1988
Field studies of HT behaviour in the environment
I. Dispersion and oxidation in the atmosphere
Fusion Technology, 14: 1165-1169

Brown R.M., Ogram G.L and Spencer F.S., 1989
Field studies of HT oxidation and dispersion in the environment
II. The 1987 June experiment at Chalk River, Canadian Fusion Fuels technology
Project
Report No. CFFTP-G-88007

Burnham C.D., Brown R.M., Ogram G.L. and Spencer F.S., 1988
An overview of experiments at Chalk River on HT dispersion in the environment,
Fusion Technology, 14: 1159-1164

Crabol B., Edlund O. and Graziani G., 1988
Overview of the assessment of the in-field French tritium experiment with
computer codes
3rd Topical Meeting on Tritium Technology, Toronto May 1-6

Diabaté S. and Honing H, 1988
Conversion of molecular tritium to HTO and OBT in plants an soils
Fusion technology, 14: 1235-1240

Djerassi H. and Gulden W., 1988
Overview of the tritium release experiment in France
Fusion Technology, 14: 1216-1221

Dunstall T.G., Ogram G.L. and Spencer F.S., 1985
Elemental tritium deposition and conversion in the terrestrial environment,
Fusion Technology, 8: 2551-2556

Förstel H., 1986
Uptake of elementary tritium by the soil
Radiation Protection Dosimetry 16: 75-81

Förstel H., 1988
HT and HTO conversion in the soil and subsequent tritium pathway: field release data and laboratory experiments
Fusion Technology, 14: 1241-1246

Förstel H., Lepa K and H. Trierweiler, 1988a
Reemission of HTO into the atmosphere after HT/HTO conversion in the soil,
Fusion Technology, 14: 1203-1208

Förstel H., Papke H. and Hillmann I., 1988b
Uptake of tritium in the organically bound form into the biomass of the soil,
Fusion Technology, 14: 1258-1263

Garland J.A. and Cox L.C., 1980
The absorption of tritium gas by English soils and plants and the sea
Water, Air and Soil Pollution, 14: 103-114

Garland J.A. and Cox L.C., 1982
Uptake of tritiate water vapour by bean leaves
Water, Air and Soil Pollut. 17: 207-212

Guenot J. and Belot Y., 1984
Assimilation of ^{3}H in photosynthesizing leaves exposed to HTO
Health Physics, 47: 849-855

ICRP, 1979
Limits on intakes of radionuclides by workers
Publication 30, Part 1. Pergamon Press, Oxford

Jeanmaire L. et Piéchowski J. 1986
Métabolisme et toxicité du tritium - moyens de surveillance des expositions
Journées Tritium, Dijon 23-25 avril

Lafuma J., 1986
Action toxique du tritium
Journées Tritium, Dijon 23-25 avril

Laylavoix F., 1988
Personal communication

McFarlane J.C., Rogers R.D. and Bradley D. V., Jr, 1978
Environmental tritium oxidation in surface soil
Environ. Sci. Techn., 12: 590-593

Murphy C.E. Jr, 1989
Controlled environment estimates of HT uptake by vegetation in Proceedings of the Third Japan-US Workshop on Tritium Radiobiology and Health Physics, Nov 8-10, 1988, Kyoto, Japan
(S.Okada ed.) pp 64-73

Noguchi H., Matsui T. and M. Murata, 1988
Tritium behavior observed in the Canadian HT release
Fusion technology, 14: 1187-1192

Ogram G.L., 1982a
Atmospheric chemistry of radionuclides: a review, Ontario Hydro Research
Division Report No. 82-83 K

Ogram G.L., 1982b
The oxidation of molecular tritium released to atmosphere, Ontario Hydro
Research Division Report No. 82-449 K

Ogram G.L., Spencer F.S. and Brown R.M., 1988
Field studies of HT behaviour in the environment: 2. The interactions with
soil
Fusion technology, 14: 1170-1175

Ogram G.L., 1988
The Canadian HT dispersion experiment at Chalk River - June 1987: Summary
Report, Canadian Fusion Fuels Technology Project
Report No. CFFTP-G-88027

Paillard Ph., Clerc H., Calando J.P., Gros R. and B. Hircq, 1988a
Tritium release experiment in France: presentation, organization and
realization
Fusion Technology, 14: 1222-1225

Paillard Ph., Calando J.P., Clerc H., Gros R and Y. Belot, 1988b
Tritium release experiment in France: Results concerning HT / HTO conversion
in the air and soil
Fusion Technology, 14: 1226-1229

Peterman B.F., Johnson J.R. and McElroy R.G.C., 1985
HT / HTO conversion in mammals
Fusion Technology, 8: 2557-2563

Russel S.B. and G.L. Ogram, 1988
Modelling elemental tritium conversion and reemission using Ontario Hydro's
tritium dispersion code
Fusion Technology, 14: 1193-1198

Spencer F.S. and T.G. Dunstall, 1986
Molecular tritium conversion in vegetation, litter and soil
Radiation Protection Dosimetry, 16: 89-93

Spencer F.S., Ogram G.L. and Brown R.M., 1988
Field studies of HT behaviour in the environment: 3. Tritium deposition and
dynamics in vegetation
Fusion Technology, 14: 1176-1181

Sweet C.W. and Murphy Jr C.E., 1981
Oxidation of molecular tritium by intact soils
Environ. Sci. and techn., 15: 1485-1487

Sweet C.W. and Murphy JrC.E. 1984
Tritium deposition in pine trees and soil from atmospheric releases of molecular tritium
Environ. Sci. Techn. 18: 358-361

Täschner M., Wiener B. and Bunnenbert C., 1988
HT dispersion and deposition in soil after experimental releases of tritiated hydrogen
Fusion Technology, 4: 1264-1269

Täschner M., Bunnenberg C., Edlund O. and Gulden W., 1989
Maximum permissible amount of accidentally released HT derived from the 1986 French experiment to meet dose limits for public exposure
Personal communication

Wiener B., Täschner M. and Bunnenberg C., 1988
HTO reemission from soil after HT deposition and dose consequences of HT releases
Fusion Technology, 14: 1247-1252

Zettwoog P., 12 July 1988
Major increases in knowledge about the biospheric transfer of tritium
An information note on the French experiment
IPSN/DPT/SPIN/10.1.3.4.2 (163)

About the French Environmental HT Experiments - September 1986

See: "Environmental tritium behaviour. French experiment", Final report 88/07
Edited by H. DJERASSI and B. LESIGNE, Commissariat à l'Energie Atomique,
IPSN/DPT/SPIN, CEN/SACLAY, Bât. 389, 91191 GIF SUR YVETTE CEDEX, France

Participating organizations

CEA	Commissariat à l'Energie Atomique
CFFTP	Canadian Fusion Fuels Technology Project
KFA	Kernforschungsanlage Jülich
KfK	Kernforschungzentrum Karlsruhe
NET	Next European Torus
NIR	Niedersächsisches Institut für Radioökologie
OHRD	Ontario Hydro Research Division
Studsvik	Studsvik Energiteknik AB
JRC	Joint Research Center of Ispra

About the Canadian Environmental HT Experiment - June 1987

See: OGRAM G.L., 1988 "The Canadian HT dispersion experiment at Chalk River"
June 1987, Summary report, Canadian Fusion Fuels Technology Project, Report
n° CFFTP-G-88027

Participating organizations

CEA	Commissariat à l'Energie Atomique
CFFTP	Canadian Fusion Fuels Technology Project
JAERI	Japan Atomic Energy Research Institute
KFA	Kernforschungsanlage Jülich
KfK	Kernforschungzentrum Karlsruhe
LANL	Los Alamos National Laboratory
NET	Next European Torus
NIR	Niedersächsisches Institut für Radioökologie
OHRD	Ontario Hydro Research Division
PPPL	Princeton Plasma Physics laboratory
SRL	Savannah River Laboratory
Studsvik	Studsvik Energiteknik AB

APPENDIX - BASIC NOTIONS REGARDING MANAGEMENT OF THE RISK OF EXPOSURE TO
IONIZING RADIATION

INTRODUCTION

Among changing attitudes in contemporary society, it is apparent that
demands in the field of safety are increasing. The fact is that the prospects
offered by technology for improved wellbeing and quality of life are
unfortunately accompanied by harmful consequences, notably deterioration of
the environment and damage to human health. Tolerance of these harmful effects
is decreasing. In particular, there is an increasing aversion to large
collective accidents. This is already true for natural disasters. The
inevitability of destruction due to earthquakes, floods or landslides is no
longer believed. The carelessness and negligence of those responsible for
prevention and forecasting are roundly condemned.

The same is even truer for technological hazards, and the aversion to
risks is especially evident with new technologies involving forces or
materials as yet unknown to the public.

This explains why the public and safety management specialists have
consideraly different perceptions of the relative risks involved. Specialists
use a factual and quantitative approach to risk management. Risk evaluation
by the public on the other hand is subject to individual factors and
psychosociological pressures and hence involves another type of reasoning.

It is normal that, as in other fields, individual preferences concerning
risk should be expressed. Some people prefer the risks of tobacco to those of
alcohol, for instance. Others eschew both of these in favour of mountain
climbing. However, in terms of collective risk management it is necessary to
make the transition from articulation of varied individual preferences to the
expression of common determination, and it is here that the specialists must
be in step with public thinking.

It is therefore to be deplored that educational systems do not prepare
the public for objective understanding of risk management, which means that
when confronted with two alternatives, the public is generally unable to
identifiy the lower-risk option. The public is threatened by three pitfalls:

1) Irrationality. This results from an inability to understand the
mechanisms of risk generation, or more exactly from limited or erroneous
understanding. It leads individuals to adopt behaviour that has consequences
at odds with the desired aims. Militating for the protection of forests
threatened by acid rain while behaving in a way that favours production of
electrical energy by coal-burning plants rather than by nuclear power stations
is an illustration of this irrationality.

When the first motor cars reached out-of-the-way villages, local free-
range fowl instinctively recognised the associated danger. But their limited

appreciation of the risk inevitably led them to run in front of the vehicle rather than to the edge of the road. This is another illustration of irrational behaviour in the face of risk.

2) <u>Wishful thinking</u>. Many people have suffered negative consequences from a belief that a particular situation will arise simply because it is desired. This has led communities to become involved in policies that are out of touch with reality. The real situation will sooner or later become apparent, and it is then very expensive to rectify the situation.

In terms of risks to the environment, an illustration of wishful thinking is offered by the decision taken by the Swedes to shut down their nuclear power stations over a short time-scale. These nuclear facilities currently provide 50 % of Sweden's electricity requirements at competititve prices and under safe conditions.

3) <u>Ideological manipulation</u>. People who allow themselves to be swayed by ideology, whatever its persuasion, lose their capacity for reasoning. Certain groups exploit the term "Nature", with a capital N, in terms of environmental protection. "Nature" is an eminently cultural concept, which only exists in the mind, and which bears little relation to the ecosystems it is supposed to represent.

Ideological manipulation becomes apparent when "Nature" is identified with truth, beauty and wellbeing, technological progress being wrongly branded as false, ugly and bad. People are pressed as to whether they are for "Nature" conservation. The dice are cast and the discussion is closed when the argument takes a theological turn and God's displeasure at interference in His Creation is invoked.

There is however a scientific approach to the management of collective risks. Its aim is to help both the adjudicators and the public to take decisions that are:

a) rational, i.e., decisions that are consonant with community aims;
b) realistic, i.e., decisions that take available data into account;
c) free of ideology, i.e., decisions taken for themselves and not because of other unavowed aims.

The scientific approach must be completely impartial, and its success depends on the constant provision of explanations in terms accessible to the affected population, whose qualified representatives should at least be involved at decisive moments.

The approach is very general. It does apply to all collective technological risks but has only been taken into account (the nuclear industry apart) in a limited number of industrial sectors. The same safety requirements as those encountered in the nuclear fuel cycle are generally only seen in the air transportation sector, although the approach is different. This general approach is explained below. Its application will be examined in the case of the nuclear industry, with its risk of exposure to ionizing radiation.

1. SCIENTIFIC APPROACH TO THE MANAGEMENT OF COLLECTIVE TECHNICOLOGICAL RISK

This approach involes four stages:

1.1. Data acquisition

The first task falls to research scientists and involves collection of data relating to the generation and extent of the risk. In general, scientists are only able to define a few islands of certainty, which are lost in an ocean of doubt. Scientific doubt is inevitable, and difference of opinion among experts is not in short supply. Data acquisition must be conducted openly and should lead to a clear and unambiguous identification of areas of doubt and certainty.

1.2. Drawing up of basic rules

The second stage involves the drawing up of basic rules designed to limit individual risks. Here the responsibility is not that of research scientists, but rather of groups of risk management experts. Bearing in mind the areas of doubt identified by researchers, these experts must first produce a limited number of reasonably conservative risk management hypotheses. They must then establish a set of basic rules, which will complete the framework of the risk limitation system. Incomplete understanding thus gives way to procedural conviction. Those who draw attention to scientific uncertainty at the moment of application of the basic rules are out of step and have forgotten that doubt has already been evaluated and taken into account during rulemaking by the most qualified experts.

It is vital that these basic rules are unambiguous, simply understood and applicable in practice, and that their proper application can be verified.

1.3. Definition of safety arrangements in technological facilities

The third stage involves definition of technical and organisational arrangements derived from the basic rules. These will result in effective protection of the public against any detriment occasioned by the operation of technological facilities.

These arrangements relate to the design, development and operation of the nuclear facility, the interposition of various successive and interlocking barriers between the machinery and the public, and the limitation of liquid and gas effluents released to the environment, bearing in mind the ecological transfer properties of the receiving sites.

1.4. Monitoring of regulatory implementation of data feedback

The fourth stage is designed to ensure that the safety arrangements are implemented and applied at all levels, this completing the special restrictions defined by the appropriate regional authority for the nuclear facility and environs.

Data relating to all incidents, even harmless, occurring within the facility or on the site are recorded by the site-surveillance network. Real-time processing of these data is a fundamental element in the safety of the facility, and can be used to prevent degeneration of normal and temporary operational failures into irreversible incidents involving either part or all of the facility.

2. APPLICATION OF SCIENTIFIC APPROACH TO ENVIRONMENTAL RISKS RELATED TO NUCLEAR FACILITIES

Nuclear facilities release to the environment gaseous and liquid effluents containing radionuclides likely to result in exposure of the public to ionizing radiation. The general approach described above has been developed for the management of collective risks to the public due to the proximity of an operating nuclear facility. It encompasses not only routine releases to the environment, but also any unforeseen or uncontrolled discharges caused by mishaps, or accidents compromising the integrity of containment barriers.

The health risks associated with exposure to ionizing radiation have been studied in-depth, not only for the short-term effects of high doses, but also for chronic effects. Physicians have used ionizing radiation for diagnosis and therapy for the last seventy years, and have, themselves, suffered the consequences of uncontrolled use of radiation.

Radioactive contamination of the biosphere is not in itself detrimental, since it does not alter the equilibrium of ecosystems. It has, nevertheless, to be fully investigated, as it does affect food chains, from which contamination is likely to pass from plants and animals to humans.

2.1. Results provided by research scientists for risk management experts

Considerable uncertainty persists in this complex and involved subject, but it has been possible to define a core of results accepted by scientific consensus which can be used in risk management.

a) The effects of ionizing radiation on a tissue, organ or organism can in the first instance be related to the quantity of energy (joule) per unit mass (kg) incident on the tissue considered. The notion of absorbed dose is introduced, expressed in gray (Gy), which corrresponds to a dose of 1 joule per kilogram of matter.

b) When high doses are absorbed over a short period, for example when an individual is exposed to energy delivered by gamma rays of energy greater than 0.5 Gy in a few minutes, the effects soon appear in all exposed individuals, and severity increases with the absorbed dose. Reversible burns (erythema) are first seen. Doses above a few Gy lead to certain death within a few days.

c) When the doses are absorbed at low dose rates but over long periods, the immediate effects cited in b) do not develop, but long-term effects may appear, i.e., after a latency period of greater than 5 or 10 years. These effects are seen when the absorbed doses range between a few tenths of a Gy to a few Gy. Only an as yet unidentifiable fraction of the exposed individuals is involved. The severity do not depends anymore on the absorbed dose. Cancer or leukemia may be involved, or a genetic defect in future generations. The biological and organic factors that define whether an individual is affected or not at a given dose are not yet completely elucidated. This ignorance is camouflaged by declaring that the occurence of these effects is random, or "stochastic". Progress in molecular biology and immunology may allow radiosensitive individuals to be identified, and the stochastic nature of these effects to be eliminated.

In any case, it should be remembered that the effect of a "low dose" is mainly an absorbed dose-dependent increase in the probability of occurrence of a severe long-term effect. The relation linking the increase in probability to the absorbed dose is obtained through epidemiological surveys in human populations subject to occupational or medical exposure, or affected by the radiation released in the destruction of Hiroshima or Nagasaki.

Experiments in animal models can provide similar results, but with more rigorous control of mortality and absorbed doses.

It is seen that the type of radiation plays a role. The probability of an effect for a given dose (Gy) is 20-fold higher if exposure involves alpha particles or neutrons, as compared with exposure to beta particles, X rays or gamma rays. Likewise, for a given organism and dose, the probability of an effect when a single organ is irradiated depends on the organ.

d) At absorbed doses of less than a few tenths of a gray, it is known that the stochastic effects, if they exist, are too weak to be observed. This corresponds in all cases to increases in probabilities of occurrence less than 10^{-5} for a given individual. It is in fact clear that epidemiological surveys can only account for such weak effects if the study population is very large (several hundred thousand individuals), and is compared with a sample population of the same size, which comprises individuals with the same way of life, the only between-population difference being in absorbed dose. Such sample populations are not found in practice.

In addition there are numerous other oncogenic agents in the environment, and synergic factors, such as nicotinism, dominate and may mask any effects of low doses.

Another difficulty in assessing the effect of very low doses is that the doses inevitably absorbed by individuals following exposure to natural radiation (cosmic and terrestrial, as well as radiation emitted by natural radionuclides absorbed by the body) are generally in the 0.1 to 0.3 gray per year range, with spatial or temporal fluctuations greater than the absorbed doses of artificial origin being investigated. It is necessary to add here a notable source that few people escape, i.e., medical irradiation during radiography, which currently accounts for most exposure of populations.

Experiments with animal models cannot provide answers either, given the difficulties of managing large animal populations and the existence of spontaneous, random cancers seen in unexposed animals.

e) There may also be beneficial effects resulting from exposure to low doses, and this is the subject of many studies.

2.2. Laying down rules for management of risk to the public

First of all it will be considered that accidental and, even more so, routine, high-dose exposure likely to provoke immediate effects can be excluded. This assumption is valid, since even at Chernobyl no "high dose" was absorbed by affected populations. Only low-dose effects will therefore be considered, i.e., long-term stochastic effects characterized uniquely by their probability of occurrence.

Individual risk R_i is defined as: $R_i = P_{ri} \times d_i$, where d_i is the value of the detriment. For reasons of simplicity, the severity of d_i is usually assumed to be the same for cancer, leukemia or a genetic defect, and for infants and the elderly. P_{ri} is the probability of being affected by the detriment, and is the product of the probability of absorbing a given dose following a release of radionuclides and the probability of the effect if this dose is absorbed. The probability of release is unity for routine discharges.

The aim of risk management is the reasonable limitation of the collective risk represented by all individual probabilities P_{ri}, given the acceptability constraint of each P_{ri}, which amounts to defining an upper limit for this probability.

We have seen that the probability of an effect for a given absorbed dose (gray) depends, among other things, on the type of radiation and the radiosensitivity of the irradiated organ.

The fundamental assumption in radiation risk management, the socalled linear relation with no threshold assumption, is that the probability of occurrence is proportional to a magnitude called the "committed absorbed dose equivalent", H_E , which is expressed in sievert (Sv). This is a reasonably conservative hypothesis for very low doses. By simplifying and by bearing in mind the different possible long-term effects, it is possible to adopt a value of 2.10^{-2} for the probability of a long-term effect resulting from an absorbed dose equivalent of 1 Sv. Physicians have established the correspondence rules between doses (in gray) absorbed by different organs (or the whole body), as a function of the type of radiation, and the committed absorbed dose equivalents (in sievert), such that the committed absorbed dose equivalent is in fact a measure of the probability, taking into account the factor 2.10^{-2} per sievert.

The aim of radiation risk management is to limit the <u>mean</u> dose equivalent absorbed H_E by the population annually at 0.5 millisievert per year, such that the <u>mean</u> probability of an effect in the affected population is less than 10^{-5} per year of exposure.

It is clear that this probability level is no longer meaningful for a given individual, bearing in mind the variability of the individual biological resistance for such low exposure, and given the other sources of radiation.

Collectively, on the other hand, this probability level is meaningful, and the proportionality hypothesis can in fact introduce the collective risk R_c:

$$R_c = \sum_i P_{ri} \times d_i = d \times 2 \times 10^{-2} \sum_i H_E i$$

The collective dose equivalent is introduced: $H_E c = \sum_i H_E i$, which is expressed in person x sievert. The product of $H_E c$ and 2×10^{-2} represents the number of anticipated effects in the affected population, within the context of risk management hypotheses. It is this number of effects which may subsequently be identified by an epidemiological survey, individual occurrences being indiscernible because of the other etiological causes.

It is only the collective dose equivalent that needs to be controlled, as it is the measure of collective detriment.

Ecologists are able to calculate the radionuclide concentrations in the surrounding atmosphere, water and physical media, and in the food chain, following routine or accidental discharge of these radionuclides.

It is then possible to calculate external exposure of the body and internal exposure of organs following inhalation, cutaneous absorption and ingestion of radionuclides. There are tabulated data giving values for the absorbed dose equivalent corresponding to the intake of a given quantity of radionuclides, bearing in mind the type of intake and its physicochemical form. The notion of Annual Limit of Intake (ALI) is introduced, such that the intake of the ALI results in an annual absorbed dose equivalent of 50 millisieverts, i.e., a probability of occurrence of 10^{-3} per year of exposure which is the upper limit of risk for occupational exposure. From this are deduced the derived limits of mean concentrations in the air, water or food consumed over the whole year.

It is therefore possible to calculate, for routine or accidental discharges, the collective dose due to all the quantities of the various radionuclides released to the atmosphere or water at a known site. The different pathways for transfer in the environment are considered, as are the eating habits of the affected populations.

The aim is to set standards for release from the nuclear facility considered, bearing in mind the dispersivity of the environment, population density, and the existence of any other radiation sources.

It is usual to set a limit value above the committed absorbed dose equivalent for the most exposed individual, which is taken as 5.10^{-3} Sv per year for short-term exposure (probability of effect of 10^{-4} per year of exposure), and as 10^{-3} Sv per year, for exposure that may last a lifetime (probability of 2.10^{-5} per year of exposure). The idea is that on average this automatically ensures for the whole population a mean probability of less than 10^{-5} per year of exposure.

Corresponding to this condition is an upper limit for the annual release (routine discharge) and an upper limit of statistically expected accidental release (the probability of which will thus be automatically limited).

This just marks the beginning of risk limitation. In fact, the residual collective detriment to the population may still be too great. A mean risk of 10^{-5} per year of exposure for an affected population of 1,000,000 individuals still results in 10 effects per year. It may be possible to diminish this admittedly hypothetical figure through additional spending on the nuclear facility. In this case it is necessary to evaluate the spending costs in the light of the anticipated reduction in health effects.

The notion of optimization of radiological protection, and hence of optimization of the release, can be introduced here. This means that the costs involved in further reducing risk would no longer be justified, given the limited health benefit gained. These costs would in any case be more effectively used for other health benefits gained.

The risk management strategy involves, first, individual elimination of risk levels unacceptable to the populations concerned, ensuring, in the case of an accidental release, a probability of effect less than 10^{-4} for the most exposed individual, by establishing an upper limit of the release. Release is then optimized so that the collective risk is as low as possible bearing in mind the technical and economic constraints.

Certain population groups constantly demand reductions in the release limits defined on the basis of exposure of the most exposed individuals or even of the optimized release. Apart from the fact that seeking risk probabilities of less than 10^{-6} or 10^{-7} per year no longer has any behavioural meaning, these groups do not realize that their aims lead to allocation of resources that runs counter to stated aims (overshoot of the optimal collective risk level).

2.3. Regulatory safety arrangements

The relevant authorities of each state define regulatory and organisational arrangements which allow application of the basic rules, bearing in mind the results of analysis of accident scenarios.

2.4. Safety maintenance

Experience gained in the management of nuclear risks shows that the safety of a nuclear facility depends upon strict sharing of responsibilities. The operator alone is liable for damages following any detriment to the population. The authorities are responsible for drafting rules and ensuring compliance. A body of experts relying on powerful safety research methods provides all concerned with the understanding required for coping with constraints and unforeseen events and for making use of data feedback. The whole process must be conducted openly and in the public eye in order to gain public acceptance.

3. THE CASE OF FUTURE CONTROLLED FUSION REACTORS

These reactors will routinely release only tritium as far as radioactivity is concerned. In the event of an accident, it will also be necessary to consider scenarios (fire followed by increased pressure within the buildings for instance) describing the release of aerosols of high specific activity produced by erosion of the first walls.

The above considerations are strictly applicable to future prototypes.

In the case of routine releases of tritium, the limit of 1 mSv per year for the most exposed individual may be not easy to respect if the tritium is in HTO form or contains a significant part of organically bound tritium. This limit could be exceeded by more than one order of magnitude in the case of an accident with releases of HTO quantities in the order of 100 g and with an unfavorable meteorological situation (see Fig. 4).

For production reactors, which will only be developed in the period after 2050, it may well be asked whether the sociological context will remain unchanged, since we have seen that it strongly influences the safety level demanded by the population. It may be thought that over this period the populations concerned will have come to a more realistic appreciation of the meaning of risks in general and radiological risks in particular, and that their demands will not have increased in this particular field. In addition, progress in the medical sciences will probably help in curing efficiently most of the sanitary effects. It follows that the constraints communicated to reactor designers by radiation protection experts today are not likely to be more severe in the future. This is fortunate since fusion technology is complex enough as it is.

It should be noted that the genuine radiological problems, for fusion reactors of the far future as well as for the next prototypes, will not be related to the environment, but to the control of the contamination of the plants by tritium and by activated erosion dusts.

If the transfer of all the radionuclides inside the operational areas is not adequately limited, this will result in too much time being spent in repair and maintenance, and in an excessive production of wastes. Both consequences mean higher costs for the production of electricity. This will also results in an increase of the collective dose to the workers.

The decisions taken in the design phase of the reactor itself and of its pephericals, and of the surrounding buildings including the use of intermediate confinments and barriers, will determine the capacity of the operators to prevent the migration of the contamination more or less efficiently inside the plant.

V. SAFETY AND ECONOMY OF FUSION PROTO-REACTORS

SAFETY AND ENVIRONMENTAL IMPACT OF ITER/NET

J. Raeder, W. Gulden

The NET Team, c/o Max-Planck-Institut für
Plasmaphysik,
Boltzmannstrasse 2
D-8046 Garching bei Muenchen, W. Germany

1. INTRODUCTION

Deuterium - tritium fusion is a nuclear energy source. Its safety and environmental (S&E) aspects are being systematically considered in all of the on-going world's major fusion programmes. There are broad efforts, of which ITER/NET are parts to develop appropriate design philosophies and procedures that recognize and properly account for the specific characteristics of the fusion energy source and the technologies that will be employed in its utilization.

In essential design principles, plants based on fusion reactions will follow safety concepts developed for now-conventional nuclear power sources, i.e., fission power plants. In detail, however, there are decisive differences between fission and fusion devices with respect to potential consequences of normal operation, including waste management, and possible accidents.

These issues will predominantly be judged upon in terms of doses or dose rates. Those, however, are determined on an isotope-by-isotope basis which accentuates the potential of fusion compared to fission.

With respect to normal operation, an important difference is the creation of radioactive fission products but no radioactive fusion products. There is residual radioactivity due to reactions of neutrons with the surroundings in both cases although not in equal amounts. With respect to safety from accident consequences, the most important characteristic of fusion is the expected self-

Safety, Environmental Impact, and Economic Prospects of Nuclear Fusion
Edited by B. Brunelli and H. Knoepfel
Plenum Press, New York, 1990

231

termination of the fusion process under accident conditions. So-called "passive safety" appears to be achievable in fusion power plants because no criticality problem exists and the energy sources that could drive accidents are low compared to fission, whereas the volumes and surface areas are large if magnetic confinement is used.

Most of the fusion device components are novel and have not yet been built at the performance level and scale finally required. Fusion now is in the process of detail evaluation of its risks and of development of its safety related guidelines, mostly due to the incentive of constructing next generation devices such as ITER/NET.

In ITER/NET the safety and environment (S+E) related part of the working programme focuses the related activities and guides them according to the following objectives:

-   ITER/NET have to comply with all relevant regulatory requirements which will be imposed by licensing authorities.
-   Harm to any member of the public under both normal and accidental conditions has to be minimum.
-   ITER/NET should develop, test and demonstrate safety technologies and safety standards applicable to future fusion technology development.

To meet these generic objectives, tasks have been defined for the S+E work. Within the European Fusion Safety and Environment Programme, for example, they are broken down as follows:

Component Related Studies
    Safety of first wall, blanket, divertor, limiter, launcher systems,
Safety of cryosystems,
Safety of superconducting magnets,
Safety of tritium cycle and storage systems,
Safety of plasma heating and fuelling systems.

For each component, the energy and radioactivity inventories are being quantified. In addition, based on analyses of accidents anticipated to the best of knowledge, safety margins (e.g. maximum temperature reached during an accident compared to critical temperatures) and/or the source terms resulting from an accident (e.g. amount of radioactivity released into the next containment) will be determined.

Plant Related Studies
    Radioactivity source terms inside the plant,
Environmental impact of tritium and activation products,

Waste management and disposal.

Safety Guidance and Assessment
    Safety related design guidelines,
Reference accident sequences,
Risk studies,
Assistance in preparation of safety reports.

Long Term Safety and Environment Studies
    These studies are centred around future fusion reactors
and will comprise radioactivity releases, waste management,
decommissioning, non D-T fusion processes, and environmental
impact statements.

    This paper reviews the state in the fields of
radiological design targets, inventories, accidents,
dispersion of radioactivity, waste and implications of
component handling. Most of the information has been drawn
from the European S+E and NET work as well as from the ITER
activity. There in the area of S+E the international
expertise has been collected and reviewed and radiological
design targets have been proposed.

2. RADIOLOGICAL DESIGN TARGETS

    One of the important ITER/NET engineering objectives is
to demonstrate the potential of safe and environmentally
acceptable operation of a power-producing fusion reactor. In
light of this objective, ITER has decided to adopt passive
safety as an ultimate design goal. This involves recognizing
the unique characteristics of the fusion process and making
it an integral part of the design work. This objective
implies adequate definition of radiation protection targets,
site requirements, and safety and environmental analyses to
be performed. Much of the latter work can only be done during
the design phase after the device design has developed to
some detail. During the definition phase it was most
important to establish a basic philosophy for the safety of
the machine and to make an effort of proposing radiological
working design targets as a first safety and environmental
guideline. The presently proposed working design targets for
ITER are [1]:

Normal operation effluents (public exposure)

Dose to maximum exposed individual (MEI):

atmospheric	50 μSv/a	(5 mrem/a)
liquid	50 μSv/a	(5 mrem/a)
total	100 μSv/a	(10 mrem/a)

<u>Worker exposure</u>

10 mSv/a	(1 rem/a)
25 µSv/hr	(2.5 mrem/hr) radiation workers
5 µSv/hr	(0.5 mrem/hr) non-radiation workers

<u>Maximum accident (public exposure)</u>

Dose to MEI: 100 mSv (10 rem).

International criteria for radioactivity differ significantly, especially in the area of waste management. Thus, the exact legal safety and environmental requirements cannot be defined during a conceptual design phase. For normal operation effluents, worker exposure, and maximum accidental releases to the public the targets proposed consider the range of national criteria, past fusion device design studies and the ALARA principle.

3. INVENTORIES

Basically there are two types of inventories which represent safety and environmental hazards:

(1) Energies available to initiate or drive accidents which ultimately may lead to releases of radioactive or chemically toxic materials.
(2) Radioactive materials that might harm plant workers or the public.

The following sections present quantitative information on characteristic inventories. In the context of radioactivity both the units Bq and Ci (1 Ci = 3.7 x 10^{10} Bq) are used.

3.1 <u>Energy inventories</u>

The dispersal of radioactive or toxic materials is conceivable by various sequences of events, stored energies playing an important role both as initiators and propagators. Major energy inventories in ITER/NET are (see also Raeder [2]):
-    fusion energy in unburnt fuel,
-    thermal energy in the plasma and poloidal magnetic energy,
-    afterheat due to radioactive decay in neutron irradiated materials,
-    chemical energy in protection materials of plasma-facing components,
-    thermal energy in hot graphite tiles,
-    chemical energy in breeder and neutron multiplier materials,

-       thermal/mechanical energy in pressurized cooling water,
-       energy stored by evaporation of cryogenic helium,
-       magnetic energy in coil systems.

In terms of these energies, ITER/NET can be characterized as follows:

The mass of tritium residing in the reaction volume ranges from about 0.1 g to 0.25 g which upon complete fusion burnup would release 5.6×10^{10} J to 14×10^{10} J. Complete fusion would occur within 50 s to 100 s but appears impossible due to the significantly smaller energy and particle confinement times expected.

The thermal energy of the plasma is up to 9×10^8 J. The fastest release mode for most of this energy would be a hard disruption with a duration which may be as low as 0.1 ms (see Engelmann [3] and [1], p. 261, for example). The poloidal magnetic energy is up to 6×10^8 J with a release time due to disruption of about 20 ms (see [1], p. 261, 265).

After fusion shutdown, radioactive decay heat (afterheat) is delivered predominantly to the structural material in the components surrounding the plasma. For NET the heat released within 5 days is about 9×10^{11} J for AISI-316 as structural steel. In the context of safety, the first few hours up to several days are most significant. Typical powers during this time span are 8.2 MW to 6.1 MW during the first hour, 1.9 MW after 1 day and, 1.7 MW after 10 days according to Ponti and Raeder [4,5]. These numbers are valid for NET with a first wall area of about 400 m², a neutron flux of 1 MW/m² and a fluence of about 1 MWy/m². Larger NET versions and ITER entail numbers up to a factor 2 higher.

First wall and divertor substrates may be totally covered with graphite (typical thicknesses being 2 and 0.5 cm, respectively). Complete reaction of these protections with oxygen would release energies ranging from about 5.4×10^{11} J of heat (5×10^{11} J due to the first wall, 0.4×10^{11} J due to the divertor) up to twice these values. Time scales are not yet known because this fire hazard could not yet be analyzed in depth (see section 4.2.4).

Radiatively cooled graphite tiles would store several 10^{10} J if the first wall is entirely covered by them. After plasma shutdown this energy would be radiated to the adjacent first wall within 500 to 1000 s.

No final decisions on breeder and multiplier materials have yet been taken. A material combining both functions and considered quite extensively in the past is the eutectic 17Li-83Pb. A NET fully equipped with a blanket based on it would store energies of about 2×10^{11} J (heat) and 4×10^9 J (mechanical work) if complete chemical reaction with water is assumed. Knowledge on chemical kinetics (see Kottowski et al. [6], for example) would have to be combined with design based accident analyses to quantify release time scales. At present it is doubted, however, that a significant fraction of the above energies could actually be released in an accident. On the other hand there is a significant activation hazard due to the alpha-emitter Po-210 produced by neutron irradiation of lead (see section 2.3).

Until recently, water at 10 bar and about 80°C was considered as reference coolant in most cases. The enthalpy difference between these conditions and depressurized water at 20°C (after a cooling circuit rupture, for example) is 2.5 x 10^{10} J per cooling circuit with 100 m^3 water. No significant hazard could be associated with this energy inventory because of the low temperature of the water. This situation may, however, change significantly if higher pressures would be required for normal operation (35 bar at a maximum temperature of 150°C are presently considered) or if bakeout by water at typically 235°C /35 bar would be selected. For 150°C and 235°C respectively the enthalpy differences relative to 20°C are 5.5 x 10^{10} J and 9.3 x 10^{10} J per 100 m^3 of water. The associated hazard is higher due to the significantly enhanced water temperatures. A total number of 5 to 10 cooling circuits each containing 80 m^3 to 180 m^3 water are indicative numbers.

The cryosystem for the superconducting coils will contain about 8 x 10^3 kg (64 m^3) of supercritical helium at around 4.5 K (according to Katheder [7] and S. Brereton [8]). Evaporation and heat up of this helium to ambient conditions requires 1.3 x 10^{10} J. Depending on the dynamics and the actual sequence of events during an accidental rupture of the helium containing system, this energy will be provided by the building atmosphere and/or device structures. This may lead to atmospheric underpressure during the early phase and to overpressure in a pressure tight structure at the end. This can be a hazard if the pressurized volume is small. In a volume of 1.5 x 10^5 m^3 for example, the additional pressure due to the gaseous helium would be 0.3 bar.

Magnetic energy is stored in the toroidal field (TF) and poloidal field (PF) magnet systems. The energies amount up to about 4 x 10^{10} J (TF system) and 1.5 x 10^{10} J (PF system) according to J. Miller, ed. [9].

3.2 Tritium inventories

Due to its mobility, tritium represents an important radioactivity inventory of a fusion plant. The tritium inventory and its mobility depend strongly on design concepts and on the details of the design. The main uncertainties in this respect arise from the various design solutions for plasma feed and exhaust, isotopic separation, breeding blanket test modules, thermal protection (such as graphite) of plasma-facing components, and fuel storage. Incomplete information on tritium behaviour in materials is an additional source of uncertainty.

To minimize tritium releases to the environment, a multiple-containment concept will be used. The inner primary containment consists of the equipment enclosing the tritium process. This containment is installed in a secondary containment (e.g. gloveboxes, jacketed tubing). The last barrier against tritium release into the environment may be the reactor building (if equipped with a leak-tight liner inside), tritium facility buildings, or other tight buildings. Table 1 shows estimates or indicative figures of NET tritium inventories (1g tritium corresponds to 10^4 Ci) in different systems or components mainly derived from Dinner et al.
[10].

The tritium inventory retained in graphite protections on the first wall varies strongly with graphite temperatures. Typical inventories for NET according to Wu [11], are around 3×10^5 Ci and 2×10^4 Ci in unirradiated graphite tiles operating at 1200K and 1500K, for example. The experimental results by Causey et.al. on tritium retention agree with these values. Tritium retention could, however, be significantly enhanced by effects due to damage by neutron irradiation (see Causey [12]).

The value for the tritium breeder is based on the assumption that NET contains shielding blankets in all 16 sectors. The term "mobility" used in Table 1 indicates qualitatively the potential for accidental release. High mobility implies that multiple lines of defence are required around this system.

Beryllium used in significant amounts as neutron multiplier in test blankets or as first wall protection could significantly raise the tritium inventory. This is due to the in-situ production of tritium in beryllium by neutron irradiation. Rough estimates for a full coverage blanket for example range up to 400 g [13] corresponding to 4×10^6 Ci.

Typical tritium concentrations estimated for the various cooling circuits are quite different: 30 to 40 Ci/l (aqueous salt blanket coolant), 1 to 3 Ci/l (first wall coolant), several 10^{-2} Ci/l (vacuum vessel/coil shield coolant). For the typical water volume of 100 m^3 per cooling circuit the above concentrations imply up to 4×10^6 Ci per aqueous salt blanket cooling circuit, up to 3×10^5 Ci per first wall cooling circuit, up to several 10^3 Ci per vacuum vessel/coil shield cooling circuit.

Table 1. Estimates and indicative figures for NET tritium inventories

System	Tritium Inventory (g)	Mobility[a]	Main chemical form[b]
Plasma chamber evacuation	150	++	G
Plasma gas transfer & processing			
Plasma exhaust transfer	20	++	G
Surge tank	40	++	G
Fuel impurity removal	50	++	G
Impurity processing	50	+	W
Tritium breeder			
Breeder and structural material (shielding blanket)	400	+	W
Blanket tritium recovery	30	+	W
Tritiated water processing	3	+	W
Isotopic separation	150	0	G
Atmosphere processing	1	+	W
Fuel storage	1000	0	HY
Fuel preparation	150	++	G
Fuel delivery	150	++	G
First wall and divertors			
Structural material	250	+	G
Protection material	50	+	G
Waste			
Solid waste storage	100	0	G
Tritiated waste treatment	5	+	W
Primary coolant systems (shielding blanket water excluded)	30	+	W

[a] O= low mobility, + = moderate mobility, and ++ = high mobility
[b] G = HT, T2, DT; W = HTO, DTO, T20; HY = hydride.

3.3 Inventories of neutron-induced radioactivity

In all fusion devices based on the D-T reaction, neutrons will activate the surrounding structures. The plasma-facing components will build up the major fraction of the total neutron-induced radioactivity. The bulk of the activation products is trapped in the solid structural material and cannot as such be dispersed. However, small amounts of radioactive dust inside the torus (being produced by erosion due to plasma-wall interactions) need special attention during maintenance and under accidental conditions.

Table 2 shows calculated values of the radioactivity inventory (RAI) as a function of time after shutdown according to Ponti [14]. The numbers may amount up to twice these values depending on the ITER/NET option considered The major contribution stems from the structural materials AISI-316: at shutdown about 50% of the total RAI is concentrated in the first-wall steel. This figure increases to 85% to 90% if the breeding blanket structural material is included.

Going from 10 to 100 years of decay time, the RAI decreases substantially (by a factor of 350). During this time interval, the contributions of the radioisotopes ^{55}Fe, ^{60}Co, and ^{54}Mn, which dominate in the first period, become negligible and the residual RAI resides in long-lived isotopes (^{99}Tc, ^{93}Mo, ^{94}Nb, ^{59}Ni, and ^{63}Ni). It is important to note that in many cases of practical relevance (handling, waste management, recycling) the contact dose rate is the more important parameter. Contact dose, however, can be dominated after 30 to 70 years (depending on the type of structural steel) by the presence of "commercial impurities" such as Ag, Nb, Os, Bi, rare earths even if their concentrations are only on the ppm level or below (see G.J. Butterworth and L. Giancarli [15, 16]).

The relative contribution of the breeder material, in this calculation assumed to be the eutectic 17Li-83Pb (the lead acts as neutron multiplier), is already modest after less than 1 year subsequent to shutdown (see Table 2). Special attention, however, has to be paid to the alpha-emitter Po-210 (produced by neutron irradiation of lead and the lead impurity bismuth). In a typical blanket front row module its specific activity at shutdown is about 5×10^4 Bq/cm^3 (according to Ponti [17]) but its internal dose conversion factor with about 10^{-7} to 5×10^{-7} Sv/Bq (depending on the type of dose) is high.

Tungsten potentially protecting the divertor substrate becomes strongly activated. The typical value 540 Ci/cm^3 according to Ponti [14] at fusion shutdown (valid for a first wall neutron fluence of 1 MWy/m^2) implies a total inventory of about 3×10^8 Ci.

Table 2. Radioactivity inventories (in Bq) of NET components as functions of time subsequent to fusion shutdown

Materials	Decay time (yr)							
	0	1	10	100	300	1000	10^4	10^5
AISI-316 in:								
First wall	2.6E19[a]	8.0E18	6.7E17	1.6E15	5.4E14	1.6E14	5.1E13	1.3E13
Breeding blanket	1.8E19	4.3E18	8.1E17	2.0E15	6.5E14	2.0E14	6.0E13	1.5E13
Shield	1.6E18	2.0E17	3.1E16	7.4E14	2.1E14	4.0E13	1.6E13	5.0E12
TF coil	1.3E14	7.4E12	9.6E11	1.3E10	4.4E9	-	-	-
17Li-83Pb	5.5E18	3.3E15	7.1E14	3.8E12	3.0E12	2.6E12	2.6E12	2.6E12
coil mixture (except AISI-316)	2.8E14	2.6E12	2.6E11	1.3E10	7.8E9	4.4E9	3.2E9	1.5E8
Total	5.0E19	1.2E19	1.5E18	4.3E15	1.4E15	4.1E14	1.3E14	3.3E13
1 kg tritium	3.7E17	3.5E17	2.1E17	1.3E15	1.7E10	-	-	-

[a] Read as 2.6×10^{19}

The inventory of activated metallic corrosion products in the coolant is estimated to 10 Ci up to several 10 Ci per circuit containing 100 m³ of water.

More detailed information on induced specific radioactivity can be found in section 6 of this paper.

4. CONCEIVABLE ACCIDENTS

Deterministic analyses and a few experiments have been performed for next step devices of the worldwide fusion programme. The work for NET was performed in European laboratories under the Safety and Environment Programme and within the NET Team .

In the European Fusion S+E Programme also a significant activity on probabilistic safety assessment has been launched. This OPAS ("Overall Plant Accident Scenarios") activity devotes the major fraction of its resources to this topic (see section 4.3).

4.1 Major conceivable initiating events and accidents

Broadly the following cases are being considered:

- Loss of coolant accident (LOCA),
- Loss of coolant flow accident (LOFA),
- Loss of vacuum accident (LOVA),
- Loss of plasma confinement (LPC),
- Magnet system failure.
- Tritium system failure

4.1.1 Loss of coolant accidents

This type of accident implies a fast and virtually complete loss of coolant from components such as divertor substrates, first wall panels, blanket and vacuum vessel/coil shield segments or from the device as a whole. A direct consequence of a LOCA is loss of the capability to remove heat by normal coolant flow or natural coolant convection. Significant further consequences of a LOCA may be temperature transients, pressure loads, chemical reactions and dispersion of radioactivity.

4.1.2 Loss of coolant flow accidents

In this context mainly accidents have been considered which are initiated by a loss of active coolant flow but with natural coolant convection supported by the design. Of special importance is the case of a LOFA without prompt active fusion shut-down as there is not immediate impact on the plasma which could cause a quench of the fusion reactions.

4.1.3 Loss of vacuum accidents

This type of accident would be initiated by failures of vacuum vessel elements, vacuum ducts and pumps, heating and fuelling devices. In many cases also a subsequent failure of the coil cryostat must occur for an actual accident to happen. The most immediate consequence would be an ingress of the surrounding atmosphere entailing pressure loads, chemical reactions (depending on the type of atmosphere) and dispersion of radioactivity.

4.1.4 Loss of plasma confinement

At present the most important manifestations of this type of event seem to be major disruptions. They have been considered in various design contexts (see Salpietro et al. [18], for example) but not yet systematically with respect to safety implications. Major direct consequences of safety relevance expected are mechanical loads due to magnetic forces to components and thermal loads to plasma facing components.

4.1.5 Magnet system failures

An analytical approach to magnet safety has only been initiated quite recently. At present insight is still limited and mainly stems from systematical experiments (see Juengst [19]) with the comparatively small toroidal coil system TESPE (bath cooled, NbTi) and from a few experiments with LCT (partly forced flow cooling, partly bath cooling, 5 NbTi coils, 1 Nb_3Sn coil). Individual failures were triggered on TESPE such as current asymmetry, coil short circuit, loss of cryostat vacuum, loss of coil cooling, hot spot in a winding, propagating quench, electric arc across the coil terminals.

4.1.6 Tritium system failures

The individual inventories in the systems components cannot be totally isolated from each other since tritium must move from the breeding blanket and vacuum pumps to the processing system, the fueling system, and into the torus, for example. As a result, a single component failure could release combined inventories into the next confinement. Location and design of confinement/containment barriers separating inventories may isolate them to a certain extent but have to consider common mode failures, propagating accidents such as fires or explosions and credible external initiators such as seismic events that can affect multiple subsystems.

4.2 Key results

The analyses and experimental investigations have led to

a significant amount of information. Because of variations within the individual designs and due to basic differences between the planned next step devices many results are specific and generalization may be debated. In the following sections an attempt is made to extract from the existing material general insight as far as possible.

4.2.1 Major component LOCA's inside the vacuum vessel

A major break (guillotine type) of a first wall or divertor coolant pipe could lead to pressurization of the vacuum vessel by steam within typically 200 s to about 1 bar (10 bar cooling water pressure at about $80^{O}C$) according to Mazille [20]. Condensation on the accessible surfaces of the vacuum vessel has been accounted for. Condensation could be impaired if the accident leads to the generation of non-condensable gases such as H_2 and CO (see Raeder [21] and Jahn [22]).

A coolant water pressure of 35 bar (at about 200°C) subsequent to a major pipe break would pressurize the vessel to about 2.5 bar by steam (condensation taken into account) and gases produced. This seems to be close to the limit set by design problems.

The tritium inventories (see section 3.2) of the cooling circuits concerned range from several 10^5 Ci (first wall, divertor) to several 10^6 Ci (shielding blanket). The corrosion product inventory is estimated to be about 10 Ci/circuit. These inventories could be released by major LOCA's into the vacuum vessel and into the next containment should the vacuum vessel fail due to the accident.

Shutdown of the fusion process within a few seconds or its quench by inherent mechanisms would limit first wall and blanket steel temperatures to maximum values ranging from $400^{O}C$ and $700^{O}C$ if at least one component (first wall or blanket or vacuum vessel/coil shield) remains cooled (see Klippel, Renda and Raeder [24,26,27]). The temperatures are driven by the afterheat powers, mainly by those released in the steel. Between the components, thermal radiation (even at "natural' emissivities of 0.3 to 0.4) is the dominant heat transfer mechanism. The time scales of the temperature transients range from about half an hour (first wall) to several hours (blanket components).

Radiatively cooled graphite protections (typically at 1500°C to 1800°C) on the first wall substrates can lead to a water/steam - graphite reaction. It is true that the cooldown time of about 150 s to non-reactive temperatures below about

800°C (see fig. 1) of the graphite for immediate fusion shutdown is rather short due to thermal radiation but it is sufficient for allowing the production of a few up to a few tens kg of H_2 and the corresponding amount of CO according to Raeder [21], Mazille [30] and Jahn [22]. Delayed fusion shutdown would lead to hydrogen production rates up to about 0.25 kg/s. The water-graphite reaction models used in the studies are, however, still under debate particularly due to conflicting evidence from recent experiments at temperatures from 900°C to 1700°C.

Rupture of a vacuum vessel window during the course of the accident (at an internal pressure of 1 bar) and interaction of the vessel atmosphere with air seems to entail no gas mixtures prone to deflagration or explosion (according to Jahn [31]).

Hot tungsten reacts with water/steam by predominantly producing the highly volatile oxide WO_3. Similar to the case of hot graphite protections, the reaction may occur during a tungsten temperature transient (see fig. 2) down to non-reactive temperatures (below about 1000°C) even if fusion is shut down. The tungsten reaction rate steeply increases with temperature, a factor 2.5, for example, when going from 1200°C to 1400°C. According to an analysis (using rather specific assumptions) by Mazille [30] for the tungsten protecting the divertor substrate of NET, about 300 g of WO_3 could be volatalized. The starting temperature assumed is 2500°C (on 10 m^2 surface area) to account for a preceding plasma disruption. With a specific radioactivity of 540 Ci/cm^3 according to Ponti [14] this would lead to volatilization of about 7×10^3 Ci.

Beryllium which may potentially be used, coming into contact with water could lead to an exothermic reaction (about 3.9×10^7 J/kg Be according to Wu [32]) producing BeO and H_2. The relevance of this potential rests with the radioactivity in beryllium (tritium produced in situ by neutrons; see section 2.2) and chemical toxicity (permissible concentration in air is around 10^{-6} g/m^3).

4.2.2 Major LOCA outside the vacuum vessel

This event is meant to describe a loss of coolant from the whole device due to a major failure in central parts of the cooling system outside the vacuum vessel.

A fusion process not shut down or inherently quenched could lead to consequences for divertor, first wall and blanket similar to those described in section 4.2.1, i.e. melting on the time scales given there.

Shutdown of the fusion process limits the temperature transients to a slow heatup of the whole device by afterheat. If the design supports thermal radiation between all

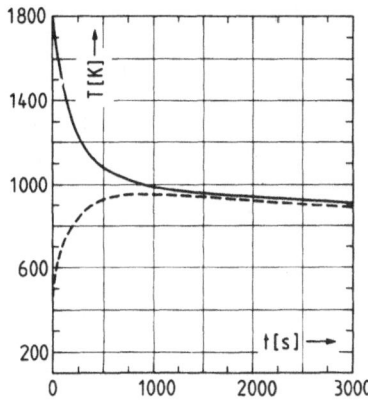

Fig. 1.
Temperatures of graphite protection and first wall (solid and dashed curves respectively) vs. time subsequent to fusion shutdown with no first wall cooling. The effective emissivity is 0.739 (corresponding to physical emissivities of 0.85 on both surfaces).

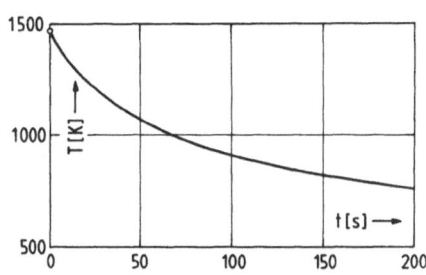

Fig. 2.
Temperature of a divertor tungsten protection vs. time subsequent to fusion shutdown. The effective emissivity between tungsten and substrate is 0.5.

components surfaces (by effective emissivities beyond typically 0.7) no gross melting of components seems to occur even if no further cooling mechanisms (such as atmospheric convection) would become operational according to Raeder and Renda et al. [27,33]. Hot spots, however, cannot be excluded in particular in the divertor region by the analyses performed. Figure 3 shows temperature distributions across NET with time subsequent to shutdown as parameter. In an austenitic steel structure maximum temperatures of 900°C (in the inboard shield region) would be reached after about 20 days, subsequently the temperatures begin levelling down (reaching about 750°C after 100 days). In the short term (after 3 days) the maximum temperatures do not yet occur in the inboard shield region (about 400°C there) but in the inboard and outboard first walls (about 600°C). If 'natural' emissivities (0.4 for steel, 0.85 for graphite) are assumed, maximum (inner shield) temperatures of 1200°C are reached after about 25 days. These analyses neglect the inboard legs of the TF coils and the central solenoid which would act as significant heat sinks if coupled radiatively to the inner shield and to each other. Adopting this assumption leads to temperatures of about 400°C and 600°C in the inboard shield and in the first walls respectively after three days. The maximum temperature (inboard shield) remains at about 900°C but is reached only after about 27 days according to Raeder [34].

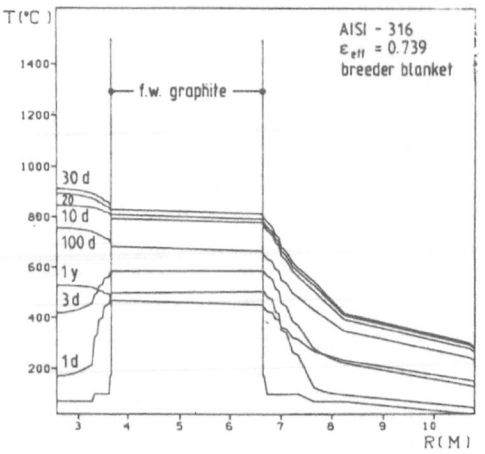

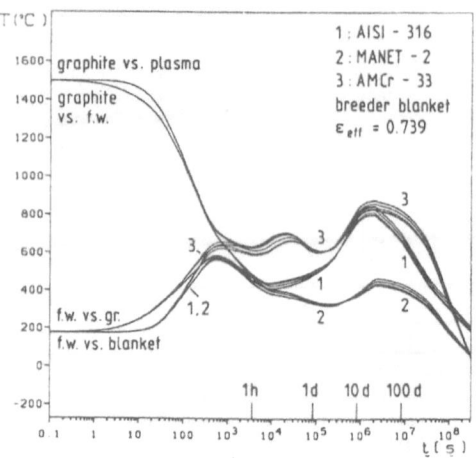

Fig. 3.

Radial distributions of component temperatures
of NET under complete device LOCA; parameter
is the time subsequent to fusion shutdown;
all radiating surfaces have a physical emissivity of 0.85 yielding an effective emissivity
of 0.739.

Fig. 4.

Graphite and first wall front and rear
temperatures (inboard and outboard) vs.
time subsequent to fusion shutdown;
same assumptions (LOCA, emissivities etc.)
as for fig. 4. The three steels AISI-316,
AMCr-33, and MANET-2 are compared.

A very significant assumption has been made in all
cases: the outer surface of the vacuum vessel/shield remains
cooled under the LOCA event by some passive means. Thermal
radiation has been chosen for the actual calculation but it
may be impaired in practice even at the elevated temperatures
pertaining to the LOCA by thermal shields and intercoil
structures housed by the coil system cryostat.

A martensitic steel (MANET-2) structure according to an
analysis of the same situation as before (coils not taken
into account) yields significantly lower temperatures.
Maximum temperatures (inboard shield) of 500°C are reached
after about 30 days according to Raeder [27]. After 3 days
the inboard shield and first wall temperatures are 200°C and
350°C respectively. The lower temperatures are due to the
much lower afterheat power densities (because of low Ni
content) in martensitic steels as compared to austenitic
steels. The reduction is by a factor 0.85 at fusion shutdown
and by about 0.30 after 1 day up to 30 days according to
Ponti [4,35]. A comparison of temperatures in inboard and
outboard graphite and first walls is shown in fig. 4 for the
steels AISI-316 and MANET-2. Figure 4 also shows the
temperatures which develop if the manganese modified
austenitic steel AMCr-33 (Ni content below 0.1%) is used in
the structures. The temperatures between about one hour and
one day are up to 240°C higher than those for the case of
AISI-316. The steels of the AMCr type are discussed due to

expected long term merits in the waste context (lower induced radioactivity, contact dose rate and afterheat power in the longer term).

To reduce the inboard coil shield thickness or to make it more effective at constant thickness, it has been proposed to incorporate a major amount of tungsten into the shield. In an otherwise austenitic steel structure according to estimates this may lead to melting of the steel in the device centre. Again the assumption of immediate fusion shutdown together with a design supporting thermal radiation has been made. It is the high afterheat power density in the tungsten (see Ponti [14]) which causes the problem. It may be reduced to a smaller concern by keeping the tungsten in the rear part of the shields only but this remains to be analyzed in more detail.

The thermal energy stored by radiatively cooled graphite protections on first walls (see section 3.1) under LOCA/fusion shutdown conditions could be radiated to the adjacent first walls within 500 to 1000 s. This would lead to fast transients (see fig. 4) of the first wall temperature (typically 2°C/s) up to about 600°C prior to the temperature maximum due to afterheat (see Raeder and Renda et al. [27,37]). The design implications of these transients are being analyzed.

Significant reduction of activated masses and irradiated volumes around the device is an attractive objective. Therefore a concrete biological shield (typically 1.6 m thick) close to the device has been proposed (see Salpietro et al. [38]). The inner surface lined by steel could serve as part of the coil cryostat. The thermal conduction time constant of the thick concrete shell is around 35 days, hence the shell is an effective thermal barrier. Estimates by Raeder [5] for the whole device under total LOCA/afterheat conditions (AISI-316 steel structure) show that the shell's impact on inner component temperatures (first wall, blanket, shield) is moderate for 3 to 5 days (temperature increment below 100°C compared to the case with a more conventional cryostat) and increasingly significant beyond 5 days. The inboard shield would approach the melting temperature of steel after about 20 days. Even the moderate temperatures during the first few days may be of concern for concrete integrity (100°C - 200°C at its inner surface after 1 to 2 days), hence further passive heat removal mechanisms like natural coolant and/or atmospheric convection should be supported by the design.

The temperature increase of the components due to afterheat can be terminated on the timescale which is characteristic of re-establishing cooling of inboard and outboard vacuum vessel/coil shield. Only a moderate fraction of the design cooling power (of the order 20 MW) is required to this end because afterheat power is only 1% to 2% of the nuclear heat deposition during normal operation.

4.2.3 Major component LOFA's with natural coolant convection

Should the fusion process proceed under LOFA conditions (without natural coolant convection) as active shut-down fails, a divertor substrate (made of TZM) would begin to melt after 20 up to a few tens of seconds (according to Zolti [23]). The corresponding times for the steel in first wall and blanket are significantly larger and lie in the range of 100 - 400 s (according to Klippel [24] and Jakeman, Autrusson, Mezola [25], for example).

Most LOFA's which have been analyzed are characterized by the onset of natural convection of the cooling water and by the assumption that the secondary cooling circuit remains operational.

Fusion shutdown (within typically 10s) together with natural coolant convection in all components limit the temperature increments to about $10^{o}C$ above the coolant temperature according to Klippel and Renda [24,37]. The coolant flow decays to natural convection within typically several minutes. The correspondingly reduced heat transfer is sufficiently high for afterheat removal.

Delay by 150 s of fusion shutdown leads to component temperatures which at maximum lie $40^{o}C$ above the coolant temperature (see Klippel [24]). Again the heat transfer characteristics of natural convection are sufficient for afterheat removal subsequent to fusion shutdown.

Total loss of coolant in one out of two first wall cooling cycles and natural convection in the other one allow fusion to proceed for about 200 s before shutdown is required. After that time heat transfer via nucleate boiling conditions has dropped to the unacceptably low level characteristic of film boiling (see Klippel [24]).

The broadly benign behaviour of first wall and divertor under LOFA conditions has been confirmed by a refined analysis (performed by Klippel [39]) based on modelling by the ATHENA code.

It has been observed during the analyses that complex geometry, "hot spots" and lack of space may render design for natural coolant convection difficult (but perhaps possible) at some locations. The major example for a difficult situation is the divertor region.

The ultimate heat sink has been assumed intact in all these analyses. Its total loss (by drainage from feedwater, for example) requires fusion shutdown within typically 100 s

(80 s according to an actual calculation by Klippel [24]) to avoid leaving the regime of nucleate boiling.

4.2.4 Loss of vacuum accidents (LOVA's)

The analyses known are not yet based on elaborate event sequences. Hence the results up to now are of an indicative nature, mainly based on energy and radioactivity inventories and basic design features such as volumes and port diameters.

Pressure loads which may be produced by heating the entering air by hot surfaces of in-vessel components or by burning protection materials (such as first wall/divertor graphite tiles) seem to be moderate. Both cases have been treated in the context of broken ICRH launcher seals by Jones and Rocco [40]. The resulting pressure increments are around 0.1 and 0.2 bar respectively (radiatively cooled first wall graphite tiles operating at 1550°C but not burning in the first case, burning tiles in the second one).

Shock waves due to a sudden air ingress have also been considered. A one-dimensional analysis by Deleanu et al. [41] shows that one break would lead to ambient pressure in the vacuum chamber, whereas two breaks with subsequent confluent shock waves at maximum can entail an overpressure of 1 bar.

Chemical reactions between air and hot in-vessel components could manifest themselves as fires. The actual ignition and burn temperatures significantly depend on the convection pattern which would establish under accidental conditions.

Roughly speaking fires are conceivable for graphite, tungsten, and beryllium beyond the following materials temperatures: 1000°C (graphite), 1200°C (tungsten), 1200°C (beryllium).

Radiatively cooled graphite tiles (typically at 1500°C) in contact with a flow of air therefore could burn in a selfsustained way. Maximum graphite temperatures due to a chimney effect between about 1700°C and 1900°C have been measured for this case (see Mazille [20]), the corresponding combustion gas temperatures ranging from 1000°C to 1300°C. Simple model calculations yielded gas temperatures up to about 2100°C even without accounting for chimney effects due to a potential second break (see Jones and Rocco [40]). These figures can only be indicative as in-depth analyses have not yet been made. The problem is complex and strongly depends on design details (chemical kinetics, boundary layer effects, convection patterns have to be included) and ultimately may require experiments on an adequate geometric scale. Computer model analyses have been specified and will be performed in the near future.

Conductively cooled graphite tiles (typically at 800°C) are much less prone to combustion with air. Under extreme conditions there still exists a fire hazard under loss of cooling conditions and if the graphite has been highly irradiated by neutrons, if it is covered by metallic erosion dust (Fe, Co, W, for example) or graphite powder or if hydrocarbon compounds are present (see Mazille [20]). Thermal conduction to the coolant is very effective in preventing a fire.

Tungsten fires could be driven by an exothermic (about 1.8×10^6 J/kg W) oxidation yielding the highly volatile WO_3 (see Mazille [20]).

Beryllium beyond about 800°C can be oxidized to BeO by a strongly exothermic reaction (about 6.3×10^7 J/kg Be). Self-sustained burn is possible beyond 1200°C, volatalization of BeO beyond about 2500°C but such high temperatures are hardly conceivable in the present context (see Mazille [20]).

The relevance of the chemical reactions with air rests with the radioactive and/or chemically toxic inventories residing in the potentially burnt materials.

Graphite fires could release radioactivity in the form of tritiated water (HTO) produced from the tritium residing in graphite protection tiles and vacuum vessel atmosphere. Typical inventories are around 3×10^5 Ci and 2×10^4 Ci for tiles operating at 1200 K and 1500 K respectively and may be enhanced by neutron damage effects to the order of 10^7 Ci (see section 3.2).

Chemical reactions of tungsten could release activated tungsten as WO_3 from the radioactivity inventory of about 3×10^8 Ci (see section 3). It is yet unknown as in the case of graphite which fraction of this inventory could be volatilized in an actual accidental fire. Again thermal conduction to the coolant is very effective in preventing a fire so that for a major hazard to exist the LOVA has to be combined with LOCA or LOFA events.

Tritium inventories in steel and beryllium are estimated to be in the order of 10^6 Ci but no release analyses are known for actual LOVA events.

Beryllium has to be judged also from the chemical toxicity point of view because of the low permissible concentrations in air (in the order of 10^{-6} g/m^3).

Dust will be produced by erosion of plasma facing

components. Various types of plasma-wall interactions contribute to this phenomenon. At present the main source for dust characterization is the JET experiment. According to Charruau et al. [42] the dust (Fe, Ni, Mn, Co, Cr, C) is fine (diameter about 10^{-6} m on average), hence a certain fraction may become airborne under LOVA conditions. The safety relevance is due to the activation level of around 60 Ci/cm^3 (first wall, AISI-316) according to [4] and due to tritium retained by graphite. Dispersion of typically 1 kg of dust subsequent to air ingress is conceivable (see Jones and Rocco [40] and Wykes [43]) and small amounts (of the order of 1 g) jeopardize human access to the reacor hall according to Giancarli et al. [44]. Hence the potential contamination of the next containment structure by expelled dust is severe (of the order of 10^4 Ci). The corresponding environmental impact depends on the leak tightness of confinement buildings, detritiation systems and filters. Such measures would result in rather low consequences (at maximum of the order of 1 mrem to the MEI) according to Jones and Rocco [40]. Without mitigating features the environmental consequences could be severe too (up to the order of 1 rem to the MEI) according to the same authors and Mazille [30].

4.2.5 Loss of plasma confinement (LPC)

Rupture of the blanket box by a disruption would be a significant safety hazard, which could also affect the first wall, breeder modules and the vacuum vessel/coil shield by which it is supported. Therefore a "reference disruption" (a plasma current of 10.8 MA decays to zero within 20 ms) has been considered in static vibration mode and dynamic 3D analyses by Crutzen et al. [45]. The characteristic frequencies (about 15 to 80 Hz) indicate resonances with the current decay spectrum. Peak stresses (elastic analysis) are due to torsional box vibration modes and add to the normal operation stresses caused by the thermal and pressure loads.

These additional stresses appear to be significant but can only be determined in the design context due to sensitivity to all mechanical details.

Thermal loads to first walls immediately or in the longer term may lead to LOCA events. Therefore analyses have been performed for bare (AISI 316 steel) and coated first walls by Klippel [46]. Deposited energy densities ranging from 2 MJ/m^2 to 10 MJ/m^2 with deposition times between 1 ms and 20 ms have been considered. Depending on the combination of energy density and deposition time, melt or evaporation layers develop on bare and coated walls respectively. The layers are up to 0.2-0.3 mm thick and initiate cracks penetrating into the bulk material. To mitigate the associated LOCA risk, coatings or protection tiles are foreseen as long as a large number of disruptions cannot be avoided.

4.2.6 Magnet system failures

Most of the reactions of a TESPE type coil system to failures are mild or can be accommodated by proper design (see Juengst 19]). Questions still debated are transfer of these conclusions to ITER/NET type coils with much larger dimensions and different conductor and cooling technologies as well as the possibility of damaging confinement barriers (such as the vacuum vessel) by coil movement under accidental conditions.

The safety relevance of arcs rests with the rather high power (up to several 10^5 W are conceivable) they may deliver and with loss of coil control if the arc bridges the terminals. Depending on geometry and magnetic stray field the arc may move with velocities up to about 100 m/s. Whether this is a hazard to other components remains to be analyzed in the context of detailed design. Circuit design has to be such that a situation where an arc cannot be extinguished in due time cannot occur.

4.2.7 Tritium system failures

Accident analyses to a very large extent depend on details of systems design and selection of hardware. Therefore as yet insight into the potential hazards is only generic.

For ITER a first attempt has been made to quantify the amount of tritium whose mobilization is conceivable [47]. The inventories used for this estimate have a similar basis as Table 1. Under moderately optimistic assumptions the amount of tritium mobilizable to reach the next confinement structure is estimated to be about 300 g. Pronounced conservatism concerning inventories and mobilizable fractions lead to about 2500 g.

4.3 Probabilistic analyses

Although not establised as legal standards, several countries are moving in the direction of limiting accident probabilistic risk (occurance rate times consequence). To the extent that regulatory officials can judge on the basis of risk instead of maximum accidents, the overall effectiveness of design actions to protect against conceivable accidents might be enhanced as effort is concentrated on actualy reducing accident risk, rather than concentrating on highly improbable or even unconceivable events.

Preliminary discussions have occurred regarding a risk assessment approach for ITER. For accident analysis it was concluded that simple deterministic calculations should be done first. If these indicate significant hazard potential, then more sophisticated deterministic consequence calculations are warranted. Probabilistic risk calculations would be useful in helping to make design decisions regarding the value of possible design options or features to reduce accident frequency or consequence.

A related activity, "Overall Plant Accident Scenarios", launched within the European Fusion S + E Programme has the objectives of identification of important accident sequences, quantification of the corresponding risks, recommendations for improving the design, and provision of basic results for a safety report.

The actual analyses proceed along the following scheme:

- functional analysis,
- construction of fault and event trees for failure identification,
- quantitative evaluation of fault and event trees,
- evaluation of risks,
- recommendations to improve the safety characteristics.

At present the following systems are being analyzed:

Vacuum Vessel   Plasma Heating
Shielding Blanket   Cooling and Heat Transfer
First Wall   Control
Divertor   Diagnostics
Magnets   Shielding
Plasma Exhaust   Confinement
Tritium and Fuel Handling   Waste Storage and Treatment
Fuel Injection   Auxiliaries
                               Maintenance Equipment

The results of a complete first round are expected to be available in the second half of 1989.

5. DISPERSION OF RADIOACTIVITY IN THE ENVIRONMENT

Quantification of radioactivity releases to the environment (source terms) during normal operation, maintenance, and under accidental conditions strongly depend on design solutions. Dominating factors are the tritium inventories and their mobility inside the plant, the materials used for highly irradiated components (mainly plasma facing components) and number and quality of barriers between the inventories and the environment.

Work is under way within the European S+E Programme to quantify the source terms. To date only estimates are at hand. As a consequence conservative quantification of the environmental impact in terms of dose to a maximum exposed individual of the public (MEI) has to be based on the radioactive inventories and their level of mobility. Only limited credit is taken from chemical or physical mechanisms inside the plant which could reduce the source terms.

5.1 Routine emissions and doses to MEI

Tritium although belonging to the low radiotoxicity group of radioactive isotopes (e.g. according to the Official Journal of the European Communities [48]) is expected to be the dominating element with respect to environmental impact.

The largest releases of tritium inside the plant during normal operation will most probably occur via the coolant lines. On the basis of experience gained from Canadian fission reactors (CANDU), a water leak rate of 10 l/d per coolant loop can be expected. Assuming a tritium concentration of 30 to 40 Ci/l in the shielding blanket water and 1 to 3 Ci/l in the first wall loop, a daily release of about 300 to 400 Ci into the reactor hall has to be envisaged. Air dryers and air detritiation systems, however, will assure reduction of the activity released to the environment to the required level.

The most likely release of activation products during normal operation is due to leakage of corrosion products from the primary cooling circuits. Based on fission reactor experience, it is estimated that this results in a small release of activity: typically $4 GBq/m^3$ (0.1 Ci/m^3) of dissolved material in the coolant and a leak rate of 10 l/d result in a release of about 15 GBq/a (0.4 Ci/a). This activity will only reach the environment, if filtering at the stack is not operational.

Releases will also occur due to maintenance operations. They have been estimated by Casini, Ponti and Rocco [49] - mainly on the basis of CANDU reactor experience - to be about 12000 Ci/a of HTO and 9000 Ci/a of HT (atmospheric emissions) and ca. 5 Ci/a of activation products and 1500 Ci/a of HTO (aquatic emissions). The most relevant mechanisms in this context are unavoidable coolant losses and re-suspension of material adhering to pipe walls.

Therefore, considering both normal operation and maintenance, the total tritium released daily from a fusion device like ITER/NET is expected to be less than 1/100 g (corresponding to 100 Ci = 3.7 TBq). According to Piet [50] this release would result in a maximum inhalation dose of about 0.03 mSv (3 mrem) per year to the MEI.

5.2 Accidental emissions and doses to the MEI

Studies are being made of accident scenarios resulting

from major technical failures of the plant. If such a severe accident caused the final surrounding building to be breached (although this seems impossible), the radioactive release into the environment would be mainly tritium rather than in the form of activated structural materials. The maximum quantity of releasable tritium contained inside the NET plant is estimated to about 1 kg (see table 1). In addition another 0.5 to 1 kg will be contained mainly in the non-mobile hydride form in the fuel storage. Therefore the tritium inventory should be contained in locally separated compartments or buildings. Adopting this local separation and considering the mobility as indicated in table 1, a maximum release of about 200 g into the containment building under severe accident conditions appears to be a plausible estimate.

If these 200 g of tritium subsequently were released in the most hazardous form (HTO) from tbe building roof at 20 m height (rather than from a high chimney stack) and assuming weather conditions highly adverse with respect to acute doses at the plant boundary (Pasquill F, wind speed 2 m/s, dispersion parameters calculated by Martin-Tikvarts formulas), it would cause according to Devell [51] a maximum dose of about 50 mSv (5 rem) at a distance of 1 km from the plant to the MEI (due to inhalation and skin absorption during plume passage).

Figure 5 showing iso-dose-curves allows quantification of the ground surface area contaminated by plume passage:
50 mSv are exceeded within less than 0.02 km^2
10 mSv are exceeded within about 1 km^2
5 mSv are exceeded within about 3 km^2

Figure 6 shows the corresponding results for a ground level release (release height in the actual calculation is 2 m), all other assumptions being unchanged. In this case the maximum dose occurs inside a 1 km radius. At the site boundary (1 km from the release point) the MEI would be exposed to about 150 mSv. The iso-dose-curve for 50 mSv covers about 0.1 km^2 outside the site boundary, for 10 mSv about 1.5 km^2 and for 5 mSv about 5 km^2.

All these values are within the range of 50 to 150 mSv (5 to 15 rem) more or less accepted by the licensing authorities for abnormal events of low probability in most countries.

Doses to the MEI, however, strongly depend on the assumptions on weather conditions usually modelled by Pasquill classes and on release height. Changing from the most adverse moderately stable conditions (Pasquill F, wind

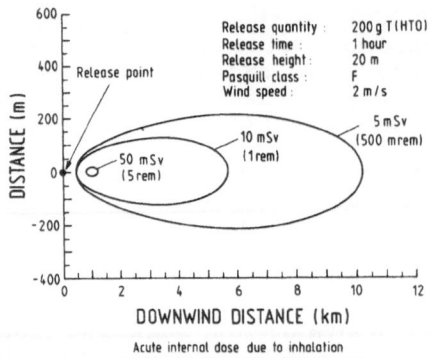

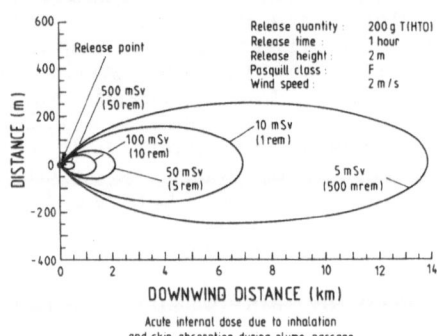

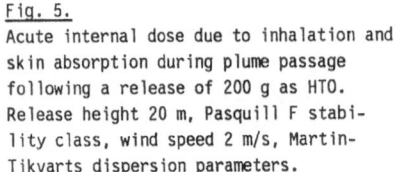

Fig. 5.
Acute internal dose due to inhalation and
skin absorption during plume passage
following a release of 200 g as HTO.
Release height 20 m, Pasquill F stability class, wind speed 2 m/s, Martin-
Tikvarts dispersion parameters.

Fig. 6.
Acute internal dose due to inhalation and
skin absorption during plume passage
following a release of 200 g as HTO.
Release height 2 m, Pasquill F stability
class, wind speed 2 m/s, Martin-Tikvarts
dispersion parameters.

speed 2 m/s, Martin-Tikvarts dispersion parameters) to
neutral conditions (Pasquill D, 5 m/s) reduces the dose to
the MEI at the site boundary and outside considerably (e.g.
11 mSv instead of 54 mSv at the site boundary). Replacing
Martin-Tikvarts dispersion parameters by Briggs dispersion
parameters reduces the dose to the MEI at the site boundary
(in this case identical with the location of maximum
exposure) from 54 mSv to 41 mSv. Increasing the release
height reduces the dose to the MEI and shifts the location of
maximum exposure away from the release point: e.g. 41 mSv at
1 km, for a release height of 20 m, 0.26 mSv at 16 km for a
release height of 100 m.

Taking into account reemission of HTO following plume
passage, the dose to the MEI is increased (assuming that the
MEI stays permanently at the same location). Calculations by
Edlund [52] show that the conservative value of a constant
reemission rate of 5%/h (defined as the emitted fraction of
the actual HTO soil activity inventory) derived from tritium
release experiments in France (according to Djerassi and
Gulden [53], Djerassi and Lesigne [54]) and Canada (according
to Brown [54]) increases the dose to the MEI by about 65%
during 8 hours and by more than 100% (to about 96 mSv) during
the first day (see fig. 7). Beyond 3 days doses due to
inhalation and skin absorption become negligible. A less
conservative reemission rate of 1%/h increases the dose to
about 60 mSv during the first day, 82 mSv during 3 days and
96 mSv during 5 days. In all these cases, however, changes
in the wind direction during these time periods will reduce
the dose considerably.

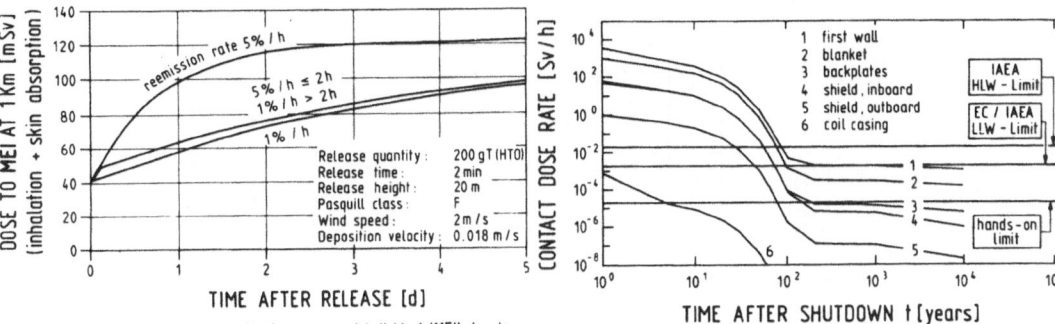

Fig. 7.

Internal dose to a maximum exposed individual (MEI) due to
inhalation and skin absorption at 1 km from the release point

Fig. 8.

Contact dose rates for NET components after 1 FPY of operation

Fig. 7.
Acute internal dose for a MEI at 1 km downwind from the release point due to inhalation and skin absorption following a release
of 200 g HTO. Release height 20 m, Pasquill F
stability class, wind speed 2 m/s, Briggs
dispersion parameters, HTO reemission rate
of 5 %/h and 1 %/h.

Fig. 8.
Contact dose rates versus time subsequent to
fusion shutdown for NET components after 1
full-power year of operation.

All doses reported above relate to tritium release in
HTO form. Most of the tritium inside the plant, however, is
in an elemental form (HT, DT) being considerably less
radiotoxic than HTO. Table 3 shows doses to the MEI
calculated by Edlund [52] for a release of 200 g of tritium
in HTO and in HT form. In both cases conservatively large
deposition velocities were assumed, $2*10^{-4}$ m/s for HT and
$1.8*10^{-2}$ m/s for HTO. For the HT release case conversion of
HT into HTO in the soil followed by reemission of HTO into
the environment was taken into account. Reemission rates of
1%/h, 3%/h and 5%/h were assumed to show the influence of
this parameter.

The results show the dominating influence of the HTO
reemission rate and clearly demonstrate that doses due to
inhalation and skin absorption for the MEI resulting from HT
releases are orders of magnitude smaller than from HTO
releases (a factor of 20,000 for plume passage, 230 after 8
hours, 160 after 1 day and more than 130 during the whole
exposure time of about 5 days, see table 3). The explanation
is the much smaller dry deposition velocity of HT resulting
in less tritium contamination of the soil.

To allow calculation of more realistic transient
reemission rates based on parameters describing the
dominating factors (e.g. weather and soil characteristics,
kind of vegetation) in more detail, dedicated theoretical
and experimental investigations are being performed within
the European S+E Programme.

Table 3: Doses to the MEI from HTO and HT releases

Integration time (= residence time of the MEI)	2 minutes (plume passage)	8 hours			1 day	3 days	5 days
Reemission rate HTO [%/h]		1	3	5	5	5	5
HTO release 200 g tritium Pasquill F, 2 m/s Briggs dispersion parameters Release height 20 m Dose to MEI at 1 km [mSv] (inhalation and skin absorption)	4	47	59	68	98	120	122
HT release 200 g tritium Pasquill F, 2 m/s Briggs dispersion parameters Release height 20 m Dose to MEI at 1 km [mSv] (inhalation and skin absorption)	0.002	0.08	0.20	0.31	0.64	0.89	0.91

Experimental results of small-scale field measurements on the kinetics of HTO reemission from contaminated soils reported by Belot [56] showed reemission rates to the atmosphere immediately after contamination of 1 to 5%/h.

The results of a large-scale experiment reported by Djerassi and Gulden [53] (release of gaseous tritium from a 40-m-high stack for 2 min) confirmed that direct atmospheric conversion is negligible. The dominant process is the conversion of HT/T_2 into HTO in the surface layer of the soil, followed by reemission of HTO. The reemission rates measured are in the range of 1 to 8%/h during the first hours after the release. As a consequence, ~30% of the gaseous tritium was reemitted as HTO during the first 21 h. During the following 5 days, only negligible amounts of HTO were reemitted from soil, due to rainfall 21 h and 3 days after the release.

Comparable results were obtained according to Brown [55] from a Canadian experiment on dispersion of tritium in the environment.

In the case of accidental releases of activation products into the environment, definition of a source term is more or less speculative at present. To get at least an answer which allows scaling of expected doses to the MEI, 8 g (about 1 cm^3) of highly activated structural material (AISI 316, 66 Ci/cm^3) was assumed by Piet and Brereton [57] to be dispersed as aerosol under the same weather condition and release height as defined in Table 3 for the tritium release. In this case a maximum early whole body dose (50 years dose commitment, 7 days exposure) for the MEI of 0.033 mSv (at 1 km) was calculated, corresponding to 4.2 mSv/kg steel dust dispersed. Similar calculations for the ITER outboard first wall (water cooled lithium oxide breeder blanket) resulted in an early whole body dose of 10.4 mSv/kg steel dust dispersed. Based on these results, about 5 to 12 kg of activation products could be released until the MEI receives 50 mSv (which is about the dose obtained during the plume passage of 200 g tritium in HTO form).

6. RADIOACTIVE WASTES

6.1 Criteria for characterisation of fusion wastes

To date, some internationally accepted criteria exist for characterisation of the radiological impact of waste disposal and waste disposal sites (e.g. dose limits for the most exposed individual of the public). The derived limits expressed in contact dose rates, specific activities or total activities, however, show considerable discrepancies because of different country-specific waste management and disposal strategies. Some tendency exists to divide waste into the 3 categories HLW, MLW and LLW (high, medium and low level waste), but big differences in the classsification are obvious from country to country.

No disposal sites for HLW are operational. However, apparently many countries plan to eventually dispose of HLW in deep geological repositories. Waste technologies for LLW and MLW exist in most countries but vary considerably. There are some shallow land burial (SLB) sites in Europe, but there is a trend towards requiring some sort of geological (deep or shallow) repositories for both MLW and LLW.

In a report on the present situation and prospects in the field of radioactive waste management in the European Community [58] low level waste in general is characterized by dose rates not exceeding 2 mSv/h (0.2 rem/h), high level waste is any waste that releases significant amounts of decay heat, which can be correlated to a specific activity of above 100 MBq/cm^3.

IAEA [59] at present propose 2 mSv/h as the LLW limit and 20 mSv/h as the lower limit for HLW to characterize fission wastes at short term (some years). Three categories for handling and transport of beta/gamma solid wastes are defined: Category 1 waste (< 2 mSv/h) can be handled and transported without special precautions, Category 2 waste (2 to 20 mSv/h) can be transported in simple concrete or lead shielded containers, Category 3 waste (> 20 mSv/h) needs special precautions for handling and transportation.

Of the small number of European sites available for SLB each has different acceptance criteria. According to present UK regulations [60], waste should contain less than 12 MBq/kg (0.32 Ci/t) beta or gamma activity to be acceptable for SLB. France apply isotopic-specific limits for each site.

In the USA isotop-specific limits apply to the acceptability of waste for SLB according to the Code of Federal Regulations 10CFR61 [61]. Some limits for Class C wastes (not decaying to safe level in 100 years, but decaying to acceptably safe level in 500 years) are relevant for steels used as structural material in fusion devices: 8 MBq/cm^3 (220 Ci/m^3) for Ni 59, 260 MBq/cm^3 (7000 Ci/m^3) for Ni 63 and 7400 Bq/cm^3 (0.2 Ci/m^3) for Nb 94. These limits lead to efforts in reducing the Ni contents and to minimize Mo and Nb in the structural materials to meet these requirements.

6.2   Characterisation of NET steel wastes

Quantification and qualification of radioactive wastes for final disposal produced by NET will strongly depend on the site chosen. Indicative data to qualify the waste, however, have been derived by comparison with some of the criteria given in the previous section.

As soon as NET begins to operate, activation products will be formed. According to Guetat [62], after the first D-T pulse NET will have produced waste requiring special handling facilities. The first wall contact dose rate will be about 50 mSv/h after one minute of exposure and a decay time of one month, and has to be measured against the EC-LLW-limit/IAEA category 1 limit of 2 mSv/h.

After about 25 full-power days of operation (the end of the physics phase corresponds to about 14 full-power days or a fluence of 0.03 MWa/m^2 for the first wall), NET first-wall and blanket structures will exceed the U.S. limits for shallow land burial (as defined in the U.S. Nuclear Regulatory Commission Regulation 10CFR61 [61] and assuming the long-lived ^{94}Nb as the dominating isotope), even after a decay time of 100 years.

Figure 8 shows the contact dose equivalent rate versus time for the relevant components of NET after 1 full-power year of operation (corresponding to a fluence of 1 MWa/m^2 for the first wall). These values can be compared with the notional "hands-on limit" of 2.5 x 10^{-5} Sv/h, which would allow the material to be worked on and machined in a controlled workshop, or with the EC-LLW-limit/IAEA waste category 1 limit of 2 mSv/h (handling and transport without special precautions). First wall and blanket structures will not meet these limits within practicable cooling times. Back plates and inner shield, however, could reach the "hands-on limit" after about 150 years, the coils after 4 years.

Comparison with EC proposed limits [58] for HLW and LLW shows that the first wall and blanket will be HLW for about 90 years, the blanket reaching the LLW limit after 100 years.

6.3 Waste management strategy for NET

LLW and MLW fusion operational wastes (wet and dry) contain tritium and activation products. They are expected to be similar to fission waste streams, having a somewhat higher volume. The waste management and disposal strategies developed for fission plants can be applied, provided that tritium has been adequately removed and/or immobilized.

Handling and treatment of dismantled first wall and blanket segments (high-level waste) will involve more complex procedures because of their larger volume, weight, tritium content and activation level.

A tentative strategy proposed for handling and disposal of wastes from NET is summarized in fig. 9. After a certain time in a short-term storage acting as a buffer or required for the decrease of activity and decay heat, some valuable materials with low specific activity may be separated by sorting. The remaining highly active materials will be disintegrated, sorted according to different materials categories, (e.g. steel, copper, tungsten, graphite, ceramics etc) and tritium contents.

All materials containing tritium above a certain concentration (typically 74 MBq/kg = 2 Ci/t) will have to undergo a detritiation process that will be specific to the chemical nature of the material. After the detritiation process, the waste has to be conditioned (which might include a tritium immobilisation treatment) for intermediate storage.

After the decay heat becomes negligible (depending on the composition of the materials involved, this lasts from a few years to some decades), the waste can be adequately conditioned for disposal in the corresponding final repositories.

The facilities required for tritium recovery from components routinely removed from inside the torus will be large. They have to be equipped for all-remote handling, due to the high gamma activity of the components.

There are many materials involved, each of which will require separate treatment methods. Examples are vacuum degassing of steel, combustion of graphite, oxidation of tungsten. Each materials treatment will produce impurities which must be taken into consideration in process design.

Tritium recovery requirements are determined by both economics and regulatory limits for disposal. While typical values of 100 to 1000 Ci/t have been derived from fusion waste management studies by Broden et al [63] for economic recovery of tritium from steel by melting, much lower limits exist in some national regulations for disposal (e.g. 74

MBq/kg = 2 Ci/t in France for the SLB site Centre de la Manche, CSM, according to [64]).

Work is under way within the S+E programme to experimentally determine actual detritiation factors which may be achieved by heating and melting. First experimental results obtained by Ochem [65] indicate that for metallic components a level of about 1 to 30 Ci/t can be reached by melting. Figure 10 shows measured detritiation factors versus initial activities. Even starting from about 1000 Ci/t results is less than 10 Ci/t after melting.

No specific or additional facilities are needed for decommissioning of the highly activated NET components. Intermediate storage, hot cells and facilities have to be provided anyway to handle replaced in-vessel components at the end of the physics phase. Their decommissioning could simply be performed as another "replacement" at the end of the technology phase.

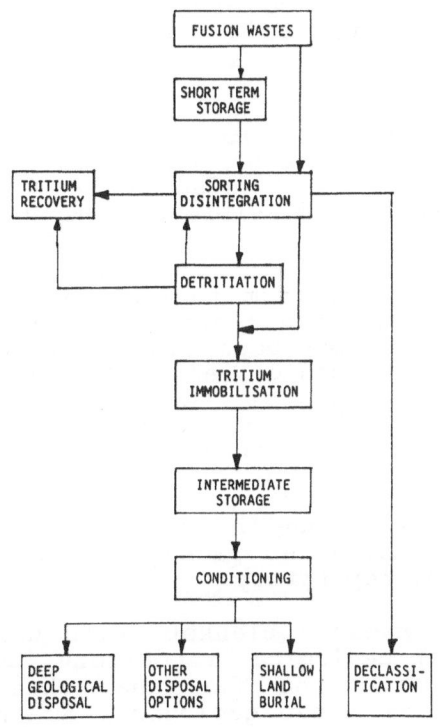

Fig.9.
Tentative strategy for handling and disposal of wastes from NET.

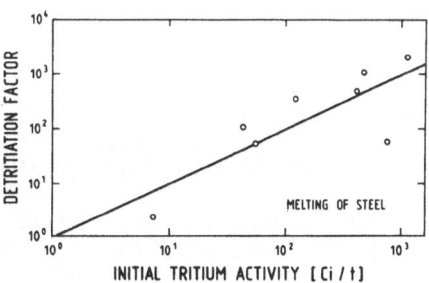

Fig. 10.
Measured detritiation factors versus initial activities obtained by melting of steel components.

6.4 Quantification of NET wastes according to a special scenario

Sweden is the only country to date having well defined specifications for final disposal of LLW, MLW and HLW waste (as outlined in the "Radioactive Waste Management Plan" [66]). A shallow geological repository for medium level waste is already commissioned, a deep geological repository for high level and long-lived wastes expecting commissioning by the year 2020. Therefore the specifications of this country have been used to obtain at least some indicative values for the volumes that will have to be disposed in final repositories.

As described by Broden and Olsson in [67] LLW and MLW waste arising from the processing systems (i.e. fuel cycle and coolant purification systems), from decontamination and maintenance operations, and packed in steel drums and/or concrete blocks, could be disposed of in the shallow geological repository, located below sea bottom with 50 to 60 m of rock cover. HLW from disassembly and periodic replacement of parts of the inner components (mainly plasma-facing components and blanket segments), could be disposed in the deep geological repository, located at a depth of about 500 m in the bedrock.

The highly active first wall and adjacent components would need about 1 year of cooling time before transport is allowed under present regulations. After a period of 10 to 30 years in an intermediate storage, final disposal of the adequately shielded packages would be possible in the deep geological repository.

Some estimates of volumes arising from replacement and decommissioning of the NET steel components and a tentative choice for NET waste repositories have been summarized in fusion waste management studies by Gulden et. al. [67], [68]. The total mass of about 7000 tonnes of the NET steel components correspond to a waste volume of about 2000 m^3 including packaging. Of this waste, 80% can be disposed of into the shallow geological repository and 20% in the deep geological repository. Each replacement of first wall, blanket structure, and back plate during the lifetime will produce additional waste of 400 m^3 after packaging, to be disposed of in the deep geological repository.

In summary, first wall and breeder blanket units will have to be disposed of in the deep geological repository, the back plates of the blanket segments and the inner shield could be disposed of in either the deep geological or the shallow geological repository, the outer shield and the TF coils (steel and other materials) in the shallow geological repository.

6.5 Low activation materials

One of the key questions for fusion waste management is, whether the wastes emerging from spent first wall and blanket segments need to be qualified and treated as HLW, MLW or LLW. It is generally agreed that fusion waste management could benefit from modification of the structural materials, using mainly Mn instead of Ni and replacing Mo as far as possible by W. A typical example for this type of low activation material is the austenitic steel AMCr 33 containing less Ni and Mo and more Mn as compared to the presently used austenitic steel AISI 316.

To quantify the difference from the waste management point of view, activation calculations were performed by Ponti [69] for the outboard first wall of NET-DN (shielding blanket option, 0.8 full power years of operation) for both AISI 316 and AMCr 33. The results - contact dose rate and specific activity as a function of time - are presented in fig. 11 and 12.

Figure 11 compares the contact dose rates with the notional "hands-on limit" and with IAEA and EC/IAEA disposal limits for high and low level waste. About 100 years after shutdown, when the contribution of Co 60 becomes negligible the contact dose rate is dominated by Nb 94. In the case of AISI 316 (containing 2.4 weight % of Mo) this isotope is mainly generated by the n,p reaction on Mo 94. The contribution of the low Nb concentration of 0.2 ppm in both steels plays only a minor role, at least for AISI 316. As a consequence of the small amount of Mo in AMCr 33 (0.06 weight %) the EC/IAEA-LLW-limit (2 mSv/h) is reached at less than 100 years after shutdown (the required time was about 10^4 years for AISI 316). The "hands-on limit", however, will be reached by AMCr 33 only after more than 10^4 years.

Using the specific activity as a measure for SLB qualification, AMCr 33 satisfies the US isotopic specific limits (Ni 63, Ni59, Nb 94) a few years (< 10 years) after shutdown whereas AISI 316 will not do so before 10^3 to 10^4 years (fig 12). Neither AMCr 33 nor AISI 316 wastes, however, would be qualified for SLB under UK regulations (10^5 Bq/cm^3).

As no European SLB site exists with similar limits as 10CRP61 [61] and as European countries tend to prefer geological repositories in the future, it may be debated whether low activation steel using mainly Mn instead of Ni and replacing Mo as far as possible by W should be promoted for use in fusion plants if final disposal is the only argument. Similar conclusions were drawn by Butterworth [15], Giancarli [16] and Ponti [70] who investigated the influence of specific elements concentrations on meeting the 10CFR61 [61] requirements for SLB and UK limits for SLB, including the impact of impurities.

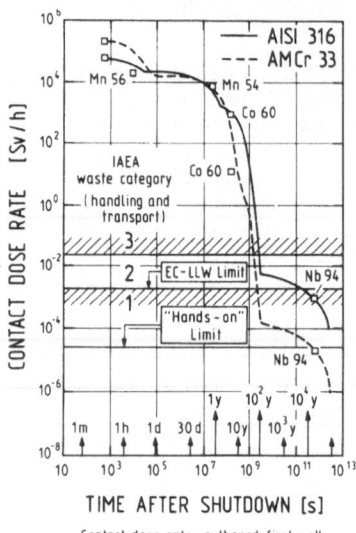

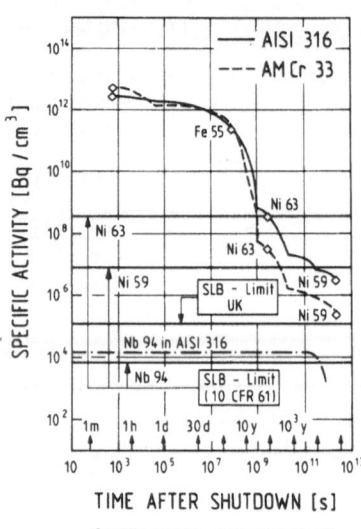

Fig. 11.
Contact dose rate for AISI 316 and AMCr 33
outboard first wall versus time subsequent
to fusion shutdown. NET-DN (shielding blanket)
after 0.8 full-power years of operation.

Fig. 12.
Specific activity for AISI 316 and AMCr 33
outboard first wall versus time subsequent
to fusion shutdown. NET-DN (shielding blanket)
after 0.8 full-power years of operation.

7. IMPLICATIONS OF COMPONENT HANDLING

Maintenance will include disassembly, removal and transportation of large torus components containing considerable amounts of radioactivity. The basic maintenance and handling requirements emerge from the necessity to meet legal requirements for radiation protection of workers and the public. As the maximum dose equivalent rate permissible for 'hands on' operations is 2.5×10^{-5} Sv/h, the data shown in fig. 8 confirm that the in-vessel components can be handled only remotely.

First wall and divertor bakeout for detritiation is highly desirable before vacuum chamber opening to keep tritium contamination levels below maximum permissible concentration. This operation should begin immediately after shutdown. Its duration will vary according to the baking temperature.

As tritium releases from the first wall and divertor depend on temperature, torus components during their maintenance should be kept at low temperature. To achieve this goal adequate cooling has to be provided to ensure afterheat removal during maintenance.

The presence of activated erosion products inside the vacuum vessel (mainly dust of first wall and divertor surface materials) requires adequate consideration in the development of maintenance strategies.

The same holds for graphite tiles and graphite dust due to the relatively high quantities of tritium retained (depending on temperature, guesses range from a few to a few hundred grams), the large volumes of graphite, and the short component lifetime expected at present.

The use of intervention containments with inert gas or vacuum during maintenance operations where the vacuum vessel has to be opened can reduce or avoid the contamination of the reactor hall by activated dust and tritium as well as the access of air and moisture to the vacuum vessel.

8. SUMMARY AND CONCLUSIONS

Radiological design targets for ITER have been proposed as a first safety and environmental guideline after having considered the range of national criteria, past fusion device design studies and the ALARA principle.

The energy inventories of ITER/NET that could be released are low by comparison if the large volumes and surface areas involved in potential accidents are taken into account. All inventories (afterheat; thermal, mechanical, chemical, electromagnetic energies) lie in the 10^8 to 10^{12} J range. A representative total is 10^{12} J, dominated by afterheat which is characterized by a slow release time scale.

The tritium inventory residing in the various device components and the tritium in storage both are estimated to be of the order of 1 to 2 kg. Tritium related design guidelines require subdivision of inventories to keep releases into the surrounding containment by a single failure to below about 150 g. An inventory carefully to be considered in accident analysis is the several 100 g in the cooling water of an aqueous salt shielding blanket.

The major part of the neutron induced radioactivity (80 to 90%) is concentrated in the first wall and blanket structures if a conventional austenitic steel such as AISI-316 is used. In tungsten, potentially protecting the divertor substrate, about 3×10^8 Ci may be induced and have to be considered mainly in the context of loss-of-vacuum accidents (LOVA). Further relevant topics are Po-210 (if lead is used as a neutron multiplier) and activated dust due to eroded in-vessel components.

Substantial information on accidents has been collected. For several reasons, however, not the entire domain of potential accidents has been covered uniformly, rather there are topics hardly adressed or not at all. Obviously these deficiencies have to be corrected as soon as allowed by design detail and available S + E resources.

On the other hand, the general insight gained appears sufficient to draw up a list containing those initiating events which seem to be most significant in the context of the ultimately relevant consequences - radiation doses to public and workers:

- major rupture of divertor or first wall cooling pipes inside the vacuum vessel (LOCA type),
- major failure of vacuum vessel elements, vacuum ducts and pumps, heating and fuelling devices (LOVA type),
- major cooling system failure outside the vacuum vessel (LOCA type),
- major disruption (LPC type),
- major tritium system failure,
- major rupture of blanket cooling pipes (LOCA type),
- major magnet system failure (current lead break, quench, cryosystem blow-down)
- major LOFA of the whole device.

The events according to the above list and their consequences should be taken into account by the design work and accomodated by passive mechanisms to the largest extent possible.

The events and their consequences should be analyzed to the depth justified by the design detail achieved at every step. These analyses at least have to be performed in a deterministic way. Probabilistic analyses should be added if they can assist design decisions and if adequate failure data are available.

Some consequences of the events are expected to jeopardise predominantly the device. But there are also consequences which could impact on the public and environment unless mitigating means such as confinement structures are provided.

External initiators have to be accounted for according to the specifications for potential sites. Earthquakes and aircraft impact certainly belong to the initiators which have to be considered.

The total tritium released daily from NET is expected to be <3.7 T Bq (100 Ci). This release would result in a maximum dose of ~ 0.03 mSv/yr (3 mrem/yr) to the MEI of the public, being well below the limit imposed by the current regulations for fission reactors.

Release of radioactivity to the environment under accidental conditions would be mainly due to tritium and to a lesser extent, to activated structural materials.

For analyses purposes the conservative assumption of a maximum release of 200 g of tritium into the containment building under severe accident conditions (design guideline: 150 g) has been made. Releasing these 200 g in the most hazardous form (HTO) to the environment would cause a maximum dose of about 50 mSv (5 rem) to the MEI of the public (plume passage). This value is within the limits accepted by the licensing authorities for abnormal events of low probability.

The contaminated ground surface area is small, covering about 3 km^2 where the dose to a MEI would exceed 5 mSv, 1 km^2 out of this area exceeding 10 mSv, and 0,02 km^2 reaching about 50 mSv.

Using HTO reemission rates derived from experiments in France and Canada and application of the most adverse weather conditions, increases the dose to the MEI by factors of about 1.5 to 3 during 5 days. This increase, however, is only reached if the wind direction is constant during the whole time period.

Doses due to inhalation and skin absorption for the MEI resulting from HT releases (including conversion of HT into HTO in soil and subsequent reemission of HTO) are orders of magnitude smaller than from HTO releases: a factor of 20,000 during plume passage, 230 after 8 hours, 160 after 1 day and more than 130 during the whole exposure time of about 5 days.

The management and disposal strategies developed for fission waste can be applied for low- and medium-level fusion waste after sufficient tritium recovery/removal and/or immobilization has been carried out.

Repositories originally designed for high- and medium-level fission waste appear suitable for the disposal of high- and medium-level wastes from NET.

After the first D-T pulse, NET will have produced waste requiring special handling facilities.

After about 25 full-power days of operation, NET first wall and blanket structures will pass the limit of shallow land burial (10CFR61).

At decommissioning, the steel components in NET will lead to a waste volume of about 2000 m^3 including packaging. Of the steel waste, 80 % can be disposed of into a shallow geological repository, 20% needs a deep geological repository.

Each replacement of first wall, blanket structures and back plates will produce about 400 m^3 of waste after packaging and will require disposal in a deep geological repository.

Using the specific activity as a measure of SLB (shallow land burial) qualification, AMCr 33 (as an example for a low activation steel) satisfies the US isotopic specific SLB-limits (10 CRF61) a few years after shutdown whereas AISI 316 will not do so before 10^3 to 10^4 years. Neither AMCr 33 nor AISI 316 wastes, however, would be qualified for SLB under UK regulations.

As no European SLB site exists with similar limits as 10CFR61 and as European countries tend to prefer geological repositories in the future, it may be debated whether low activation steel using mainly Mn instead of Ni and replacing Mo as far as possible by W should be promoted for use in fusion plants if final disposal is the only argument.

In conclusion the following may be stated:

. During normal operation radioactivity releases on the level of conventional nuclear technology seem to be achievable but dedicated efforts are required to achieve this goal.

. Operational wastes are expected to be similar in quality and quantity to those from conventional nuclear reactors. Replacement and decommissioning waste masses are rather high compared to fission but this feature is mitigated by the lower radiotoxicity of fusion wastes as is shown by waste characterization on an isotope-by-isotope basis. Geological waste disposal seems to be a general trend for the future which would also cover tritium concentrations in wastes at the level presently expected after partial detritiation.

. The stored energies which could drive accidents are moderate by nuclear power plant standards, the same being true for most off-normal powers too. Combined with big masses (i.e. heat sinks), large surface areas (i.e. heat exchange areas) and the expected self-termination of fusion under accident conditions this feature leads to expecting rather benign behaviour and good prospects for safety due to passive, natural processes. Hence it may even be possible to base designs

on worst cases. Certainly this approach will require
strong efforts but finally would be a convincing
argument
for fusion.

Acknowledgement

The authors gratefully acknowledge the work and the
support of their colleagues contributing to the worldwide S+E
activities, the computing assistance of B. Esser and H.
Gorenflo and in particular all the typing and editing of G.
Boekbinder Mulder and K. Piller.

References

1. ITER Concept Definition, Report ITER-1, Oct 1988

2. W. Gulden, J. Raeder.
    Safety Aspects of the Next European Torus.
    Fusion Technology, Vol. 14, No. 1, July 1988.

3. F. Engelmann, M.F.A. Harrison, R. Albanese,
    K. Borrass, O. De Barbieri, E.S. Hotston,
    A. Nocentini, J-G. Wegrowe, G. Zambotti.
    Next European Torus Physics Basis.
    Fusion Technology, Vol. 14, No. 1, July 1988.

4. C. Ponti.
    Activation calculations for NET-DN.
    Tech. Note No. 1.87.116, JRC Ispra, Oct. 1987.

5. J. Raeder.
    On NET afterheat under loss of active cooling
    conditions - transient analyses.
    NET Internal Note, NET/88/IN-011, April 1988.

6. H. Kottowski, V. Renda and G. Kuhlboersch.
    State of the Art of Water-Cooled LiPb Fusion Reactor
    Blanket Safety.
    Proc. 4th European Nuclear Conf., Geneva, Switzerland,
    June 1-6, 1986.

7. H. Katheder.
    NET Team, private communication, Sept. 1988.

8. S. Brereton
    Cryogen release; part of the memo
    Passive safety in the ITER reactor vault and building
    design, personal communication, June 1989.

9. K. Koizumi, J. Minervini, A. Kostenko,
    N. Mitchell, J. Miller (ed.),
    Report ITER-EL-MG-1-9-U-1; LLNL-ITER-89-020, p.17.

10. P. Dinner, M. Chazalon, D. Evans, F. Fauser, M. Iseli
    and C.H. Wu.
    Next European Torus Plasma Exhaust and Fuel Processing
    Systems.
    Fusion Technology, Vol. 14, No.1, July 1988.

11. C.H. Wu.
    Graphite and analysis of critical problems.
    NET Internal Note, NET/85/IN-079, October 1988.

12. R.A. Causey.
    The Interaction of Tritium with Graphite and Its
    Impact on Tokamak Operations.
    8th PSI, Juelich, May 1988.

13. ITER shield and blanket work package report.
    ANL/FPP/88-1, June 1988, p. 7-11.

14. C. Ponti.
    Activation calculations for NET IIIA and FCTR.
    Technical Note 1.05.B1.85.123, JRC Ispra, Sept. 1985.

15. G.J. Butterworth.
    Low activation structural materials for fusion.
    Invited Lecture, 15th SOFT - Utrecht, The Netherlands,
    September 1988. To be published in this issue of FED.

16. L. Giancarli.
    On the radiological behaviour of first wall fusion
    structural materials.
    Culham Report, CLM-R275, Feb. 1987.

17. C. Ponti.
    JRC Ispra, private communication, March 1988.

18. E. Salpietro, R. Albanese, E. Coccorese, R. Martone,
    N. Mitchell, G. Rubinacci.
    Next European Torus Operation Cycle.
    Fusion Technology, Vol. 14, No. 1, July 1988.

19. K.P. Juengst.
    Ergebnisbericht ueber Forschungs- und Entwicklungs-
    arbeiten 1987.
    Institut fuer technische Physik, KfK 4395, March 1988,
    p. 24 ff.

20. F. Mazille.
    First wall and blanket analysis.
    Report CEA/IPSN S+E 4.1.1.1/RE1/88.01, March 1988.

21. J. Raeder.
    On hydrogen production by hot graphite reacting with
    water/steam.
    NET Internal Note, NET/88/IN-02, March 1988.

22. H. Jahn.
    Reaction of graphite with water/steam - modelling of
    combustible gas production, distribution and
    consequence analysis, Interim Report No 2,
    Gesellschaft für Reaktorsicherheit, GRS-A-1517,
    Jan 1989.

23. E. Zolti.
     Thermal and mechanical behaviour of CFC/TZM divertor
     elements under loss of coolant conditions, Report
     ITER-IL-PC-8-9-E-15, June 1989.

24. H.Th. Klippel.
     Analysis of thermal hydraulic transients of the water
     -cooled eutectic LiPb blanket of NET.
     Fusion Engineering and Design, 6, 2, May 1988, p.79.

25. R. Jakeman, B. Autrusson, M. Mezola.
     NET first wall and mock-up thermomechanical analyses,
     Report ITER-IL-PC-6-9-E-12, June 1989.

26. V. Renda, L. Papa.
     Thermal effects due to residual power on the reactor
     internals.
     Private communication, Jan. 1987; for a similar case
     (5 outboard blanket modules) see Technical Note
     I.87.50, JRC Ispra, April 1987.

27. J. Raeder.
     On NET afterheat under loss of active cooling
     conditions - revised transient analyses.
     NET Internal Note, NET/88/IN-44

28. H. Th. Klippel.
     Numerical Analysis of the Dynamic Response of Water-
     Cooled Liquid-LiPb Breeder Blankets to Coolant Tube
     Rupture. Nucl. Eng. Design/Fusion, 3, 4, July 1986.

29. V. Renda.
     JRC Ispra, privately communicated calculations.

30. F. Mazille, H. Djerassi.
     Analysis of LOCA/LOFA Risks for the Water Cooled First
     Wall of a NET Type Fusion Reactor.
     15th SOFT- Utrecht, The Netherlands, September 1988.

31. H. Jahn.
     Reaction of graphite with water steam-modelling of
     combustible gas production, distribution and
     consequence analysis, Final Report, Gesellschaft für
     Reaktorsicherheit, GRS-A-1517, July 1989.

32. C.H. Wu.
     NET Team, private communication, July 1988.

33. V. Renda, A. Soria.
     JRC Ispra, privately communicated calculations on
     LOCA/radiative energy transfer, July 1988.

34. J. Raeder.
On NET afterheat under loss of active cooling
conditions - transient analyses including the TF coils
as heat sinks.
NET Internal Note, NET/89/IN-016

35. C. Ponti.
Activation calculations for a MANET-2 first wall (NET-
DN), JRC ISPRA, private communication, October 1987.

36. W. Gulden, J. Raeder, H.H. Gorenflo:
On NET afterheat under loss of active cooling
conditions - transient analyses for the case of a
tungsten containing shield.
NET Internal Note, NET/88/IN-017.

37. V. Renda.
JRC Ispra, privately communicated calculations,
May 1987.

38. E. Salpietro, F. Casci, F. Farfaletti-Casali,
F. Fauser, H. Gorenflo, L. Ingala, T. Kaltner,
G. Malavasi, J. Minervini, N. Mitchell,
R. Poehlchen.
Next European Torus Basic Machine.
Fusion Technology, Vol. 14, No. 1, July 1988.

39. H.Th. Klippel.
On safe decay heat removal from NET shielding blanket
by natural convection cooling,
IAEA Technical Committee Meeting on "Fusion Reactor
Safety", Jackson, USA, April 1989.

40. A.V. Jones, P. Rocco.
Air Ingress Accidents in Tokamaks.
15th SOFT - Utrecht, The Netherlands, September 1988.

41. L. Deleanu, H. Djerassi, A.V. Jones, P. Rocco.
Safety Analysis Related to the Possible Release of
Activated Erosion Dust in Fusion Reactors.
14th SOFT, Avignon - France, September 1986.

42. J. Charuau, H. Djerassi.
First Experiment on Erosion Dust Measurement in a
Tokamak.
15th SOFT, Utrecht - The Netherlands, September 1988.

43. M.E.P. Wykes.
Safety Analysis of Torus Vacuum Breach.
JET Report, JET-P(88)37, 1988.

44. L. Giancarli, H. Djerassi.
Radiological Problems Related with the Plasma Induced
Erosion of the NET First Wall.
15th SOFT - Utrecht, The Netherlands, September 1988.

45. Y.R. Crutzen, M. Biggio, F. Farfaletti-Casali.
Electromagnetic effects on the NET first wall caused
by a plasma disruption event.
14th SOFT, Avignon - France, September 1986.

46. H. Th. Klippel.
Thermal behaviour of bare and coated first walls under
severe plasma disruption conditions.
ISFNT, Tokyo, April 1988.

47. D.F. Holland.
Results privately communicated, June 1989.

48. "Council Directive of 15th July 1980, amending the
Directives laying down the basic safety standards for
the health protection of the general public and
workers against the dangers of ionizing radiations,"
Official Journal of the European Communities, 23,
L246, September 1980.

49. G. Casini, C. Ponti, P. Rocco.
Environmental Aspects of Fusion Reactors, 1985.
Technical Note I.04.B1.85.156, JRC Ispra, December
1985.

50. S. Piet.
Effluent Benchmark Calculation for ITER, SJP-01-89,
Attachment 1 - Tritium Effluent Benchmark Calculations
(Task 1), Jan 20, 1989.

51. L. Devell, O. Edlund.
Environmental radiation doses from tritium releases.
Fusion Reactor Safety. Report of a technical
committee meeting on fusion reactor safety, Culham 3-7
November 1986, IAEA-TECDOC-440.

52. O. Edlund.
Radiation Doses from Atmospheric Pulse Releases of
Tritium as HT and HTO. Studsvik Report NP-88/72,
June 1989.

53. H. Djerassi, W. Gulden.
Overview of the Tritium Release Experiment in France.
Proceedings of the Third Topical Meeting on Tritium
Technology in Fission, Fusion and Isotopic
Applications. Toronto, Canada, May 1-6, 1988.

54. H. Djerassi, B. Lesigne.
Environmental Tritium Behaviour, French Experiment.
Final Report, NET Contract No.85-074/GSA, Issue 88/03.

55. R.M. Brown, G.L. Ogram, F.S. Spencer, C.D. Burnham.
An Overview of Experiments at Chalk River on HT
Dispersion in the Environment. Proceedings of the
Third Topical Meeting on Tritium Technology in
Fission, Fusion and Isotopic Applications. Toronto,
Canada, May 1-6, 1988.

56. Y. Belot, H. Clerc, J. Guenot, H. Djerassi, W. Gulden.
Assessment of the Environmental Impact of a Tritium
Gas Release: Resuspension of HTO from Soil Surface.
Presented at the IAEA Technical Committee Meeting on
Fusion Reactor Safety, Culham, U.K., November 3-7,
1986.

57. S. Piet, S. Brereton.
Status of U.S. Contributions to ITER Safety and
Environment, SJP-37-88, Attachment 2 - Activation
Product Release Benchmark Calculations (Task 2),
December 6, 1988.

58. "Present situation and prospects in the field of
radioactive waste management in the European
Community", Documents of the Commission of The
European Communities, COM (87) 312.

59. Standardization of Radioactive Waste Categories,
IAEA - Technical Reports Series No. 101 (1970).

60. Radioactive Waste, Vol. I, first report from the
Environment Committee of the House of Commons, Session
1985-86.

61. USNRC, 10 CFR Part 61: Licensing requirements for land
disposal of radioactive waste. Final Rule. Federal
Register (47 FR 57446), 27 December 1982.

62. P. Guetat, "Fusion Reactor Wastes: Technical and
Radiological Aspects for the Management of Wastes from
NET and a Commercial Reactor," presented at the IAEA
Technical Committee Meeting on Fusion Reactor Safety,
Culham, U.K., November 3-7, 1986.

63. K. Broden, A. Hultgren, G. Olsson, H. Djerassi,
P. Giroux, J-L. Rouyer, Fusion Waste Mangement
- Safety and Environment Studies 1983-84 - European
Fusion Technology Programme, EUR-FU/XII-361/85/35
(1985).

64. Prescriptions techniques relatives a l'exploitation du site de stockage de la Manche. Annexe II a la lettre SIN A 693/85 du 06.02.85. Service Central de Surete des Installations Nucleaires, Ministere de l'Industrie et de la Recherche, France (1985).

65. D. Ochem, Metallic Tritiated Waste Reprocessing - a Compendium of Know How Gained in Valduc, CEA Progress report. To be published

66. SKBF/KBS
Radioactive Waste Management Plan, Plan 82, Part 2, Facilities and Costs, SKBF/KBS Technical Report 82 -09:2 (1982).

67. W. Gulden, C. Ponti, P. Guetat, D. Ochem, K. Broden, G. Olsson, J. Butterworth.
"Fusion Waste Management - Safety and Environment Studies 1985-86, "EUR-FU/XII-80/87/72, Commission of the European Communities (1987).

68. W. Gulden, C. Ponti, Ph. Guetat, K. Broden, G. Olsson, G.J. Butterworth.
Waste Management for NET. 15th SOFT - Utrecht, The Netherlands, September 1988.

69. C. Ponti.
Comparison of low activation materials and AISI 316. JRC Ispra, 1988, privately communicated calculations, August 1988.

70. C. Ponti.
Fusion Reactor Materials To Minimize Long Living Radioactive Waste, Paper presented at the International Symposium on Nuclear Technology, Tokyo, April 10-15 (1988).

COST ANALYSIS OF NEXT STEP DEVICES AND THE IMPLICATIONS FOR REACTORS

W.R.Spears

The NET Team
c/o Max Planck Institut für Plasmaphysik
D-8046 Garching bei München

INTRODUCTION

Costing of next-step devices such as NET or ITER is at a very early stage. The design is mostly conceptual, and still evolving rapidly. In general, design work is concentrating on key components at the forefront of technological advance, and there has been little attempt to date to design or cost "conventional" plant components in detail. For the key components the database for cost estimation is weak or non-existent, and cost estimation for these items therefore has a wide uncertainty.

Despite these difficulties, cost estimates for the next-step device, and beyond, are demanded by those inside and outside the fusion programme. This paper therefore summarizes the status of cost estimation today and, while pointing out how the cost estimates even for the next step device are still subject to large uncertainties, to indicate what are the implications for future power reactor costs. The intention is to be clear about the quality of today's cost estimates, so that no more credibility is given to them than they deserve.

There is an alternative to "detailed" cost estimation of fusion. An attempt can be made to scale with the reactor core power density from costs of existing power sources, such as fission, by comparing its size or volume[1] in a one-line expression. This can easily run into trouble, because a fusion reactor is not a fission reactor of much larger volume, and cannot therefore be costed on the same cost/unit volume or cost/unit mass basis. In particular, the cost of the high-quality items, which both these plants need, scales only weakly with their volume or mass (i.e the quantity of material), and much more strongly with the man-hours spent in manufacture and quality control. These labour costs are much weaker than linearly related to size. The only way out of this difficulty, with a one-line equation, is to quantify this weaker scaling, or to calibrate the scaling law more reasonably[2,3] so that a larger proportion of the plant cost (30-50%) scales with core volume, using a lower core unit cost. As shown later, this latter approach is justified, as the reactor core influences a larger proportion of the cost than the fraction it itself contributes.

An obvious improvement on the above approach, with a smaller likelihood for error, is to replace the one equation, with its uncertain scaling dependence and range of applicability, with a number of more well-founded rela-

Safety, Environmental Impact, and Economic Prospects of Nuclear Fusion
Edited by B. Brunelli and H. Knoepfel
Plenum Press, New York, 1990

tionships. These equations usually scale the cost of the main hardware items according to major technical driving parameters such as size, power level, mass, area, etc. This makes such costing equations useful as part of scoping calculations for reactor parameters where cost minimisation is the goal, so they are often referred to as "systems-code algorithms".

The derivation of such equations requires input from those who are expert in building each group of hardware items for which a cost equation is written, and estimates of unit cost (e.g $/kg) based on expert judgement and/or a survey of the existing database. For many fusion items, the latter is the main problem. The database is either:

- non-existent, and "similar" items have to be used to evaluate a unit cost;

- not extensive enough in range, with fusion components often required to be larger or more massive than current experience; (In this case the scaling can suffer from the same criticism levelled at crude one line equations, mentioned above.)

- weak in number, where few similar items have been built, the data scatter is large, and each point carries anecdotal evidence of why it was much cheaper/more expensive than the norm.

Thus although useful attempts have been made to correlate the existing database[4], in the end it is the expert's judgement on unit costs which carries most weight.

Cost estimates carried out for real manufacturing work are made on the basis of breaking down the work into packages, and costing off-the-shelf items, raw materials and labour costs, in a so-called "bottoms-up" approach to costing. It is not easy to incorporate this approach into a systems code, since the man-hours for novel jobs needs particularly careful assessment. This approach is also only possible when the design has progressed so far, and NET is only now at the stage where it is possible to get meaningful estimates from industry for some of the NET components worked out on this basis. These estimates unfortunately can still only be considered as illustrative, as manufacturers are not being asked to bid to <u>build</u> the item concerned, and this cannot fail to colour their estimate. Furthermore, because the dimensions of next-step devices are still not fixed, some extrapolation from the bottoms-up estimate to a changed set of dimensions has to be made and, if the extrapolation is significant, is only a little better than the systems-code algorithm approach previously outlined. The design status can have a strong influence on cost uncertainty, since although it is possible to estimate costs for conventional equipment, this can only be done when the full requirements for such equipment have been worked out. For novel equipment, item costs are more clearly known only when the manufacturer has completed mock-up, sample or prototype manufacture;

There are a number of additional factors which will affect the extrapolability and applicability of a given cost estimate. These include:

- project organisation - a large centralised project team with staff mainly loaned by governments at the project's expense, sharing design development and thus responsibility with the manufacturer, can significantly reduce the manufacturer's contingency allowance; despite the extra costs of the project team staff, for novel designs this arrangement can reduce the total construction costs (i.e. the sum of indirect and direct costs) compared to a turnkey project with industry;

- prestige - the next-step reactor is likely to attract manufacturers wishing to participate for company prestige reasons, and encourage them to quote prices with negligible profit margins;

- licensing/nuclear class - the cost of a high-quality components is significantly affected by the man-hours spent in manufacture and quality control, and these will be much greater if nuclear code quality assurance is required;

- series production - although a single next-step device cannot take advantage of series production in reducing engineering man-hours by both central project team and manufacturer due to past experience, this "learning" effect can be applied to power reactors; however, the strength of this effect for a power reactor cannot easily be estimated before the man-hour figures are better known;

All these factors will affect any cost estimate, and are distortions often not clearly recognised or quantified in the existing cost database. Care must therefore be exercised in using this database for costing.

These issues will show up again in the following, which describes the costing done in the past for NET and now underway for ITER. The predictions and use of systems codes for costing power reactors is also discussed.

COSTING OF NEXT-STEP REACTORS

Up to now, costing of next-step reactors has been based on increasingly sophisticated systems code algorithms[4,5,6]. To use the EC experience as an example, parameters for NET have been scoped out using the SUPERCOIL[7] systems code, which was used to evaluate cost impacts using first a crude model[8], then moving on to more sophisticated models using a wider database, which eventually grew into the SCAN-2 cost model[9]. SCAN (System for Cost Analysis of NET) was put together based on experience of unit costs and cost scaling laws on existing experiments within the fusion community, plus industrial experience of constructing nuclear plants. NET, after all, has many of the trappings of today's fission plants. This model was expanded, reviewed and modified by Motor Columbus Consulting Engineers Inc., a company that has considerable experience in the planning, management and execution of nuclear power plant construction projects worldwide. A brief synopsis of the direct cost scalings of the SCAN model is given in Table 1. Studies[10] (albeit for a power reactor) show that, neglecting spare parts and contingencies, the direct costs in the SCAN model are roughly 15% fixed, 20% power related, and in the remainder driven strongly by the TF coil outer dimension.

Subject to the uncertainties mentioned in the introduction, the SCAN model is as accurate a model as can be produced by simple scaling laws which are thought to be valid over a wide range of parameters. That is not to say that it is always accurate in its detail, but rather that where the equation for one system underestimates an item cost, another will overestimate, so that the expectation is that in the end the total cost arrived at will have a reasonable credibility. In any case, the advantage of a multi-equation cost model is that any distortion caused by incorrect scalings will not dominate the result.

More recently, the SCAN-2 cost model has been rearranged according to the ITER Cost-Centre Breakdown[11]. The result of using this model for the direct cost of the NET-2 device[12] is shown in Table 2. Costs are quoted in 1984 Ecu, and exclude spare parts and contingencies. The results are shown graphically in Figure 1.

TABLE 1 SYNOPSIS OF SCAN-2 COST MODEL FOR DIRECT COSTS

CODE	ITEM	MAIN SCALING	COMMENT
1	SITE & IMPROVEMENTS	%, item	frac. of direct cost + elec. connections
21	Nuclear Buildings	volume	related to tokamak outer dimensions
22	Heat Transport Buildings	volume	ditto, plus power scaling for power reactor
23	Cooling System Buildings	waste heat	
24	*Electrical System Buildings	fixed	using JET and PWR experience
25	PIC Buildings	volume	related to tokamak outer dimensions
26	Plant Auxiliary Buildings	volume	based on current fission/fusion experience
27	Additional Buildings	volume	ditto
28	Site Service Civil Works	vol + fixed	ditto
2	BUILDINGS	[≈volume]	
31	First Wall	mass	assuming a specific composition
32	Divertor	mass	ditto
33	Limiters	-	not itemised separately
34	Blanket	mass	includes feed-through region and its shield
35	Shield	mass	usually zero, since integrated in vac. vessel
36	Vacuum Vessel	mass + area	includes ports, pump ducts and machine supports
37	TF Magnet System	cond. length	plus mass-related and fixed installation costs
38	PF Magnet System	cond. length	ditto
39	Overall Cryostat	mass	
3	TOKAMAK	[≈mass]	
41	Vacuum Pumping	alpha power	only plasma exhaust pumping in this item
42	Central Cryoplant	peak power	sized by peak magnet cooling demand at 4K
43	Heating and Current Drive	absorbed power	RF systems assumed unless stated
44	Fuelling	item	default is 1 gas puffer, 1 pellet injector
45	Fuel Handling	T throughput	plus building-volume-related HVAC costs
46	Heat Transport	cooling power	numerous circuits, for main and aux. systems
47	Power Supplies	peak power	only specialised power systems (see also 51)
48	PIC Systems	no. systems	plus diagnostics and safety systems
49	Maintenance Equipment	fixed	based on current fission/fusion experience
4	TOKAMAK AUXILIARIES	[≈power]	
51	Electric Power Distribution	power demand	various components and scalings
52	Waste Handling	fixed, %	non-nuclear waste is an overhead (%)
53	Cooling Water System	heat load	includes turbogenerator in power reactor
54	Fluid Supply System	%, heat load	
5	PLANT AUXILIARIES	[≈power]	

*PIC = Protection, Instrumentation & Control

TABLE 2 SCAN SYSTEMS CODE EVALUATION OF DIRECT COSTS OF NET-2

CODE	ITEM	1984ME	%
D.1	SITE & IMPROVEMENTS	36	1.0
D.2	BUILDINGS	778	21.9
D.3	TOKAMAK	1255	35.3
D.4	TOKAMAK AUXILIARIES	1276	35.9
D.5	PLANT AUXILIARIES	210	5.9
D	DIRECT COST	3555	100.0

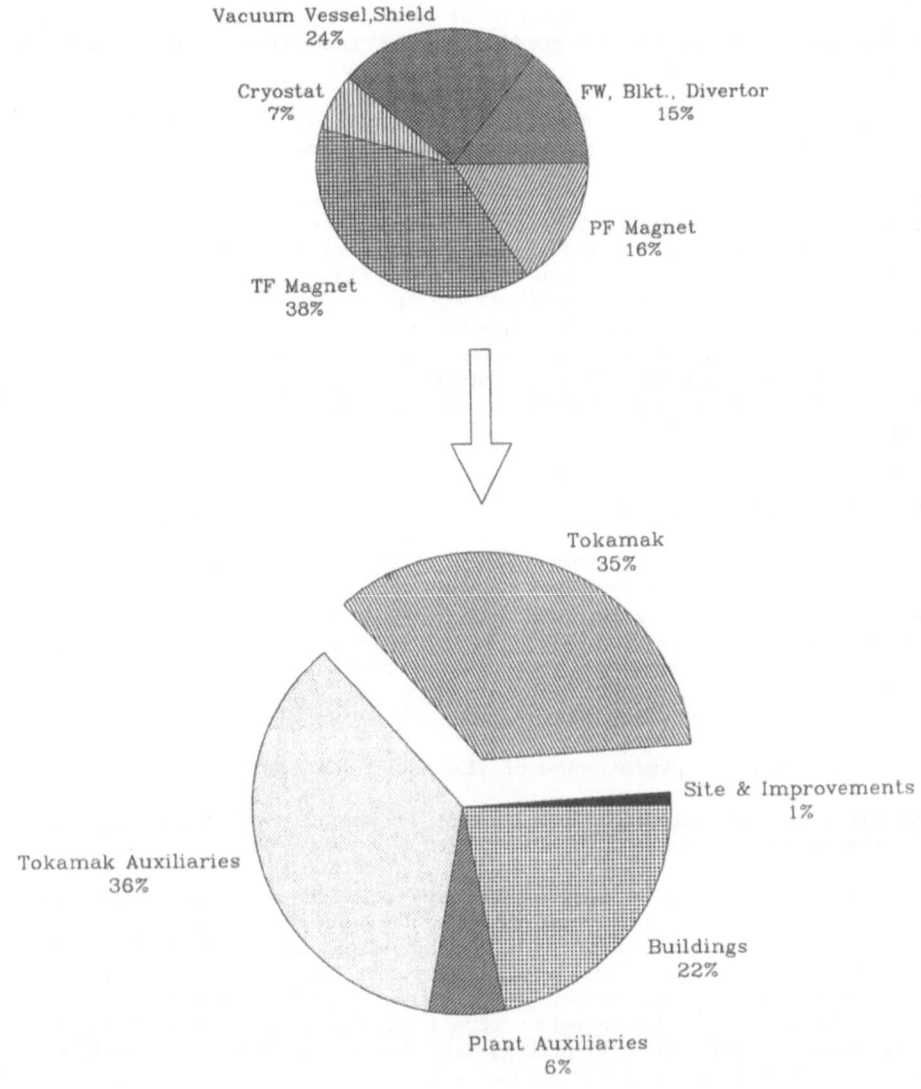

Figure 1. NET-2 cost breakdown as calculated by SCAN code

Although it is expected that the estimates provided by systems codes will give a reasonable approximation to the total cost, this cannot be strongly defended on close examination of each system. The assumption, that the scaling laws of the cost model are valid for a particular design point, requires a reexamination of their validity at that point in relation to a well-developed design. The designs of next-step devices are now beginning to reach the stage where the predictions of the systems code equations can be checked by the designers.

Existing plans are for designers to produce a first estimate of the direct construction costs of ITER this year, with more accurate estimates including operation costs next year. At this stage the total cost derived in this way is not an estimate of what ITER will eventually cost, but is rather an input to focus design decisions that will have to be made before entering a detailed design phase in 1991. The efforts of the designers will provide an excellent snapshot of the state of the art in costing each system, indicating the most reliable algorithms and unit costs available today. This should enable new systems-code algorithms to be written for next-step machines, giving a better estimate than today's models about exactly how to minimise costs when it comes to choosing final parameters. At the very least, it will enable a good check to be made of the results of today's cost models.

IMPLICATIONS FOR POWER REACTORS

As pointed out in the introduction, costing of devices beyond the next step is a necessary factor in judging the potential of turning today's favoured approaches into an eventual power reactor. As also pointed out above, any estimate made today for the cost and parameters of such a device will be quite crude, because it is not yet possible to accurately predict:

■ the cost of the next-step device;

■ the plasma physics performance of the power reactor;

■ the state of technological development at the time of designing and building the power reactor;

With this state of affairs, two ways forward are usually followed. The first assumes that the cost scalings for next-step devices are accurate, and extrapolates them to the reactor scale assuming plasma physics is fully understood and unit costs for materials around the reactor are as found for today's materials. The second again uses a scaling approach, often of a less detailed nature, is somewhat more adventurous in its assumptions about plasma physics and the technology available, and assumes reduced costs brought about through the "learning" effects of series production. Both of these are systems code approaches to costing, and obviously therefore the results have a wide uncertainty given the large extrapolation, the long way ahead before implementation, the fact that the details of the design are unknowable, and that advances are likely in the next half century before the first commercial-sized tokamak power reactors become available, and certainly in the longer period before the most commercially acceptable fusion power reactor design is obtained.

The first approach merely requires an extension of the SCAN cost model to apply to reactors. This involves the application of discount rates to the lifetime project costs right up to decommissioning, the inclusion of interest to be paid during construction, and the use of a scenario for electricity generation over the plant operating life at an assumed availability. All these items lead to the calculation of the "levelised" generation cost of electricity[13] (sometimes also known as the "over-night" cost of electricity). All the cost scalings and unit costs are kept as for next-step devices. This approach should therefore give a reasonably good estimate of "one-off" or "first-of-a-kind" devices. An additional assumption, much harder to justify, is that the design assumptions implicit in the hardware items of the cost code will apply to the power reactor.

The second approach is exemplified by the GENEROMAK code[14], as used for the ESECOM study[15]. This is based mainly on component costs derived for the STARFIRE reactor design[16], adjusted to 1986 unit costs with improved information on some unit costs (e.g. magnet unit costs increased about 2.5 times). Component masses and volumes are calculated assuming a series of nested toroids of the required thickness and density. Direct capital costs are calculated based on simple mass scaling of the main fusion power core components (blanket, first wall, shield, vacuum vessel, structure, magnets) plus linear power scaling for auxiliary plasma power requirements, less-than-linear (0.60) scaling for steam generators and other balance of plant, and less-than-linear (0.67) scaling of all building costs with the "nuclear island" volume. These parts of the scaling are normalised to STARFIRE costs where appropriate. The direct cost is then added to lifetime costs assuming certain discount and tax rates, to come up with the generation cost.

Major differences in the cost calculation in the SCAN and GENEROMAK models are summarised in Table 3. Note particularly that although superficially the direct costs for the two approaches give similar unit costs overall for the fusion power core, GENEROMAK assumes a vanadium-based structure (partly $400/kg, partly $50/kg), whereas SCAN assumes steel ($50/kg according to reference 15). Also, because the calculations of the fusion power

TABLE 3 COMPARISON, ON THE SAME BASIS, OF GENERATION COST CALCULATION BETWEEN SCAN AND GENEROMAK MODELS

COST ITEM	SCAN (1984 ME)	GENEROMAK (1986 M$)
Direct Costs (excluding contingency & including spare parts)	For details see text and ref. 9 -typical cost/unit mass of fusion power core ≈ 45 E/kg (SCAN def. of FPC) (Stainless steel struct.)	For details see text and ref. 15 -typical cost/unit mass of fusion power core ≈ 48 $/kg (GENEROMAK def. of FPC[1]) (Vanadium structure)
Contingency	9.8% of direct costs	20.6% of direct costs
Indirect Costs	31.8% of direct costs	37.5% of direct costs
Interest During Construction	25.2% of direct costs (5% real net interest rate assumed, on borrowing to cover direct, indirect costs and contingencies, over a construction period of 8 years, with a typical S-shaped expenditure profile)	13.5% of direct costs (equivalent to a real net interest rate of 3.35% assuming the same expenditure profile as in the SCAN model, but over the specified period of 6 years)
Basic Operation & Maintenance	28.4% of direct costs (discounting all charges at 5% discount over assumed plant lifetime of 25 years)	43.4% of direct costs (see reference for discounting rules)
Other Operation, Maintenance & Decommissioning (including fuel)	15.1% (typically) of direct costs (made up of various items, discounting costs according to when they occur during plant life)	45.1% (typically) of direct costs (various items and discounting rules, and based on different design assumptions to SCAN model)
Net Result	108.5% of direct costs (typically)	160.1% of direct costs (typically)

[1]Typically less than half the mass of the SCAN FPC estimate, for the same device

core mass and volume are quite different in the two cases (SCAN attempts to estimate "designed" component masses), the GENEROMAK unit costs would be significantly lower if referred to the SCAN-calculated masses.

The result of running the SCAN code on a reactor based around the physics and engineering assumptions of the NET design in 1985[17] is shown in the first column of Table 4. As part of a new assessment of the environmental and economic prospects for fusion, now underway, these parameters have been updated to more closely reflect today's NET/ITER assumptions (see Reactor-1 in Table 4). This study[18] is also looking at the effect of assuming some progress in attainment before deployment of fusion reactors, as shown by Reactor-2 in Table 4. Also shown in Table 4 are the parameters from the "point-of-departure" case in the ESECOM study[15]. The cost figure quoted for this device is not that from the GENEROMAK code (53mills/kWh), but is the equivalent SCAN code value for the same sized machine according to the SCAN code design assumptions. Costing assumptions (e.g.discount rates, duration of project) are also set to the usual SCAN values for comparison.

A breakdown of the main cost components for the reactors of Table 4 is shown in Table 5, alongside a cost breakdown for a PWR[19]. The results for PCSR-E, Reactor 2 and a PWR are shown graphically in Figure 2. These results show that the generation cost for fusion is capital-cost-dominated, and thus there is considerable opportunity for generation cost reduction by clever design. However, this also means that the rather uncertain cost estimation of capital items at present makes the estimate of fusion power generation costs even more questionable.

Apart from predicting, rather uncertainly, the generation cost of future fusion reactors, systems codes such as SCAN and GENEROMAK can be used to point out how best to reduce future reactor costs. This was previously

TABLE 4   COMPARISON OF MAIN REACTOR PARAMETERS

	PCSR-E	Reactor-1	Reactor-2	"ESECOM"
Plasma configuration	single-null	double-null	double-null	double-null
Plasma elongation (null)	1.71(av.)	2.2	2.5	2.5
Plasma safety factor (qpsi)	3.0	2.9	2.9	2.1
Beta scaling coefficient,g	3.5	3.0	4.0	4.0
Fuel beta/total beta	0.74	0.67	0.67	0.91
Burn mode	pulse(5000s)	steady state	steady state	steady state
Turbine thermal efficiency (%)	35	35	40	40
Blanket/Shield thickness (m)	1.34/1.54	1.36/1.57	1.38/1.59	1.54/1.54
Average availability (%)	73	75	75	65
Constr./Operation Time (yrs)	8/25	8/25	8/25	8/25
Major radius (m)	9.30	7.07	5.31	5.89
Minor radius (m)	2.39	2.03	1.40	1.47
Plasma current (MA)	16.64	22.35	16.62	15.81
Toroidal field: coil/axis (T)	11.33/6.36	13.46/6.44	14.85/6.22	10.00/4.29
Total beta (%)	3.83	5.14	7.62	10.00
Current drive power (MW)	-	216	91	78
Current drive efficiency (A/W)	-	0.10	0.18	0.20
Mean neutron wall load (MW/m^2)	2.22	3.10	4.15	3.36
Burn fusion power (MW)	3578	3902	3051	2870
Total thermal power (MW)	4166	4933	3779	3618
Recirculating power (MW)	211	524	302	297
Power sent out (MW)	1247	1203	1209	1164
Fusion Power Core mass (t)	41500	35200	22800	26500
Mass power density (kW$_{so}$/t)	30.0	34.2	53.0	43.9
Mass expenditure (t/MW$_{so}$)	33.3	29.2	18.8	22.7
Generation cost (E$_{84}$/MWh)	109	105	78	88

TABLE 5 COST BREAKDOWN OF REACTORS (1984 MEcu)[*]

		PCSR-E	Reactor-1	Reactor-2	"ESECOM"	PWR[**]
D.1	Site and Improvements	57	55	40	37	–
D.2	Buildings	1239	1032	826	788	–
D.3	Tokamak	1863	1617	1045	1046	–
D.4	Tokamak Auxiliaries	2041	2189	1610	1380	–
D.5	Plant Auxiliaries	541	576	472	457	–
D.S	Spares	133	131	94	87	–
D.C	Contingencies	574	576	399	371	–
I.1	Basic Indirect Costs	904	861	628	583	–
I.2	Launching Costs	969	922	673	625	–
C	Total Construction Costs	8332	7929	5788	5375	2257
Z	Interest During Construction	1483	1411	1030	956	697
M	Operation and Maintenance	2071	2174	1806	1730	646
F	Fuel	105	107	104	102	1003
R	Decommissioning	381	363	265	246	34
T	Total Project Costs over Life	12372	11983	8993	8409	4637
	Actual energy production over life (TWh)	200	198	199	166	194
	Discounted energy production over life (TWh)	114	114	115	96	110
	Generation cost (E_{84}/MWh)	109	105	78	88	43

[*] Using SCAN model rules only for fusion cases

[**] Averaged from single unit station costs with unit sizes from 1100-1300 MW (5 European, 4 USA) from NEA/OECD Study of reference 19

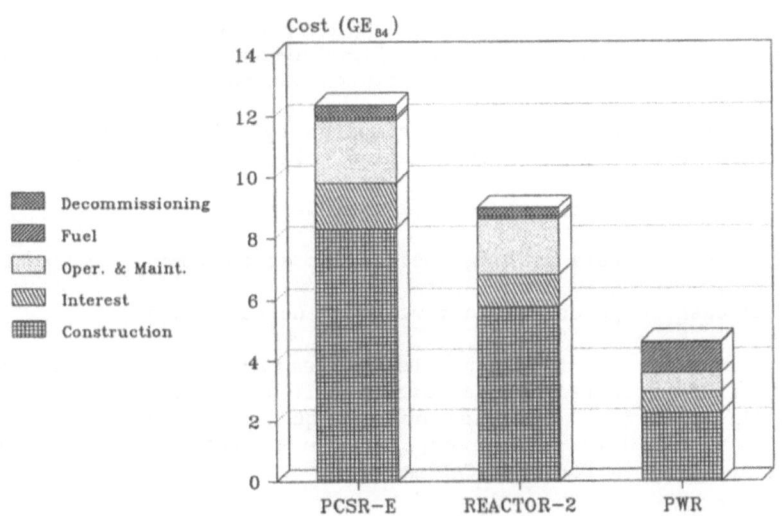

Figure 2. The main generation cost contributors for fusion and fission plants

287

investigated based on the early NET design assumptions which led to PCSR-E[20], but the same general conclusions still apply today. The generation cost parameter space is typically as shown in Figure 3. Here generation cost is shown as a function of "mass expenditure" on the fusion power core, i.e. the total mass of the fusion power core (see above) divided by the net electrical power generated after deduction of all circulating power within the plant, the "power sent out". This mass expenditure is the inverse of the "mass power density" used in other studies[21]. For a fixed plasma shape, temperature, etc., fixed assumptions about breeding blanket thicknesses around the plasma, and fixed assumptions about availability, etc., the generation cost is essentially determined by choosing the required power sent out and the beta scaling coefficient, g (g~$\beta[\%]$a[m]B[T]/I[MA] for plasma total beta, β, plasma minor radius, a, plasma current, I, and toroidal field on axis, B). (In fact there is some small flexibility with respect to the choice of wall load, but this can be neglected here, especially as any change in wall load would increase costs.)

These results show that reduction in generation cost can be achieved by significant increase in the beta scaling coefficient at constant power sent out. However, as Figure 4 shows, this can only be achieved by managing to handle higher wall loadings (note also that in SCAN the unit costs for first wall materials are not increased with power handling capability, as would in reality be likely). Also there is a diminishing return with going to higher g-values, due to stress limits in the magnets of the smaller, higher-g, devices. Another possibility to improve generation costs is to increase the power sent out. Again, this may be problematical, especially for geographically spread out networks not under central government control, as for example in North America. A third possibility is to reduce the unit cost of the fusion power core by clever design favouring the most economic design solutions. This is likely to be possible since the design of all systems are not optimised even for next-step devices, let alone for power generation. The way costs are estimated at the moment, for instance, the fusion power core contributes typically only 30% to the direct costs, but its design influences, for example via building dimensions, about 70% of the direct costs. About 60% of the non-direct costs are then calculated in terms of the direct costs. So although the FPC only contributes about 15% of the generation cost, its design and cost influences 65%, a vital factor when formulating simple scaling laws, as pointed out in the introduction.

The above requirements can be compared by noting that a 15% reduction in generation cost from the PCSR-E design point can be achieved by:

∎ fusion power core unit cost reduction by 50%;

∎ a 60% increase in beta scaling coefficient at constant power sent out;

∎ a 40% increase in power sent out without increase in g;

FPC unit cost reduction would be more easily achieved if account can be taken of "safety credits". These are factors for reducing hardware costs by avoiding the need for satisfying nuclear quality controls (often known as "N-Stamping"). Typical factors applied previously[15] would lead to the following safety credits for the SCAN model:

Site & Buildings............0.68
Tokamak....................0.60
Tokamak Auxiliaries...0.85 (PCSR-E), 0.94 (Reactor-2)
Plant Auxiliaries..........1.00

(The different Tokamak Auxiliary values arise because PCSR-E includes considerable thermal storage since it is a pulsed-plasma device.) Generation costs would drop to 83 E_{84}/MWh for PCSR-E and 64 E_{84}/MWh for Reactor-2.

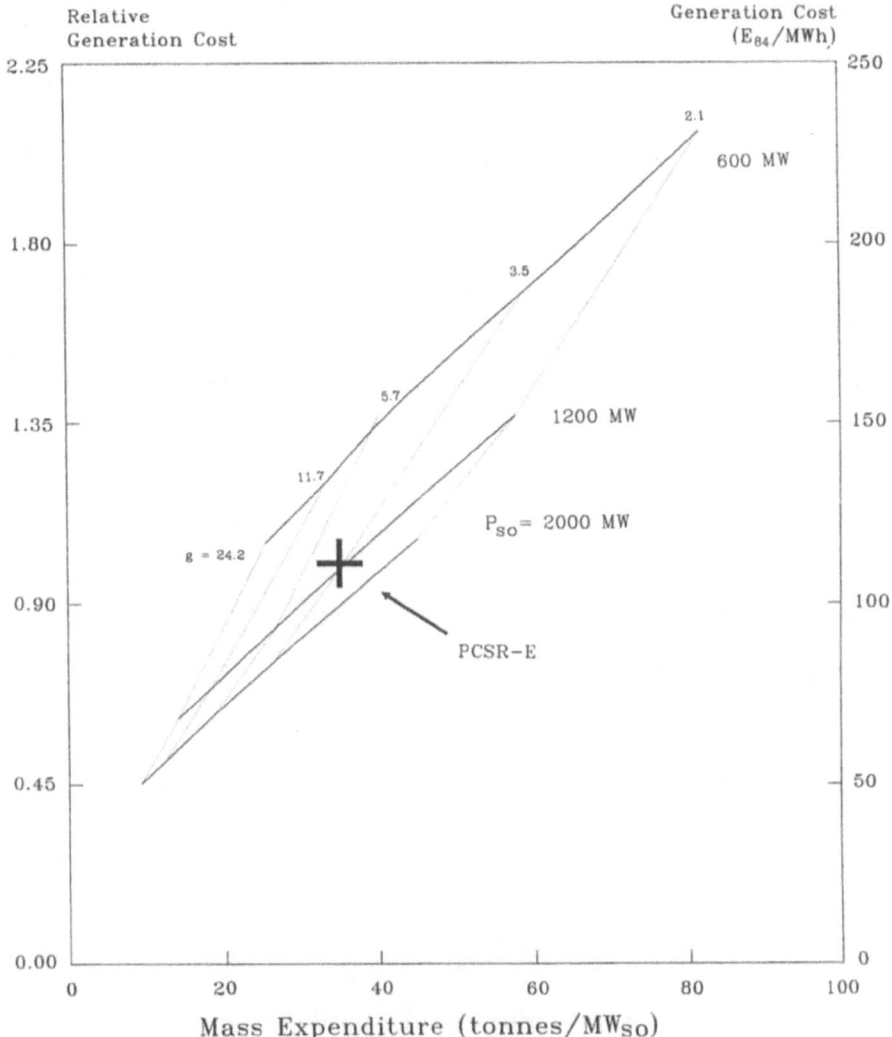

Figure 3. Correlation between generation cost and mass expenditure
for minimum-cost devices at given values of the beta
scaling coefficient and the power sent out.

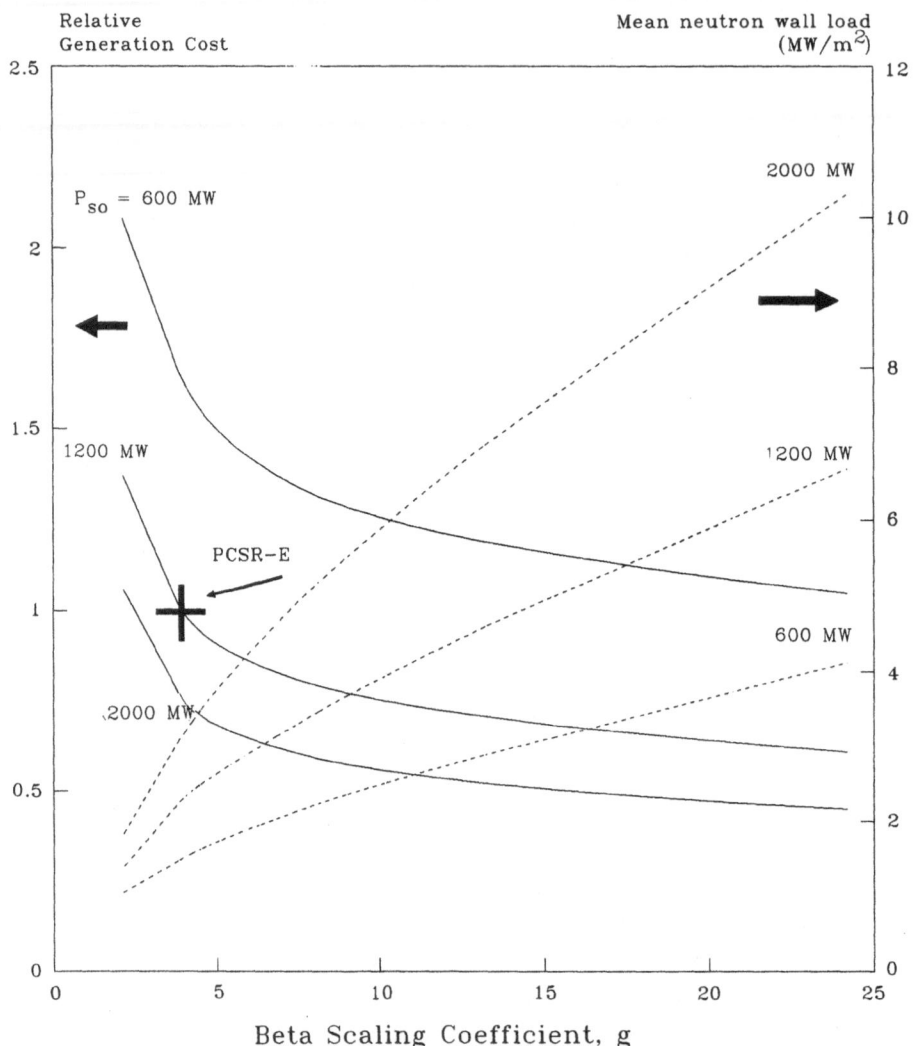

Figure 4. Generation cost of minimum-cost devices as a function of the
beta scaling coefficient at different values of power sent out,
and the corresponding wall loading levels required.

All the reactors described in Table 4, and Figures 3 and 4, are costed as if they were first-of-a-kind devices. The effect of learning on bringing costs down is hard to estimate, but assuming the following arbitrary construction cost changes:

Site & Buildings............-15%
Tokamak....................-30%
Tokamak Auxiliaries.........-20%
Plant Auxiliaries...........-15%
Indirect Cost fraction......-50%
Construction time...........-25%

would reduce generation costs to 77 E_{84}/MWh for PCSR-E and 56 E_{84}/MWh for Reactor-2. Putting this together with the safety credits, noted previously, would lead to generation costs of 60 E_{84}/MWh for PCSR-E and 47 E_{84}/MWh for Reactor-2.

It is also worth noting from the analysis of figure 3 that target values for economic viability of a fusion device, which have been previously estimated as requiring mass power densities of 100kW$_{so}$/tonne[21] (mass expenditure 10 tonnes/MW$_{so}$), are virtually unattainable. This is not because fusion reactors cannot become economically viable, but rather that the fusion power core mass estimation underlying the fusion power core calculation in figure 3 (SCAN-like) and behind the above value (GENEROMAK-like) differ by at least a factor of two, as pointed out previously. A more realistic figure in SCAN terms would be a mass expenditure around 20 tonnes/MW$_{so}$ (50kW$_{so}$/tonne).

CONCLUSIONS

Costing of next-step devices is only now starting to be done based on developed designs of components. There is therefore insufficient data yet available to make a reliable estimate of the cost of such a device, and it is certainly not possible to make a truly justifiable statement on the generation cost of a future fusion reactor 50 or more years hence. This is particularly so since the plasma physics attainable in such a device can only be guessed at.

However, given the above caveats, it is mathematically possible to make such estimates based on component cost scaling laws that have a fair basis in reality, and usually reflect past experience in both fission and fusion. These scaling laws at least are capable of giving a "ball-park" estimate for cost. They may even turn out to be correct in the end.

Assuming that such cost estimates are accurate, the next-step device is likely to have a construction cost around 3500-4000 ME (or M$). The most expensive hardware cost items are presently perceived to be the tokamak itself (35%), its auxiliaries (36%) and all the buildings (22%). Similar percentages also apply for the power reactor. Since these items (particularly the first two) are heavily dependent on the fusion power core design, there is clearly room for cost minimisation provided certain directions are followed in future development.

First, the future power reactor design must exploit as many "inherent safety" features of fusion as possible, in order to avoid having to satisfy nuclear component quality control wherever possible, and thereby reduce costs. Secondly, crude, robust and simple design solutions are needed to minimise unit costs. Thirdly, attempts must be made to increase the attainable beta scaling coefficient. Here it is not worth pushing too hard, but some improvement over today's expectations would be welcome. If pursuing these lines falls short of making fusion economically attractive, even when its full impact on society concerning its special safety features are considered, there is still the alternative of making the unit size large to bring unit costs down, although for some countries this must be considered a last resort.

For the future, as the design and construction of a next-step device proceeds, the cost estimates for the device and for future reactors will become clearer. Although present estimates indicate fusion costs to be on the high side of the costs of today's fission plants, their cost can be more easily changed by innovation in design and improved understanding of plasma behaviour, compared to fission where design limits are now well-known and where fuel costs are not negligible. It is therefore reasonable to expect, based on what is known about the cost of fusion power today, that fusion power can be made into an economically attractive power source in the future.

REFERENCES

1.   Pfirsch, D., Schmitter, K.H., "On the Economic Prospects of Nuclear Fusion with Tokamaks", Fusion Technology, 15, 1471-1484, (1989)

2.   Spears, W.R.,Wesson, J.A., Nuclear Fusion, 20, 12, 1525-1532, (1980)

3.   See Chapter IV of "ITER Concept Definition", Report ITER-1, IAEA Vienna 1988

4.   Thomson, S.L., "Systems Code Cost Accounting", FEDC-M-88-SE-004, available from Fusion Engineering Design Centre, Oak Ridge, Tennessee (1988)

5.   Meldianov, A.I, Muravyev, E.V., Orlov, V.V., "Analysis of Possible Parameters for a Tokamak Research Reactor: I - Description of Model Calculations", Voprosy atomnoi nauli i tekhniki, Ser. Termoyadernyi sintez (Problems in Atomic Science and Technology, Series Thermonuclear Fusion), 4, 3-8 (1984)

6.   Mizoguchi, T., Sugihara, M., Shinya, K., Kobayashi, T., Miki, N., Nakashima, K., Saito, R., Yamada, M., Fujisawa, N., Yamamoto, S., Iida, H., Nishio, S., and FER Design Team, " Development of Tokamak Reactor Conceptual Design Code 'TRESCODE'", JAERI-M-87-120 (1987)

7.   Borrass, K., Söll, M., "SUPERCOIL: A Model for the Computational Design of Tokamaks", Nuclear Engineering and Design/Fusion, 4, 2, 21-35 (1986)

8.   Borrass, K., "Normal Conducting versus Superconducting TF Coils in Next-generation Tokamaks, a Competitive Study", Max Planck Institut für Plasmaphysik Report IPP 2/267 (1983)

9.   Spears, W.R., "The SCAN-2 Cost Model", NET Report EUR-FU/XII-80/86/62, Commission of the European Communities (July 1986)

10.   Spears, W.R., "Reactor Cost Driving Items", in "Fusion Technology 1986", Proceedings 14th Symposium, Avignon, 387-392 (1986)

11.   See Chapter VI.4.4 of "ITER Concept Definition", Report ITER-1, IAEA Vienna 1988

12. Toschi, R., "The NET Project - An Overview", Twelfth International Conference on Plasma Physics and Controlled Nuclear Fusion Research, Nice, France, October 1988, Paper IAEA-CN-50/G-1-1

13. Expert Group of OECD/NEA, "The Costs of Generating Electricity in Nuclear & Coal Fired Power Stations", OECD, Paris, 1983

14. Delene, J., Krakowski, R.A., Sheffield, J., Dory, R.A., "GENEROMAK - Fusion Physics, Engineering and Costing Model", Oak Ridge Report ORNL/TM-10278, June 1988

15. Holdren, J.P., Berwald, D.H., Budnitz, R.J., Crocker, J.G., Delene, J.G., Endicott, R.D., Kazimi, M.S., Krakowski, R.A., Logan, B.G., Schultz, K.R., "Summary of the Report of the Senior Committee on Environmental, Safety, and Economic Aspects of Magnetic Fusion Energy", University of California Report, UCRL-53766-Summary(Revised), September 1987

16. STARFIRE Design Team (Baker, C.C., et al.), "STARFIRE - A Commercial Tokamak Fusion Power Plant Study", Argonne National Laboratory Report ANL-FPP-80-1, September 1980

17. The NET Team, "NET Status Report December 1985", EUR-FU/XII-80/86/51, Commission of The European Communities (December 1985)

18. Cooke, P.I.H., Hancox, R., Spears, W.R., "A Reference Tokamak Reactor", Culham Report, in preparation (1989)

19. Jones, P.M.S., et al. "Projected Costs of Generating Electricity from Nuclear & Coal-fired Power Stations for Commissioning in 1995", Expert Group Report of the OECD, Paris (1986)

20. Spears, W.R., "Reactors Beyond NET", in "Fusion Reactor Design and Technology", Proceedings Yalta Conference, 1986, International Atomic Energy Agency, Vienna, 1, 285-301 (1987)

21. Magnetic Fusion Advisory Committee Panel X (Conn, R.W., et al.) "Report on High Power Density Fusion Systems", UCLA, Los Angeles, May 1985

PULSED VERSUS STEADY-STATE REACTOR OPERATION IN VIEW OF SAFETY AND ECONOMY

Rolf Buende

The NET Team
c/o Max-Planck-Institut fuer Plasmaphysik
D-8046 Garching bei Muenchen
Federal Republic of Germany

ABSTRACT

The impact of applying non-inductive instead of inductive current drive for tokamak reactors in future fusion power plants is assessed. The criteria for that assessment are those which in general serve for appraising the attractiveness of power plants: the impact on environment (plant safety, exhaust heat and waste materials), plant availability and cost of electricity. Apart from minor beneficial effects on various components going to the non-inductive scheme could improve reliability and, hence, safety and availability and could reduce radioactive waste but would increase the exhaust heat and possibly the costs of electricity.

1. INTRODUCTION

The transformer principle of the tokamak implies that the time span for inducing the plasma current is limited in accordance with the available flux swing thus causing pulsed reactor operation. To avoid pulsing the plasma current could be driven continuously by devices like neutral beam injectors and radio frequency systems previously foreseen only for plasma heating. Up to which extent it is worthwhile for technical reasons to put emphasis on the non-inductive current drive largely depends on its impact on the attractiveness of fusion power plants for energy supply. Hence, the technical worth of a transition from the inductive scheme, as well understood from todays experiments, to the non-inductive one must be assessed in view of the energy supply system as expected by the middle of the next century, this being the earliest date for which a usable fusion power plant may be imagined. In the following, firstly, the criteria for this assessment will be derived, secondly, the areas of influence will be defined for the transition from reactors with inductive to those with non-inductive current drive and, thirdly, the advantages and disadvantages of this transition for the overall plant and for its subsystems and components will be considered and weighted. The entire consideration focusses on technical impacts thus excluding physics reasons like profile control which could require non-inductive current drive devices irrespective of all technical consequences.

2. CRITERIA

The criteria for the assessment will be based on an energy supply scenario as it is expected in the middle of the next century in industrially developed countries (only these will then realistically be capable to handle the various technologies involved). In comparison with todays energy supply system generic changes are not expected because the overall energy consumption will probably not increase anymore and the lifetime of installations is counted in decades. Hence, the primary energy carriers will remain the same, only the pattern of their use and the structure of the consumption of useful energy will vary although the heat supply may keep its dominating role. The main reason for changes is the increasingly accepted necessity of strongly reducing the impact on environment. The further exploitation of fission energy will remain limited (presently: fission technology suffers from public acceptance because of increased perception of accidental and environmental

Safety, Environmental Impact, and Economic Prospects of Nuclear Fusion
Edited by B. Brunelli and H. Knoepfel
Plenum Press, New York, 1990

impact on environment. The further exploitation of fission energy will remain limited (presently: fission technology suffers from public acceptance because of increased perception of accidental and environmental risks; only limited construction of new LWR plants; some countries have moratoria, one decided to phase out nuclear; no need for FBRs because of ample Uranium resources; practically no further HTR development and, hence, no further development of Helium cooling technology; application practically only for electricity supply because of limited possibilities for cogeneration of electric power and heat on useful temperature level). Fossil fuels will keep their important role inspite of the now widely acknowledged problem of CO_2 enrichment of the atmosphere (fuels well suited for transportation and heating, proven technology for cogeneration). Renewable energies will be used increasingly but will probably not gain a dominating percentage. As a consequence to reduce the impact on environment requires to reduce the consumption of useful energy and to improve the efficiency of energy conversion thus reducing the necessary amount of primary energy even more.

The most effective and in the next decades most likely applied measures of achieving this target is to increase the price of energy by direct and indirect taxes and by progressive tariffs in order to give energy costs a higher weight in economic calculations. This is not unrealistic because the percentage of energy costs in goods only in rare cases exceeds some percent (Federal Republic of Germany [1]). Especially indirectly imposing progressive taxes on routine environmental loads like exhaust heat, exhaust gases and all kinds of liquid and solid wastes as well as imposing fines on safety related occurrences (depending on the absolute risk and its increase) would create a higher appreciation of the value of energy and all risks associated and consequently would make technology improvements profitable. Present such tendencies in the Federal Republic of Germany led to increased application of, for instance, cogeneration (overall efficiency up to 90%) and of combined gas and steam turbine processes (net electric efficiency up to 50%) [2], [3].

Fusion power plants on the basis of magnetic confinement promise in general higher nuclear safety than fission plants. However, they are technically more complicated due to the unavoidable mutual penetration of nuclear-thermal and electric-cryotechnic components in the reactor and the additional presence of plasma heating systems, fuel (Tritium) handling and cryoplant, their installation costs are expected to be higher. From the above considered trends in energy supply follows that the impact on environment is a first order criterion to appraise the attractiveness of a plant. With respect to the integration into an energy supply system (second order criterion) the reduced prospects of Helium cooling technology (high gas temperatures; possibly in combination with closed-cycle gas turbines) practically rule out an efficient cogeneration so that fusion will be used just for electricity supply. Especially the higher complexity of the reactor and the additional existence of heating and fuel handling systems require the implementation of specific 'high-tech' reliability assurance measures in order to achieve the same level of plant availability as in fission plants. The costs of electricity are seen as a third order criterion due to the above described necessity for anyway increased energy prices. Of main influence are the installation costs, the operating costs in the sense of spare parts for replacements, the plant load factor and anticipated charges on the impact on environment .

The attractiveness of utilising fusion is measured by means of these criteria. Accordingly, also the transition from inductive to non-inductive current drive will be appraised by considering the improvements by this transition with respect to the following items:

1. Impact on environment: Plant safety (occurrence rate of accidents and their consequences)
Exhaust heat (depending on net plant efficiency)
Waste materials (radioactive)

2. Plant availability: Reliability of components and subsystems

3. Costs of electricity: Installation costs
Operating costs
Net plant efficiency
Plant availability.

3. BASIC ASSUMPTIONS ON THE POWER PLANTS

The following items are valid for both the inductive (I) and the non-inductive (NI) case (overall plant and especially reactor parameters are those used in [4]):

---

* The net electric power output from the plant is the same in both cases and is constant at rated operation.

* The average plant load factor (in accordance with OECD/NEA studies [5], [6]) is 75% over 25 years calendar time (about 19 full power years). Fusion power plants must achieve this in order to be acceptable for an electricity supply system. Assuming base load operation this load factor can be illustrated by an average of 3 months outage per calendar year or by an outage of two months in four years each and a down time of 7 months every 5th year. In the latter case dividing the total outage per year into one month each for scheduled and unscheduled outage requires a load factor of 91% during scheduled operation. At a mean plant down time of 70h per outage [7] this corresponds to a plant failure rate of $1.4 \cdot 10^{-3}$ per hour (i.e. on average one failure after every 700h of operation). Accordingly in its lifetime the plant will see about 250 start-up/shut-down procedures.

* The reactor is cooled by pressurized water. Its thermal energy is transferred to a secondary wet steam cycle which is equipped with intermediate dewaterisation and reheating by live steam. The thermodynamic efficiency of this process is 35% [4]. In the I-case thermal storage is installed such that the steam turbine does not see any changes in live steam data and throughput due to the pulsing. Likewise auxiliary power changes due to pulsing are balanced by adequate storage.

4. MAIN AREAS OF INFLUENCE OF CURRENT DRIVE METHOD

The transition from I- to NI-current drive changes the energy balance of the overall power plant and of the reactor and reduces the fatigue damage to various components which in turn could improve their reliability.

The overall plant energy balance is changed because the auxiliary energy to operate the continuously running current drive systems is considerably higher than for operating them in a pulsed mode this causing a decrease of net plant efficiency. In addition the distribution of the energy input to the plasma facing components is changed towards a higher thermal load on the divertor plates. Thermal energy storage to bridge the dwell time is not required anymore; electric energy storage for PF coils supply is reduced.

The reduction in the number of transients influences the reliability of components. It relaxes the fatigue strength requirements for which the mechanical components have to be designed and reduces the number of demands for subsystems which otherwise work intermittently.

In summary the areas of influence are:
1. Energy balance of
   - Overall Plant
   - Reactor
2. Reliability of
   - Mechanical components
   - Subsystems

The impact of changes in these areas will be considered with respect to the criteria desribed in Sec. 2. Not all areas of influence are relevant for all criteria. The items actually considered are listed in Tab. 1.

Table 1.   Areas of influence and criteria for comparing inductive and non-inductive current drive

Areas of influence / Criteria	Energy balance		Reliability	
	overall plant	reactor	mech. comp.	subsystems
Plant safety			X	X
Waste heat	X			
Waste material	X		X	
Plant availability		X	X	X
Installation costs	X	X	X	
Operating costs			X	

5. ASSESSMENT OF INFLUENCE ON ENERGY BALANCE

5.1 Overall Plant Energy Balance

Table 2 shows data for a plant based on the 'prototype commercial-sized reactor' PCSR-E [8] for the I-case and corresponding data based on a recent assessment of requirements for steady-state operating tokamaks [4] (NI-case). The latter data are typical for a device where the bootstrap current contributes 30% to the total current, the current drive figure-of-merit is about 0.1 A/W and the electrical efficiency is 50% ([4], [9]) the latter value presently being regarded as only achievable in the future [10].

Table 2. Power data for plants with inductive and non-inductive current drive

		Inductive (PSCR-E [8])	Non-Inductive ([4],[9])
Fusion power (cycle average) P_{FUS}	[MW_{th}]	3513	4122
Reactor power, thermal P_{th}	[MW_{th}]	4169	4976
Energy multiplication in blanket	[-]	1.178	1.178
Gross electrical power $P_{el,g}$	[MW_{el}]	1459	1742
Mean auxiliary power	[MW_{el}]	215	498
Net electrical power $P_{el,n}$	[MW_{el}]	1244	1244
Exhaust power $P_{th,EX}$	[MW_{th}]	2925	3732
Efficiency, thermal ETA_{th}	[%]	35	35
Efficiency, gross electric $ETA_{el,g}$	[%]	35	35
Efficiency, net electric $ETA_{el,n}$	[%]	30	25

Exhaust heat. The exhaust power, i.e. difference between thermal reactor power and net electric power, is discharged to the environment as thermal energy. Because of the considerably higher demand for auxiliary power the reactor power in the NI-case has to be larger in order to keep the net power unchanged. The increase of exhaust heat by about 30% is a disadvantageous impact on environment. It will practically be impossible to reduce this exhaust heat because its temperature level is too low to allow any further use. The only way out would be to considerably improve the current drive efficiency. In absolute terms a net efficiency of 30% seems already to be at the lower limit of what might be accepted by mid of next century, lower values seem hardly to be acceptable in view of what was discussed in Sec. 2 on environmental impact.

Waste material. The amount of radioactive waste material will increase proportional to the increase of reactor power, i. e. by nearly 20%.

Installation costs. Following the cost breakdown for the PCSR-E [11] about 50% of the total direct costs refer to torus system, magnets system and buildings. Previous calculations [8] have indicated a cost increase of 12%, when the device size is allowed to reoptimize under the current drive condition. The costs for cooling system and energy conversion were taken for constant which is valid under the assumption that the cost increase for the higher thermal power just balances the cost reduction from omitting the thermal storage in the I-case; anyway the cost percentage of cooling system and energy conversion in the I-case is in the range of 10% so that minor changes in their costs can be neglected. The same is true for the heating devices which are given to contribute only 3% to the total direct costs; any cost changes of these devices will scarcely influence the overall. In summary the trend is for higher installation costs in the NI-case, however, the amount of that increase cannot be seen as important.

5.2 Energy Balance of Plasma

Three effects are considered, two of them concerning the distribution of power output from the plasma, one referring to the energy of the particles which hit the divertor plates:
--- The additional current drive energy increases the thermal load on the plasma facing components. As these components will anyway be highly loaded this addition is disadvantageous.
--- For getting a higher ratio of driven current to power absorbed by the plasma the plasma density in the NI-case is desired to be lower than in the I-case. As a consequence, the load on the divertor plates will be somewhat more peaked. As the divertor anyway is the highest loaded component and may also suffer more from off-normal operating conditions this loadpeaking is disadvantageous. The effect cannot be quantified now because it depends strongly on the real operation of the current drive system which even for NET/ITER is not settled [12].

--- The energy of the particles hitting the divertor plates will be substantially larger so that sputtering yields from the surface of these plates may be drastically increased.

Plant Availability. The trend in the first two effects is towards a higher load on the divertor plates, a component which already in the I-case is highly loaded and where no reasons are seen which would change that situation in principle. As it will be anyway difficult to get a sufficiently high reliability of the divertor the increase of the divertor load must be compensated by enlarging its size to keep the specific load unchanged. The third effect could increase the erosion rate and thus reduce the lifetime of the plates such that the frequency of plates replacements becomes unacceptably high.

Installation Costs. To enlarge the divertor plate surface and the distance between null point and striking points leads to a larger torus and correspondingly larger coils and buildings thus affecting three items which account for about 50% of the entire direct costs (see Sec. 5.1). Hence, the cost increase could be significant, however, the available information is not sufficient to assess any amount. The third effect (higher particle energy) can be mitigated somewhat by increasing the null to strike point distance but it is unlikely to be solved with by design changes.

6. ASSESSMENT OF INFLUENCE ON RELIABILITY

6.1 Reference plant

The plant with inductive current drive is characterised by the following features: designed for 250 start-up/ shut-down procedures and 10^5 pulses (75% load factor over 25 years at 5000 s burn time and 100 s off-burn time per pulse); the average outage time of 3 months per year is sufficient for all necessary inspections, maintenance, repairs and replacements; necessarily not redundant mechanical components with lifetimes shorter than plant lifetime can be replaced sufficiently fast to keep the average failure rate down to the level which corresponds to the 75% load factor; these components are specifically monitored for early detecting and localising failures to keep failure consequences low in terms of safety-relevant events and length of down times; sufficient reliability of subsystems is assured by redundancy of critical components. In total the plant will have a practically constant failure rate over its 25 years service life (according to [5]: 4000, 5000, 6600 h/a in year 1, 2, 3 to 25).

The degree of influence of changing to non-inductive current drive on the overall plant will be assessed on components level taking into account their relative contributions to failure rate and outage risk as presently valid for NET [7] (Tab. 3). The use of this breakdown will, however, overestimate the benefit from non-inductive current drive because the step from NET to a mid-next-century fusion power plant anyway will have required improvements which are most effective for the largest unavailability contributors thus, however, reducing their margin for further improvement.

Table 3.   Breakdown of failure rate FR and outage risk contribution
GM (GM = FR * MDT; MDT = mean plant down time)

		FR [%]	MDT [h]	GM [%]
A	HEAT SYSTEM	(39.6)		(85.9)
AA	TORUS SYSTEM	(2.0)		(27.7)
AAB	FIRST WALL	0.3	600	3.4
AAC	BLANKET	0.3	600	3.2
AAD	VACUUM VESSEL	0.2	2160	7.1
AAE	DIVERTOR	0.9	100	7.4
AAF	LIMITER	0.3	600	3.4
AAG	SHIELD	0.1	2160	3.2
AB	TF MAGNET SYSTEM	0.5	1400	13.2
AC	PF MAGNET SYSTEM	0.1	4200	13.2
AE	OVERALL CRYOSTAT	0.7	100	1.3
AF	RF SYSTEM	7.2	50	7.5
AG	NBI SYSTEM	13.2	50	13.2
AM	FUELLING SYSTEM	7.2	10	1.5
AP	PLASMA VAC. PUMPING S.	2.1	100	4.2
AQ	HEAT TRANSPORT SYSTEM	6.6	30	4.1
C	REACTOR AUX. SYSTEMS	0.2	50	0.2
D	FUEL HANDLING SYSTEM	4.6	50	4.6
K	CRYOPLANT	1.6	100	3.2
L	ELECTRIC POWER SUPPLY	2.2	50	2.2
P	AUXILIARY SYSTEMS	3.6	1	0.1
R	BUILDINGS	0.0	8760	0.0
W	ASSEMBLY & MAINT. EQUIP.	2.0	50	2.0
Y	PLT PROT., INSTR.&CONTR. S.	46.1	2	1.9

6.2 Mechanical Reliability

A simple consideration from fracture mechanics illustrates the development of a mechanical failure as a function of operating time (Fig. 1). In this figure the size of a real crack (bottom thick curve; starting from a

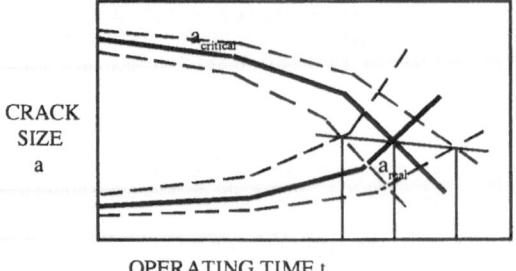

CRACK
SIZE
a

OPERATING TIME t

Fig. 1   Critical and real crack size over time

certain flaw size) is shown growing over time, its growth depending among others on the number of cycles in case of pulsed loads. The upper thick curve shows the critical crack size i.e. that size of a crack which represents if achieved the occurrence of a failure. The intersection of the two curves designates the operating time at which a failure occurs. On both curves distributions have to be superimposed to take into account a certain scatter of real flaw sizes at start of operation and uncertainties in material properties causing a scatter in critical crack size (dashed lines). Accordingly the time to failure shows also a distribution. The impact of omitting the pulsing on the time to failure is assessed for two catagories of components:

(a)   Components which to replace is planned because the lifetime is expected to be shorter than the service life of the plant (consideration valid for all in-vessel components).

(b)   Components which to replace is not planned - although not impossible - because it would be too time consuming so that they are designed for a lifetime which exceeds that of the service life of the plant with high probability (Semi-permanent components; here: PF magnets system).

6.2.1 Components with planned replacements

The present knowledge on the structural material behaviour for in-vessel components in about 60 years from now is too poor to allow to provide the required critical crack size curve. Furthermore it is not even clear which material would be used because the stainless steel AISI 316 L foreseen for NET/ITER was shown to be inadequate already for DEMO for which a martensitic steel was recommended [13]. From the same reason a real crack size curve cannot be provided. Therefore only a very rough assessment can be made. For that purpose the

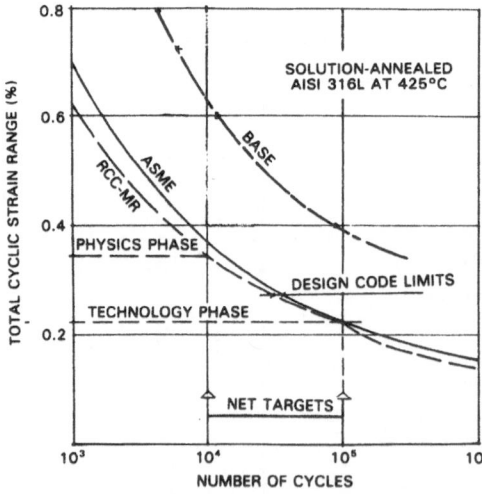

Fig. 2   Fatigue curve for AISI 316 L [13]

fatigue curve which has been used for the NET first wall (Fig.2, taken from [13]) is considered as a representative example. The mechanical reliability of a component under cyclic loads can be characterized by such a curve in which the permitted strain range at a certain temperature is represented depending on the number of cycles. The base material curve designates the 50% probability of a sample to break under a given cyclic strain range after a certain number of cycles. The design codes prescribe to reduce the base material values down to actually permitted values by using the minimum of either dividing the number of cycles by a factor of 20 or by dividing the strain range by a factor of 2. The design curve represents a failure probability of about 0.1% [14]. Following the design curve in Fig. 2 the reduction in number of cycles by about 4 orders of magnitude would allow to increase the cyclic strain range by about a factor of 4. In principle, two different conclusions could be drawn from that:

(1) Thermal stresses can be higher allowing for higher surface heat flux at constant fatigue damage and thus constant failure probability due to thermal cyclic load. Assuming that this does not change the lifetime of the components by other effects, e.g. increased erosion, the criteria (Tab. 1) would be affected as follows:

* Plant safety: Accidents occurrence rates as well as consequences in terms of impact on environment remain unchanged.
* Waste material: Occurrence rate of replacements either unchanged or increased if higher heat flux leads to reduced lifetime, e.g. due to increased erosion rate. Higher heat flux would allow for machine size reduction and thus reduced masses of radioctive waste material.
* Plant availability: Failure rate and mean plant down times unchanged.
* Installation costs: If machine size reduced then correspondingly lower costs which would enter the total direct costs with a factor of 0.5 (see Sec. 5.1; [11]).
* Operating costs: Reduced costs for smaller size spare parts and for waste material handling due to reduced mass of material.

(2) The actual load remains constant so that thermal stresses are considerably below the design curve. Other lifetime influencing effects like erosion are again assumed unchanged. The effects are:

* Plant safety: Reduced loads lead to lower failure rates (see plant availability) and accordingly lower accidents occurrence rates; consequences of a failure remain unchanged.
* Waste material: Occurrence rate of replacements only reduced if thermal cycling is lifetime limiting. Amount of waste material per replacement unchanged.
* Plant availability: Reduced loads have the effect of derated operation which is known to be extremely beneficial for reducing the failure rate. If thermal cycling were lifetime limiting, thus determining the re placement intervals, the reduced loads mean less often down times for replacement.
* Installation costs: According to unchanged size no influence.
* Operating costs: Less number of replacements reduce the financial effort for spare parts and waste material handling.

The above consideration is valid for all mechanical components which see a pulsing thermal load in the reactor: divertor, first wall, launchers and windows of heating systems. However, the assessment strongly suffer from the lack of information on the in-vessel components in a mid-next-century fusion power plant so that it is hardly more than an illustration of the generally known fact that reducing the number of thermal cycles on mechanical components in general is beneficial. The replacement planned also with respect to lifetime limitations not due to the pulsing anyway reduces the importance of the difference between pulsed and non-pulsed operation. However, if at all a conclusion were to be drawn (2) seems to be preferable because of the possibility of reduced failure rates (if thermal cycling determines the lifetime).

6.2.2 Semi-permanent Components

Components of this type must be designed for a lifetime which is longer than the service life of the plant. A characteristic component is the PF magnets system which according to Tab. 3 contributes about 13% to the overall plant outage risk. Within this system the lifetime of the central solenoid will probably be determined by the fatigue due to pulsing. Following the NET/ITER design principle it will be vertically prestressed in order not to loose its integrity when the forces go from peak positive to peak negative load for more than 10^5 cycles [15]. For the already above used reference pulsed reactor case [8] it is assumed that the same design solution will allow the cyclic strain range actually be borne by more than the envisaged 10^5 burn cycles. The fact that the major radius in that case is by about 50% larger than that for NET/ITER allows to assume that the stresses in the central solenoid are lower thus ensuring the fatigue lifetime to be sufficiently larger than the plant service life.

As in the previous section the reduction in number of cycles could be used in two different ways: Either the stresses and the strain range could be increased which allows to reduce the machine size or the size could be kept with then lower loads thus leading to a higher reliability. In the sense of Fig. 1 the (real) crack growth curve (at bottom) would remain either the same or would be lower thus reducing the failure probability.

For the consideration here it is assumed that the reduction in number of cycles is used to permit higher mechanical loads keeping the failure probability unchanged. This will not affect the criteria of plant safety, waste material, plant availability and operating costs. Only the installation costs of the reactor could be lower as the reduction of the installed volt-seconds at unchanged forces allows for a smaller solenoid diameter and consequently a smaller machine size. Previous studies [16] have shown that typically the installation cost reduction attainable by reducing the central solenoid is about 10% and is limited by the TF coil stresses from reducing further. A better quantification would require detailed parametric studies for the overall plant taking into account also the plant and reactor energy balance changes which are incurred by the transition to the NI-case.

6.3 Reliability of Subsystems

The subsystems mainly affected by the transition from the I- to the NI-case are the heating and current drive systems, the heat transport system and the electric power supply which altogether contribute nearly 30% of the overall plant outage risk. Assessing the influence of the transition results in the following:

--- Heating and current drive systems (RF system outage risk contribution about 8%, NBI system about 13%). Due to the early stage of the design of these systems the reliability analyses at present are not sufficiently detailed to allow an assessment of the influence on the various components. In general the reduction of the number of demands on the components (e.g. high voltage switching) and the transients in any other component may decrease the systems failure rate. However, for the critical components in the pulsed case sufficient redundancy was assumed to be installed so that the step towards continuous operation just means a relaxation of these redundancy requirements. So, the transition to the NI-case may result in a reduction of either cost or failure rate. As the share of heating devices in total direct costs is only 3% whereas the contribution to outage risk is about 20% the reduction of failure rate is chosen as the more effective benefit.

--- Heat transport system (outage risk contribution about 4%). Omitting the thermal energy storage reduces the complexity of the primary cooling cycles where, for instance, the control valves of the storage system may have required specific effort to achieve an acceptably low failure rate. Hence, in view of the small outage risk contribution it would not be adequate to assume a reliability increase.

--- Electric power supply (outage risk contribution about 2%). Reducing the number of demands on the supply system components may allow for a reduction of redundancy or of scheduled replacements. As also the direct cost share is only about 2% it is not worth to consider the effects on either reliability or cost although in principle the transition to continuous operation could be beneficial.

In summary, of the various criteria only plant safety and plant availability are affected by means of reduced failure rates. This might be of some importance only for the safety of the heat transport system (occurrence of coolant related accidents) and availability of the heating and current drive systems.

7. CONCLUSIONS

The consideration of the influence of the step from inductive to non-inductive current drive for a mid-next-century fusion power plant has shown that for most of the components at maximum only a slight benefit can be expected. For appraising the desirability of that transition four items are seen to be important:

(1) Fatigue damage in the central solenoid system: The reduction of this may allow to reduce the machine size considerably this being accompanied by a decrease of installation costs (maybe in the order of 10%). Alternatively an increase of reliability is possible thus reducing the plant outage risk.

(2) Fatigue damage on components with planned replacements: Its reduction may lead to an increase of reliability with consequently improved safety and to less often components replacements thus reducing the amount of radioactive waste material (valid only if thermal cycling actually influences the time between two replacements) and improving plant availability.

(3) Plasma energy balance: The increased load on the divertor plates may only partly be balanced by increasing the machine size which in turn leads to higher installation costs. The question of sputtering (higher particle energy) remains to be solved to allow non-inductive current drive.

(4) Plant energy balance: The increase of the auxiliary energy demand reduces the overall plant efficiency (by about 15% of its reference value) down to a value which scarcely can be accepted especially in view of the corresponding increase of waste heat. This decrease in efficiency causes an increase of electricity generating costs of about the same percentage.

In summary: As far as electricity generating costs are concerned the reducing effect of (1) may just be balanced by that of (4) so that in total (3) may cause an overall increase. Alternatively the effect of (1) could be turned into an availability increase at unchanged costs. With respect to the impact on environment somewhat higher reliability (availability and safety) and reduced radioactive waste (2) is combined with an increase of exhaust heat (4). Hence, in total improvements in availability and impact on environment are accompanied by an increase in thermal pollution and in costs.

As is obvious from the previous sections that this result is based on various assessments performed practically without quantification. To quantify and to take into account all considered effects together would require (a) to determine a reference power plant layout for both current drive schemes by means of parametric studies and (b) to determine the availability by means of simulating the plant operation this, however, requiring reliability analyses of the various components involved to an degree of detail which actually allows the distinction between the two current drive methods. Although for both tasks the models are at hand additional basic information is required, for instance, for the heating and current drive systems. However, given the various counter-directed effects it remains doubtful whether a more detailed analysis would result in an answer which clearly favours one method.

ACKNOWLEDGEMENT

The author gratefully acknowledges the very helpful comments and discussions by F. Engelmann, M. Harrison, J. Raeder, W. Spears, R. Toschi, J.-G. Wegrowe and E. Zolti.

REFERENCES

[1]   Garnreiter, F.: Zur internationalen Wettbewerbsfaehigkeit energieintensiver Industriezweige in der Bundesrepublik Deutschland (On the international competitiveness of energy-intensive industries in the Federal Republic of Germany). Brennst.-Waerme-Kraft 34 (1982) 297

[2]   Lezuo, A., K. Riedle, E. Wittchow: Entwicklungstendenzen steinkohlebefeuerter Kraftwerke (Tendences in the development of hardcoal-fired power plants). Brennst.-Waerme-Kraft 41 (1989) 13

[3]   Ewers, J., H. Mertikat, W. Guenster, J. Keller: Gas-Dampfturbinenkraftwerk mit integrierter Braunkohlevergasung nach dem HTW-Verfahren (Gas-Steamturbine power plant with integrated lignite gasification using the HTW process). Brennst.-Waerme-Kraft 41 (1989) 23

[4]   Harris, G. H. and W. R. Spears: Requirements for Steady-State Operation of Tokamak Power Reactors. NET Report EUR-FU/80/89-93, 1989

[5]   Jones, P. M. S. (Chairman of Expert Group): The Costs of Generating Electricity in Nuclear and Coal-fired Power Stations. OECD/NEA Report (1983)

[6]   Moynet, G. (Chairman of Expert Group): Electricity Generation Costs Assessments made in 1984 for Stations to be commissioned in 1995. UNIPEDE Study Report 1985 (available from OECD/NEA)

[7]   Buende, R.: Reliability and Availability Issues in NET. Fusion Engineering and Design 11 (1989) 139

[8]   Spears, W. R.: Reactors beyond NET. Proc. 4th IAEA Technical Committee Meeting and Workshop on Fusion Reactor Design and Technology, Yalta, 1986

[9]   Spears, W. R. (NET): personal communication

[10]   Wegrowe, J.-G. (NET): personal communication

[11]   Spears, W. R.: Reactor Cost Driving Items, Proc. 14th Symp. on Fusion Technology, Avignon, 1986

[12]   Harrison, M. (NET): personal communication

[13]   Chazalon, M. et al.: Next European Torus In-Vessel Components. Fusion Technology 14 (1988) 82

[14]   Debray, J. (Novatome): personal communication

[15]   Mitchell, N.: Operational Flexibility and Fatigue Life of the Central Solenoid. ITER-IL-MG-4-9-1 (1989)

[16]   Spears, W. R.: DEMO & FCTR Parameters. NET Report EUR-FU/XII-361/85/41, 1985

VI. PANEL DISCUSSIONS AND CONCLUSIONS

INTRODUCTION

The Seminar was concluded by thorough discussions organized within the structures of the three panels. Prior to each panel, all the participants were given the opportunity to present short, ten-minute contributions to the discussions.

This last chapter contains four short contributions that were submitted as manuscripts by the authors. In addition and under the sole responsibility of the editors, a selection of the most significant discussion points made at the three panels, based on tape recordings and notes by the Chairmen and panel members, is also presented. The guidelines for the selection were mainly conciseness and novelty of opinions and concepts (those not already presented in the papers). Although a certain subjectiveness cannot be excluded, the editors made an effort to render the spirit of the overall opinions.

SHORT CONTRIBUTIONS TO THE PANEL DISCUSSIONS

G.J. Butterworth, E.C. Brolin, P. Rocco, M. Snykers

MATERIALS SELECTION FOR FUSION

G.J. Butterworth

Euratom/UKAEA Fusion Association, Culham Laboratory

Abingdon, Oxon, OX14 3DB, UK

The safety and environmental implications of reactor materials are mainly related to their neutron-activation and their propensity for tritium permeation and release. The opportunity offered by materials selection and development, and especially the use of low activation materials, for optimising the safety and environmental advantages of fusion energy has been alluded to many times. The realisation of these benefits is seen as an essential component of the materials development programme, though further studies are needed to establish specific radiological objectives and quantify the potential advantages.

For some reactor components the engineering property requirements will allow little scope for materials selection. For instance, the use of copper may be mandated by requirements on thermal or electrical conductivity or the use of tungsten because of its refractory characteristics. In other cases, such as the first wall-blanket structure which, although of relatively small mass, represents the bulk of the induced radioactivity inventory, substantial reductions in medium and long term activity could be achieved through the use of low activation materials. Whilst components further from the plasma may be less highly activated or tritiated and have longer service lives, it is worth noting that the quantities of material in these components may be very much greater.

The advantages to be gained through materials selection involve a wide range of decay times and material properties and are, moreover, dependent on reactor design. Pending the outcome of more detailed study, a few general remarks must suffice. Of all the potential radiation exposure scenarios, those most likely to attract public attention are releases of radioactive products in major reactor accidents and radioactive waste disposal. In the first event, radionuclides having relatively short half lives (minutes-years) are likely to dominate whilst in the second case relevant half lives lie in the range decades to millenia. The materials employed for components close to the plasma are likely to be particularly relevant to accidental releases and the design criteria for low activation compositions for these components will need to take into account, e.g., through probabilistic risk analysis, the potential consequences of such low-probability events so as to maintain the right balance between short- and long-term radiological requirements, where these are in competition, and consistency with the engineering property specifications.

In relation to waste management, low activation materials have been reviewed as a means of widening the range of available management options. The US fusion programme, for example, has adopted shallow land disposal as its objective for low activation materials development. It is, however, unclear whether this objective is sustainable on either economic or radiological grounds. In order to meet the present requirements based on the 10CFR61 rule, first wall material for a power reactor would need to satisfy quite stringent compositional limitations which would incur a considerable increase in cost relative to conventional materials. More-over, international trends in radiation protection regulation will almost certainly result in a future tightening of the current rules. Overall, it is likely that geological disposal would not prove significantly more expensive and would be much safer radiologically than shallow land burial. In Europe there is a trend toward geological disposal of essentially all radioactive decommissioning wastes and, given the very large safety margins offered by this method of disposal, there appears to be no signi-ficant advantage for low activation materials with respect to permanent disposal. Thus disposal is not a driving force for low activation ma-terials design. So far as public perception is concerned, the technical details of radioactive waste, e.g., the degree of activity and the way it is disposed of, may be of less interest than the quantities of waste. One of the ways in which the amounts of waste might be significantly reduced is through the reuse or recycling of used material. With this in mind, a primary aim of low activation materials design should be the development of materials that would permit reuse or recycling whenever technically feasible and economically justified. Furthemore, the limitation of dose rates to levels that would allow recycling to be undertaken after a suitable storage period could be expected to confer radiological benefits in other operational areas involving the handling of activated materials.

LESSONS FROM FISSION

A PERSONAL PERSPECTIVE

E.C. Brolin

PPPL, Plasma Physics Laboratory, Princeton University
Princeton, N.J. 08543, USA

My observations on the future of fusion reactors are based on twenty-five years of experience with military and civilian applications of fission reactors in the United States. From that perspective I believe that the United States, and probably the majority of the industrialized world, must convince the public that fusion reactors are safe and environmentally superior to all other viable alternatives if we are to be allowed to construct and operate them.

Public acceptance will be dominated by the following considerations:

° Safety, in particular whether or not there is any possibility of catastrophic accident
° Environmental impact, with primary interest in the amount of radioactive pollution as compared to alternative fission power sources
° Economics
° Fusion community credibility. We should not promise more than we can deliver.

We must listen carefully to the public if we are to maximize the probability of public acceptance. An irrational fear of radiation dominates public actions. This fear is most apparent in public reaction to the possibility of catastrophic accidents in nuclear power plants and in the more general objection to radioactive waste disposal.

We must also listen to our customer. In the United States that customer is most likely to be a private utility. These organizations, even more than government utilities which are more prevalent in Europe and elsewhere, must minimize financial risk. Financial risk includes:

° Construction cost
° Construction schedule -- directly related to cost
° Operating and maintenance cost

Finally, government regulations can greatly affect both the cost and feasibility of fusion power plants. Fusion power has advantages over fission power. The fusion community should work now with national and international regulatory agencies to begin to develop regulations governing fusion which are technically consistent with the reduced risk from fusion power plants as compared to fission. As part of this activity we should attempt to avoid the adversarial relationship with regulatory bodies which is typical of fission power in the United States.

DOSE LIMITS FOR FUSION REACTORS

P. Rocco

Commission of the European Communities
JRC, Joint Research Center
Ispra, Italy

RADIOLOGICAL PROTECTION DATA

In connection with the discussions at the Seminar, it seemed useful to briefly recall some concepts and definitions of relevance to the safety and environmental impact of fusion reactors.

The following definitions derive from the Recommendations of the International Commission on Radiological Protection (ICRP) and from the dose calculations practice.

Dose equivalent H

$$H = DQN$$

H = dose equivalent. The unit of dose equivalent is the sievert (Sv):
1 Sv = 1 J/kg = 100 rem
D = absorbed dose
Q = quality factor which takes into account how the absorbed energy is distributed in the human body
N = a factor which takes into account the dose rate and fractionation.

The expression indicates that the assessment of the severity and probability of deleterious effects due to an absorbed dose requires both the irradiation conditions and the characteristics of the radiation to be specified.

The factor N is usually taken to be 1.

The factor Q is taken to be a function of the collision stopping power L_{00} (keV m^{-1}) of the radiation in water. The value of Q ranges from 1 to 20 for stopping powers from 3.5 to 175. In a first approximation, x rays, γ rays, and electrons have a Q value of 1; for neutrons and protons Q = 10; α particles have Q = 20.

The various "types" of doses, which are defined to assess radiological effects, take into account the quality factors, so that they are dose equivalent, even if not explicitly indicated. Some of them are given hereunder.

Whole body dose

The dose computed as if the whole body were irradiated uniformly.

Total body dose

The mass-weighted average of the dose to each body organ.

Effective dose (equivalent)

Weighted average of the doses to certain organs. The weighting factors represent the proportion of the stochastic risk to the total risk when the whole body is irradiated uniformly.

The ICRC recommends values of the weighting factors ranging from 0.25 (gonads) to 0.06 (bone surfaces).

Prompt dose

Used to quantify acute health effects from accidental releases. It is defined as the committed dose to a particular organ by the cloudshine during plume passage and by seven days of groundshine and the dose commitment over the critical time from inhalation during plume passage.

The critical time, which is organ specific, is the time during which a dose must be delivered to produce evident acute effects on the organ. For bone marrow, which is the most important organ for acute effects, this time is seven days.

Early dose

The 50 years' dose commitment due to early exposure (seven days). It includes irradiation (cloudshine and groundshine) and inhalation.

Chronic dose

The dose commitment due to irradiation and inhalation for a continuous (50 years) residence at the site.

Ingestion dose

The dose commitment due to the eating of local food and drinking water for a continuous (50 years) residence at the site.

Long term dose

The long term dose, which is composed of early dose + chronic dose + ingestion dose, is used to assess effects which are not immediately evident.

Collective dose

This is used to assess the detriment to an exposed population. The collective dose is the sum of the doses committed to each member of an exposed population.

Mitigating actions

Possible actions to mitigate the projected dose following an accident are:

- Evacuation: the movement of people away from the release point.
- Sheltering: the people remain indoors, shutting the ventilation systems.

The previous actions have the purpose of reducing short term effects (prompt dose, early dose). The long term effects can be mitigated/avoided by:

- Relocation: movement of population outside an interdicted area for a long period of time;

- Food interdiction: disposal of the food which is contaminated beyond
  regulatory levels.

SAFETY DESIGN OBJECTIVES FOR FUSION PLANTS

In the present situation the public will accept new energy sources
only if it can be reasonably confident that the related health and social
effects are minimal. The main safety items to be considered are:

- The maximum effects produced by "possible" accidents in nuclear
  plants compared with those of conventional plants. Generally,
  the concept of "possible" does not take into account the event's
  probability and is supported with statements on the likelihood of
  "unexpected events" and "human errors".
- The public is not prepared to compare the hazards of nuclear and
  conventional energy plants including all steps of the fuel cycle
  (mining, coal and oil shipments).
- The impact of nuclear waste. It can also be stated that the
  problems of the waste arising from conventional energy plants
  are not perceived.

The safety design objectives of fusion plants should take into
account the statements indicated above.

The long term doses to the public in extreme conditions (very low
probability events: $P < 10^-$ per year) must be "sufficiently low" without
mitigating actions which have disruptive effects on the public.

These dose limits can be set to be near the actual dose threshold
values (adopted in some states) to trigger evacuation, which are in the
range of 50 - 100 mSv (5 - 10 rem).

On this basis:

- The CIT design objectives, namely, 50 mSv for low probability events,
  are a sound limit, provided that this "maximum public dose" is a long
  term dose achieved without intervention on the population.

In particular:

- Evacuation or relocation of population should be avoided
- Interdiction of local food should be avoided or reduced to a minimum
- The ITER design objectives for accident conditions, namely, 100 mSv of
  "early dose" can be accepted, provided that this value is not significantly increased by the chronic and ingestion dose. A 50% increase,
  obtained without intervention on the population, is probably acceptable.

The design objectives described above could possibly be achieved with
"passive safety features" in the fusion reactors. However, beside
thorough safety and environmental analyses, they require:

a) Reduction of the neutron-induced radioactivity (development of low
   activation materials);
b) reduction and subdivision of the tritium inventory (progress in the
   tritium process systems);
c) reduction of the releasable radioactive amounts in extreme conditions
   (limitation of pressure/temperature transients).

It has to be noted that a) should include the reduction of the long
lived radionuclides (minimisation of the waste environmental impact).

PUBLIC ACCEPTANCE OF NUCLEAR FUSION

M. Snykers

Nuclear Research Centre, SCK/CEN

2400 Mol, Belgium

Lessons to learn from fission

The production of electricity by nuclear power is strongly hampered by the negative reactions from the public towards fission.

Improving public acceptance of nuclear power has become an important issue in the main nuclear power countries [1-4]. It is considered as an action with a long-term effect. It aims at "building up confidence in nuclear power" by means of :

- supply of information;
- gaining credibility in the nuclear industry.

Information has to be given in a language, understandable to the public, by third party specialists, able to gain credit from the public. The largest part of the public (about 90%) has a fragile, very low-profile level of knowledge on fission. Those persons have to be addressed before they will take a position based on incomplete information. This needs a coordinated action, with sufficient financial means, using all conventional mass media systems. Information has to cover facts, background knowledge, given in an objective way, and also the benefits for the public from nuclear power.

The nuclear industry has to become synonym of reliability. Acceptance of and confidence in the products and the people has to be established. This may lead to trust and credibility in the nuclear industry. Nuclear industry should get a human face. This task of building up confidence has to be realized by all firms involved in all industrial activities from the start (ore mining) over electricity production to the end (waste treatment) of nuclear power.

2. Perception of risk

The perception of risk is, generally speaking, strongly dependent on the personality of the individual. However, investigations have shown that there are several common factors [5].

The perception of risk, or the level of risk acceptance is strongly influenced by the involvement in the decision making process. We are generally more easily inclined to accept the risk if we have been part of the decision making process. You may decide for yourself to smoke sigarets and by doing this to increase your risk to health problems.It will be much more difficult to accept that a dangerous plant will be built in your neighbourhood, without your advice or without you seeing the benefits from this.

The perception of risk increases with increasing level of technology. Since the level of technology of our society is steadily increasing, this will, as a whole increase the level of the associated perception of risk.

The perception of the risk associated with low probability events involving a large number of casualties is higher than for high probability events with the same total number of casualties. This is clearly illustrated by the attitude of the major part of the population with respect to travelling by air or by car.

These observations show that it is possible to influence the perception of risk. It may be diminished by :

- information;
- involvement in the decision making process.

Adequate information in an understandable language, using modern mass-media systems, has to be supplied to the public. This will enable people to form a better documented opinion.
The suitable techniques have to be applied to get people involved in the decision making processes of actions which may influence considerably their quality of life.

3. Policy for fusion

In the process of research on the energy production by nuclear fusion we must take benefit from the lessons learned from fission.
Realistic information has to be supplied to the public on the future prospects of nuclear fusion. At present fusion is to the largest part of the public and of the politicians known as a clean and inexhaustable potential energy source for the middle of the 21st century.
If we want fusion to be associated with confidence, correct information has to be supplied at any time through the appropriate channels. The strong points of fusion have to be emphasized, but the weaknesses may not be forgotten.

It is too early to start building up confidence through the industries involved. Fusion is still in the stage of research and development. In the present stage confidence must be built up through on-going and planned experiments. Incidents involving important spills of radioactive materials have to be avoided. Attention has to be paid to the production of radioactive waste. This should be minimized, also for the pre-commercial reactor devices.

Building up confidence may be compared with the filling of a basket with small grains of truth. It is very hard to reach a high level of confidence, it is very easy to loose it.

REFERENCES

[1] Media and public relations of nuclear energy : is it worth the hassle?
     Carl M. Goldstein. Proceedings of the Annual Conference of the Cana-
     dian Nuclear Society / Canadian Nuclear Association, Ottawa, June
     1989.

[2] The situation of anti-nuclear movements and the concept for develop-
     ment of public acceptance strategy in Japan.
     M. Sasaki. Proceedings of the Annual Conference of the Canadian
     Nuclear Society / Canadian Nuclear Association, Ottawa, June 1989.

[3] Guided tours - The medium and the message.
     Pierre Haller. Proceedings of the Annual Conference of the Canadian
     Nuclear Society / Canadian Nuclear Association, Ottawa, June 1989.

[4] Is advertising an effective tool to reach the public on nuclear
     matters?
     Rita Dionne-Marsolais. Proceedings of the Annual Conference of the
     Canadian Nuclear Society / Canadian Nuclear Association, Ottawa, June
     1989.

[5] The psychology of preferences.
     D. Kahneman and A. Tversky, Scientific American, 160-173, vol 246
     (1982).

PANEL DISCUSSION ON SAFETY AND ENVIRONMENTAL IMPACT OF FUSION REACTORS

K. Tomabechi (Chairman), J.G. Crocker, J. Darvas, Y. Fujii-e,
J. Raeder, P. Zettwoog

1. **The Chairman introduced the discussion by addressing some main questions:**

   * How is so-called "inherent safety" or "passive safety" to be considered in fusion?
   * What are the ecological advantages of fusion?
   * How is fusion waste to be managed?
   * Which potential advantages of the safety and environmental aspects of fusion can be implemented in a next-step machine?
   * Are all the advantages of fusion, as claimed so far, real?

2. **An extensive consideration of inherent/passive safety versus engineering safety features** was presented by Y. Fujii-e, Tokyo

   These features are intended to prevent, protect or mitigate abnormal energy releases so as to assure boundary integrity.

   * Safety research is directed in principle to:

     ° finding out inherent (ISF) or passive (PSF) safety features (without ISF/PSF no device can be operated);
     ° utilizing ISF/PSF and integrating them into system designs;
     ° evaluating the design (safety) margin.

   * Characteristics of inherent (ISF) or engineering (ESF) safety features:

     ° are highly design-dependent (material selection, geometrical configuration);
     ° need simple safety principles (philosophy);
     ° are not almighty (fire, missile, pipe integrity).

   * Definition of inherent/passive safety features:

     ° Safety features without external power source (intrinsic, design independent, inherent, passive)

   * Examples for light water reactors (objective: reactor inventory containment):

     ° ISF/PSF: Doppler broadening, void formation; natural convention.
     ° ESF: Scram; forced convection; containment structure.

3. <u>Some personal considerations on the role of the ecological factor in the year 2050</u>, were given by P. Zettwoog (Fontenay-aux-Roses).

* The fusion community has always dealt with two concerns:

   ° in the short term, to find the money for the programs over the next 5 to 10 years;
   ° in the long term, to find the best way to convert the means received into a technology which should, hopefully, be able to compete in 2050 with other energy producing technologies available at that time.

* The influence of ecological factors on decision making in the energy sector is shifting from the present emotional height to levels determined by scientifically established data and facts; in the future (after 2050) the levels might depend on the capacity to master the climate and the ecosystems.

* From such general ecological considerations, it may be concluded that in 2050:

   ° The ecological factors will have introduced a major difference between nuclear energy and fossil fuel energy, in favour of nuclear energy.
   ° All competing energy sources based on nuclear energy, whatever the technology, will be in the same basket.
   ° This means that the important factor in the competition will be the price of the electricity produced .
   ° The radiological constraints inside a fusion facility may represent a substantial part of the running costs and influence repair and maintenance delays (loading factor). A good policy in radioprotection could be very cost effective and should be anticipated in the design phase.

4. <u>An extensive discussion took place during the Panel session.</u> Some of the points discussed are given in the summary presented by the Chairman, K. Tomabechi:

* Care should be taken in using the word "passive safety" in fusion because it may lead to misunderstanding. What is required is the need to develop a philosophy of passive safety as well as a clear definition of passive components and systems so that fusion reactor design can benefit from incorporating passive safety into the design.

* The ecological advantages of fusion over fossil as well as fission energies should be delineated with careful analyses of the details. Fusion has certainly advantages.

* The management of fission waste is experiencing difficulty at present in setting out strategies which can be accepted with the consensus of the public. Fusion seems to have advantages as regards the quality of the waste because the isotopes to be dealt with would become virtually known if the design of the reactor is determined, although quantities in terms of curies per unit volume are similar to those of fission.
   Careful analyses of the nature of fusion waste in comparison with fission waste should be made in order to be able to argue more definitively.

*   The development of lower activation should be encouraged to ease the problems of both accident handling and waste management of fusion reactors in the future.
    Next-step devices, such as ITER, will provide a good opportunity to demonstrate the usefulness of such an approach and accelerate the development of materials. The next step projects should pay proper attention to these points.

*   The fusion community should be careful not to overstate the advantages in the safety and environmental aspects of fusion, which would mislead the public.
    However, with regard to accidents and waste management, fusion has, perhaps, some advantages in comparison with fission.

*   An effort should be made to benefit from the present freedom in fusion in the selection of a design philosophy, and the components, systems and materials to be used to design a safer and environmentally more benign reactor.

PANEL DISCUSSION ON ECONOMIC PROSPECTS OF FUSION REACTORS

R.W. Conn (Chairman), R. Buende, J. Darvas, L. Gouni, W.R. Spears

1. The Chairman introduced the Panel session by presenting and commenting a set of questions which defined the extent of the discussion:

    * Concept of cost of fusion and use of cost estimates for fusion

        ° Why make cost estimates at all?
        ° Why make cost estimates now?
        ° Are there and should there be targets?
        ° Costs for economists (no inflation) vs financiers (role of inflation)?

    * What methodology should be used in estimating and comparing costs?

        ° For example, compared to coal, to PWR, etc.?
        ° Role of learning curves?

    * How should cost estimates be presented?

        ° One methodology?
        ° To governments? To utilities?
        ° How do we deal with uncertainties?

    * What is the role of taxes or regulation in areas of safety & environment?

    * How to "market" fusion today?

        ° Extrapolation of the near-term only?
        ° Use of longer term extrapolations only?
        ° Use of a combination of near and long term extrapolations?
        ° Fusion as an "Energy insurance policy"
        ° Role, if any, of military/defense applications?

2. We shall retain here only a few qualifying points of the general lengthy discussion, which is difficult to render appropriately because of a certain vagueness of the subject.

    * Is it meaningful to make and then defend cost estimates when there is no fusion cost data base, the physics is still uncertain, and the

device exploitation is at least 50 years away? Although I have been doing this for five years, I now think that it is better to just go ahead in developing the reactor. We shall be able to estimate the cost better when we have at least built the Next Step. (W.R. Spears)

* None of us would like necessarily to be held accountable for cost estimates now. However, the authorities have the right to ask what level of resources is to be spent on a particular project, as compared to other competing projects. We should, therefore, give some indications or whether ultimately fusion will be cost-competitive and will be acceptable by the public with regard to its safety and environmental characteristics. (various speakers)

* There are some internal reasons, as well, for doing costing of fusion devices, namely, you would like to know in which direction to spend the R&D effort in order to maximize the efficiency of your work. In this respect there can be guidance by costing studies as to where the sensitive parameters or components are. (M.S. Kazimi)

* Why not compare the cost of the fusion reactor with the cost of the fast breeder, which will probably be its competitor?.
It seems more serious to compare fusion costs with energy sources that are well understood and with large experience (such as coal, or PWR) and then apply learning cost studies to put the fusion cost into the right perspective (such studies have been made for the fast breeders in the UK and France). (various speakers)

* I can't imagine sitting in a room with utility executives and having them believe the costs that have been presented.
The kind of cost estimates that we are asked for at Princeton are not what the costs of the fusion reactor in the year 2050 would be, but what the cost of CIT is going to be by the year 1998. (E.C. Brolin)

* The approach of regulation and of taxation to control the safety and environmental features (and to affect the costs) of a particular industry is widely used (for example in the Los Angeles area which is the largest manifacturing area in the US). But we may not be able to readily quantify, in this respect, the economic benefits of fusion energy. (R.W. Conn)

* In a few years' time, hopefully, we have to cost NET or ITER, which after all are very close to a reactor. So it is inescapable that we will then have a basis, with the potential of improvement, to estimate the cost of almost a prototype reactor. (R. Toschi)

3. <u>Cost considerations on fusion energy made in Japan</u> were presented by K. Tomabechi

Some years ago, committees were set up in Japan to look at the safety, environmental impact and economics of fusion energy. These studies are still valid and I will present, in a succinct form, some of the major points of the economics study which led to the Japanese view of the feasibility of fusion as a viable energy source.

* Energy balance of fusion reactors

  ° Required energy for fusion reactors (compared with total electric-

ity produced during life time):   2 ∿ 3%
° Required energy for LWR: 4 ∿ 5%

* Available resources of special materials needed for fusion

Available Resources (t)   Required resources for fusion
                                 reactors of 3 x 10^3 GWe (t)

He 4 x 10^6 4 x 10^4
Li 8 x 10^6 5 x 10^5
Be 1 x 10^6 3 x 10^5

* Example of cost breakdown of a fusion reactor, 3200 MWt, 1000 MWe.

° Overall plant cost

Reactor plant   47% – of which: SC coils   41%
Site facilities   8%   Blanket (30% ^{6}Li) 22%
Turbine and electric   7%   Support structure   9%
Others   4%   Cryogenics   12%
Eng. facil. for construction 19%   Heating   7%
Interest during construction 15%   Others   9%

° Cost of electricity generation

Capital   87%
Operating   9%
Scheduled component replacement   4%

4. Further points raised and discussed during this panel are mentioned
    in the concluding remarks made by the Chairman:

* There is an expanding need for base-load electricity production, more
  so in the developing countries than in the developed countries. In
  Western Europe one can envisage the need to instal 5 GW (E) or more
  per year of base load plants.
* The cost of base load electricity is likely determined by the cost of
  (imported) coal and of PWR electricity, and it is prudent to assume
  that in the mid 21st century fusion must be cost competitive with
  these. The possibility of a 50% increase of coal was also considered.
* The cost credits which might be allowed for the indigenous supply and
  for environmental benefits are not likely to be large.
* The costs of generating electricity from envisaged fusion will be
  dominated by the capital component. At the present stage cost estimates are bound to be very uncertain and nearly impossible.
* Cost estimates from single-equation models (e.g., proportionality to
  volume) can be very misleading, whereas the detailed design data and
  man-power needed in manufacturing and assembly are not available.
* Present methods use the multi-equation scaling of the different
  components. An element of assurance is provided by applying this rule
  to fusion plants or to existing experimental apparatus.
* The SCAN code used in Europe indicates that fusion costs might be in
  the range of 1-3 times the best PWR experience, whilst the code
  published in the ESECOM-report has produced a range of costs which
  overlaps with the range of PWR costs.
* Both analyses reported that fusion has the potential to be cost
  competitive by the D-T tokamak route.

CONCLUDING PANEL

R.S. Pease (Chairman), R.W. Conn, V. Demchenko, Ch. Maisonnier, K. Tomabechi, R. Toschi

1. This Seminar covered nearly all the technical areas relevant to the environmental, safety and economic aspects of envisaged fusion power and had had input from all the main fusion programmes. There had been a natural tendency to concentrate on the more difficult and controversial topics, whilst other topics such as the magnitude of the fusion energy resource, the absence of chemical emission from criticality accidents in envisaged fusion power reactors had been largely omitted. The question of unwanted connections with military applications and safeguards, a strong motivation issue for some in connection with the use of fission power, had been touched upon only in connection with inertial confinement fusion. The symposium discussion contained a number of suggestions for further research and action.

2. A summary of the presentations made at the Seminar was provided by the Chairman and structured along the following main sections corresponding to the program.

* Section A: Reviews

General review of studies from reports of Euratom, USA and Japanese programmes; the following main points emerged:

° Fusion research programmes are at a level where potential benefits have to be worked out in some detail; simple generalities are not sufficient.
° Environmental benefits and foreseen economics are both required; examples are the thorough ESECOM studies in the USA and the ongoing EEF studies in Europe.
° The potential environmental benefits are seen as particularly important in Japan and the Soviet Union.
° The realisation of these benefits is foreseen for the mid 21st century.
° Such a large time scale makes for considerable difficulties in studying the right balance between caution and imagination.

* Section B: Feasibility

Magnetic fusion based on the D-T thermonuclear reaction and the

tokamak system is feasible on the envisaged time scale. This conclusion needs to be qualified in two respects, namely, the recent
important advances in experiments on the large tokamak still leave
unresolved the question of the physics that underlines the
extrapolations to the currents dimensions and alfa particle heating
powers of envisaged reactors. A clear demonstration of net fusion
power production is needed.

Studies of other magnetic fusion systems provide important supporting
physics data, but in themselves are likely to require a longer
time scale to develop into reactor systems.

Likewise, inertial confinement systems, it seems, require the
construction and operation of large 10 MJ laser systems to
demonstrate the feasibility of net power output and, therefore, may
also need a longer time scale.

The papers presented at the symposium indicate that muon and
catalysed fusion, the use of D-D thermonuclear reaction and the
provision of a source of ^{3}He for large-scale use cannot be reasonably
foreseen at present.

*   Section C: Environmental and Safety Issues

Any practical power station has to be safe and has to be seen as such
by the public. For example, a clear goal is that there should never
be need for evacuation of residential areas. In the D-T reactors now
envisaged, the tritium, which has to be used on the scale of 1 kg per
day, is a component that requires special attention because it is
volatile both in elemental and oxide forms and its radioactivity
makes it toxic when ingested in comparatively small quantities. Its
release in accidents must be limited to less than 200 g, and losses
arising from routine operations should not, it appears, be more than
50 curie/day.

The potential for accidents is roughly indicated by the potential
energy that could be released by fire. Pure liquid lithium, whose
combustion energy is of the order of 10^{12} joules, provides the least
satisfactory coolant and breeding material in this respect. Its use
is therefore avoided altogether in many designs. Another component
needing control assurance is the neutron activated blanket and the
structural materials in D-T reactors. The magnitude of the hazard
potential can be strongly reduced by the use of low activation materials. Without these, the radioactive inventory at shortly after shut
down can be of the order of 10^9 curies. Volatilisation of these
components needs to be avoided. Considerations of loss of flow and
loss of coolant accidents coupled with estimates of the afterheat
indicate that this can be avoided.

*   Section D: Proto reactors

ITER/NET will have to meet the design goal of a dose of < 10 rem at
the 1 km boundary for any accident. The assurance requires detailed
design consideration and the component tests to be completed. The
influence of steady state versus pulsed operation is not an
influential factor in these safety considerations. During discussions
on the use of low activation materials, the following specific
suggestion were made:

°   It is very important to establish the reality of the low-acti-
    vation-materials option;
°   The ITER collaboration should be extended to include international
    collaboration devoted to research on safety and low activation
    materials;
°   Specific provision should be made for ITER test modules or reactor
    components made of low activation materials, for integrated

engineering and lifetime tests in ITER.
° The benefits arising from low afterheat, from low long-lived activity, and from recycling of materials need to the assessed together;
° Cost benefit analyses of high purity and of the use of isotope separated materials need to be extended.

* Section G: Conclusions of the Panels on safety and costs

° The overall safety and environmental issues of fusion were dealt with in the special Panel led by K.Tomabechi and a summary of the discussion held can be found at the end of the Panel record.
° Similarly, some of the major points raised in the Panel on economic prospects, chaired by R.W. Conn, are mentioned at the end of that Panel.

3. In opening the Panel discussion, V. Demchenko (IAEA, Vienna) presented some general comments on the role and chances of fusion in future energy scenarii:

* The rapid population growth of the world and the enhanced energy consumption make demands on energy supplies. The expected world energy consumption from 1970 to 2050 has been estimated to be about 100 Q, where $Q=10^{21}$ J. The major commercial energy source today is fossil fuels, whose total recoverable reserves at the present time have been estimated to be in the order of 100Q. However the long-term addition of carbon dioxide to the atmosphere from the combustion of fossil fuels could have serious effects on the climate and ecology of the world. Solar electric power will take a long time to become economically attractive. The rapid deployment of solar electric power stations is limited by the requirements of large energy storage systems and a huge surface area which must be covered with collectors (100 TW thermal energy might ultimately be collected by covering 10% of the earth's desert areas with collectors). Clearly, the need for a clean and inexhaustible source of energy has never been greater.

* The overall role that fusion might play in the future energy supply depends on the characteristics of a fusion reactor. Much of the research in this area draws on comparisons of fusion technology (based on the conventional D-T fuel cycle) with fission reactor technology. In this case fusion and fission plants are comparable because both are suitable nuclear technologies for power generation and because they share some of the same environmental and safety concerns. Despite great advantages, conventional nuclear fusion cycles contain neutrons and rely on tritium-based fuels. If this energy is to be made acceptable to the public, it must be based on fuel cycles which can release energy using nonradioactive fuels and which produce no radioactive waste. Alternate fuel fusion reactors (which use proton-based reactions or D^3-He cycle, etc.) have some advantages: no tritium breeding and significantly lower tritium inventory, no radioactive waste and no need for massive shielding, high energy conversion efficiency, lower afterheat and overall high availability due to much lower wall load. Because all the energy released in neutronic cycles is carried by charged particles, such a reactor could produce electricity via direct conversion (without the release of waste heat). Since radioactive waste, heat pollution and high capital cost are now the main issues preventing the wide adoption of nuclear energy, the implications of neutronless fusion power as an alternate energy source are obvious.

* Fusion energy might be able to produce a wide variety of products other than electricity. Beneficial radioisotopes like cobalt-60, which has multiple uses including medical treatment and food preservation, fission fuel for the nuclear power plants, and synthetic fuels to replace oil could be produced in fusion reactors. Still other applications, such as space power and propulsion or radiation processing of materials may be possible. The successful commercialization of fusion energy needs the early involvement of the supposed user. That is why it is important to include early participation of the user at all phases of the fusion R&D programme. The programme objectives must be developed in conjunction with industrial executives. The fast transfer of fusion technology to industry is the key to achieving the long-range objective of commercial fusion applications.

4. In the general discussion, a number of specific points were made, of which the following can be mentioned:

* The relation between radioactivity, as measured in curies, and the potential hazard needs to be up-dated from the early IASA work.

* Cost breakdown is a very important factor determining the direction of research. For example, the development of current drive, and especially the demonstration of the reality of the bootstrap current needs to be fully explored in reactor design.

* The hybrid-reactor component of the USSR fusion programme is likely to be dropped because of environmental issues now dominant in the USSR.

* The fast breeder reactor is the natural long-term competitor of fusion. Therefore cost comparisons were reached with LMFBR in the ESECOM report.

* Material and energy accounting indicated a cost advantage to fusion as compared to the PWR (because the latter uses isotope-enriched fuel).

* Concerning, in general, the presentation of the issues of environment and safety of fusion energy to the public, the compelling need for simplicity and full information was considered.

5. The following points as a route towards attractive fusion energy were stressed by R. Toschi (NET, Garching)

* Safety and environment are the most promising and defendable features (but are hard to implement)

* Availability is the greatest concern because of reactor complexity

* Plant efficiency is a concern because of operating conditions and current drive

* Any development of the reactor concept is an improvement only if it improves also the safety, environmental aspects and economy (SEAE) of fusion energy.

* Capital cost reduction should not, at this stage, be pursued per se but only in relation to an improvement or, at least, maintainment of SEAE

* A good standard in SEAE is not only a prerequisite for global

acceptance but will have beneficial effects on economy that we may not appreciate in full today.

* Factors affecting power cost: specific mass; passive design features; reliability; lifetime; numerous options for materials, according to the different points of view; maintenance, e.g., access, simplicity.

  ° High power density affects negatively all but the specific mass, beyond 50 kWe/ton no global gain envisaged.

* The next step has to show

  ° long and fully controlled burn
  ° passive safety features
  ° environmental acceptability
  ° maintainability
  ° continuous operation

* From the next step we shall learn how

  ° to reduce complexity
  ° to reduce environmental impact
  ° to improve releability and availability
  ° to reduce capital cost

* To argue in favor of fusion, in general, it is probably best to show that:

  ° we have sufficient knowledge to built a next step with acceptable SEAE
  ° we have options for improvements over the next step (maximum credible extrapolations) to be implemented in a "first of a kind" reactor to be confirmed by next step operation and by parallel R&D.

* There are directions (derivatives) for further improvements of global acceptance (sensivity analysis):

  ° Too ambitious targets should be avoided because they reduce credibility
  ° Improvements should be balanced among physics, technology, design solutions, because pushing only basic technology is not credible

6. In conclusion, Ch. Maisonnier (CEC, Brussels) made the following points:

* Statement: fusion offers the prospect to be one of the major sources of electricity in the middle of next century.

* To substantiate this we need

  ° a clear strategy: present devices, next step, DEMO; and a good data base for the next step,
  ° to convince authorities that fusion has a good chance to be attractive from the combined environmental, safety and economic points of view.

* For safety and environment, we should follow two approaches:

  ° Long term: fusion has the potential to reach a high degree of

passive safety and to have only a moderate impact on the environment for fundamental reasons (low energy content of the reactor,
and radioactivity does not come from fusion process itself).

° Short term: we should demonstrate that we can master the safety
and environmental issues of the Next Step.

* Economy

Costs of both "conventionnal" kWh and fusion kWh in 2050 are highly
unpredictable.
The best approximation: make comparison based on present cost of kWh
and on extrapolation from next step costing and show that they are of
the same order of magnitude, as indeed they are.

VII. MISCELLANEA

Participants on the staircase to San Rocco's lecturing hall

PARTICIPANTS

Thirty-six participants, named in the following list, took part in the Seminar on "Safety, Environmental Impact and Economic Prospects of Nuclear Fusion", held in Erice, August 6-12, 1989.

A.C. BELL
JET
Culham Laboratory
ABINGDON OX14 3EA
(G.B.)

R. BUENDE
NET Team
Max Planck Institut
für Plasmaphysik
GARCHING 8046
(F.R.G.)

J. DARVAS
CEC
DG XII Fusion
200, Rue de la Loi
BRUSSELS 1049
(Belgium)

B. BONNEVIER
The Royal Institute
of Technology
Div. of Plasma Physics
STOCKOLM 100.44
(Sweden)

G.J. BUTTERWORTH
UKAEA
Culham Laboratory
ABINGDON, OX143EA
(G.B.)

V. DEMCHENKO
IAEA
Wagramerstrasse 5
P.O. Box 100
VIENNA 1400
(Austria)

A. BOSCHI
ENEA
DISP
Via V. Brancati, 48
ROMA 00144
(Italy)

G. CASINI
CEC
Joint Research Center
ISPRA 21020 (Varese)
(Italy)

H. DJERASSI
CEN
DPT/SPIN
Bâtiment 389
GIF-SUR-YVETTE CEDEX
(France)

E.C. BROLIN
PPPL
Princeton University
P.O. Box 451
PRINCETON, NJ 08544
(USA)

R. CONN
Fusion Engineering and
Physics Program
UCLA
6291 Boelter Hall
LOS ANGELES, CA 90024
(USA)

G. FIORENTINI
INFN
Sezione di Pisa
Dip. di Fisica
Università di Pisa
PISA 56100
(Italy)

B. BRUNELLI
ENEA
DISP
Via V. Brancati, 48
ROMA 00144
(Italy)

J.G. CROCKER
Fusion Safety Program
Eg & Idaho, Inc.
P.O. Box 1625
IDAHO FALLS, Idaho 83415
(USA)

Y. FUJII-E
Tokyo Institute
of Technology
TOKYO 100
(Japan)

F. GORI
Università degli Studi
di Firenze
Dipart. di Energetica
Via di S. Marta, 3
FIRENZE 50139
(Italy)

D. PALUMBO
CEC
DG XII Fusion
200, Rue de la Loi
BRUSSELS 1049
(Belgium)

M. SNYKERS
SCK/CEN
200, Boeretang
MOL 2400
(Belgium)

L. GOUNI
Electricité de France
32, Rue de Monceau
PARIS CEDEX 75384
(France)

R.S. PEASE
The Poplars
West Ilsley
NEWBURY, Berk. RG160AW
(G.B.)

W.R. SPEARS
NET Team
Max Planck Inst.
für Plasmaphysik
GARCHING 8046
(F.R.G.)

M. HEINDLER
Inst. für Teoretische
Physik
Technische Univ.
GRAZ 8010
(Austria)

C. PONTI
CEC
Joint Research Center
ISPRA 21020 (Varese)
(Italy)

K. TOMABECHI
JAERI
Naka-machi
Naka-gun
IBARAKI-KEN
(Japan)

M.S. KAZIMI
Plasma Fusion Center
MIT
CAMBRIDGE, MA 02139
(USA)

J. RAEDER
NET Team
Max Planck Inst.
für Plasmaphysik
GARCHING 8046
(F.R.G.)

R. TOSCHI
NET Team
Max Planck Inst.
für Plasmaphysik
GARCHING 8046
(F.R.G.)

H.Th. KLIPPEL
Fusion Techn. Program
Energy Res. Foundation
3 Westerduinweg
P.O. Box 1
PETTEN 1755 ZG
(NL)

P. ROCCO
CEC
Joint Research Center
ISPRA 21020 (Varese)
(Italy)

J. UHLENBUSCH
University Düsseldorf
KFA
Postfach 1913
JÜLICH 1 5170
(F.R.G.)

H. KNOEPFEL
Associazione EURATOM-ENEA
sulla Fusione
CRE Frascati
C.P. 65
FRASCATI 00044
(Italy)

E. SALPIETRO
NET
Max Planck Inst.
für Plasmaphysik
GARCHING 8046
(F.R.G.)

H. VIALLET
Centre d'Etudes
Nucleaires de
Cadarache
DRFC
ST. PAUL LEZ DURANCE
(France)

Ch. MAISONNIER
CEC
DG XII Fusion
200, Rue de la Loi
1049
BRUSSELS
(Belgium)

S. SARTO
ENEA
DISP
Via V. Brancati, 48
ROMA 00144
(Italy)

P. ZETTWOOG
CEA
Centre d'Etudes
Nucleaires
B.P. 6
FONTENAY-AUX-ROSES 9226
(France)

INTERNATIONAL SCHOOL OF
FUSION REACTOR TECHNOLOGY

The Seminar on "Safety, Environmental Impact, and Economic Prospects of Nuclear Fusion" is the 9th event held within the International School of Fusion Reactor Technology directed by Bruno Brunelli and Heinz Knoepfel. The School started in 1972 under the auspices of the Ettore Majorana Center for Scientific Culture, whose founder and director is Antonino Zichichi. An international consulting committee of the School collaborates in the definition of the program and in the choice of the invited speakers; at present its members are: D. Palumbo (chairman); R. Aymar, CEA, Cadarache; A. Gibson, JET, Abingdon; G. Grieger, MPI, Garching; G.H. Miley, Univ. of Illinois, Urbana; R. Toschi, NET, Garching; K. Tomabechi, JAERI, Ibaraki. The Courses and Seminars up to now were:

1. Stationary and Quasi-Stationary Toroidal Reactors, 1972 (Proceedings published by the Commission of the European Community, doc. EUR 4999e).

2. Pulsed Fusion Reactors, 1974 (Proceedings published by Pergamon Press).

3. Tokamak Reactors for Breakeven, A Critical Study of the Near-Term Fusion Reactor Program, 1976 (Proceedings edited by H. Knoepfel, Pergamon Press).

4. Driven Magnetic Fusion Reactors, 1978 (Proceedings edited by B. Brunelli, Pergamon Press).

5. Unconventional Approaches to Fusion, 1981 (Proceedings edited by B. Brunelli and G.G. Leotta, Plenum Press).

6. Fusion Blanket Technology, 1983 (Workshop Reprints available from the Commission of the European Community, doc. EUR-FU-Brux 12-83-571).

7. Tokamak Startup - Problems and Scenarios Related to the Transient Phases of a Thermonuclear Fusion Reactor, 1985 (Proceedings edited by H. Knoepfel, Plenum Press).

8. Muon-Catalyzed Fusion and Fusion with Polarized Nuclei, 1987 (Proceedings edited by B. Brunelli and G.G. Leotta, Plenum Press).

9. Safety, Environmental Impact, and Economic Prospects of Nuclear Fusion, 1989 (Proceedings edited by B. Brunelli and H. Knoepfel, Plenum Press).

INDEX

This index also contains, at the appropriate alphabetical locations, the explanation of the <u>abbreviations</u>, <u>acronyms</u>, and <u>radiological units</u> used throughout the text.

AB: area of blanket, 45

AC: area of containment, 45

Acceptance of risk, 315

Accidents,
    in fusion reactors, 241
    tritium release, 20
Activation,
    calculation code (ORLIB), 70
    cross section library (ACTL), 70
    of elements, 155, 239
    products in fusion reactors, 21
    structural material, 24
Active safety features, 73

ACTL: cross section library, 70

Advanced toroidal facility (ATF), 110

AECL: Atomic Energy of Canada Limited, 203

AF: area of fuel, 45

Afterheat, 191, 235

AISI: austenitic steel type, 240, 246, 265

ALARA: as low as reasonably acceptable, 234

ALI: annual limit of intake, 226

AMCr: austenitic steel type, 246, 265

American Society of Mechanical Engineers (ASME), 300

APOLLO: advanced fusion fuel reactor, 162

ARIES: advanced reactor innovation and evaluation study, 12, 175

ASME: American Society of Mechanical Engineers, 300

ATF: Advanced Toroidal Facility, 110

ATHENA:
    laser facility, 121
    thermal computation code, 248
AV: area of vacuum, 45

AW: area of waste, 45

Base-load electricity, 37

BBO: boundary of blanket area, 45

BBV: boundary between vacuum and blanket areas, 45

BCSS: blanket comparison and selection study code, 70

BDBE: beyond design basis event, 51

Becquerel: number, per second, of disintegrations with emission of particles or electromagnetic radiation (example: 1 T Bq = 10^{12} Bq $\div$ 0.0027 g of tritium), 202

Beryllium,
    chemical toxicity, 250
    exothermic reaction, 250

neutron multiplier, 150
resources, 325
Beta value,
    cost scaling, 290
    see also poloidal beta
    theoretical limit
    (Troyon-Gruber), 97, 113
    total in reverse field pinches,
    106
    total in tokamaks, 97, 103
BFO: boundary of fuel area, 45

BHP: biological hazard potential,
24

Bismuth, 151

Blanket,
    coolants, 7
    lithium-lead, 194, 235
    materials, 7, 149
    reference designs, 184
    shield, 188
BOE: boundary between containment
area and environment, 45

Bootstrap current, 7, 97, 113

BOP: balance of plant, 12, 76

Bq: see becquerel, 202

Breeder material, 184

Breeding blanket, 149

Bremsstrahlung, limit, 96

BVO: boundary of vacuum area, 45

BWO: boundary of waste area, 45

Canadian heavy water cooled and
moderated reactor (CANDU), 254

CANDOR: advanced fusion fuel
reactor, 165

CANDU: Canadian heavy water cooled
and moderated reactor, 254

Carbon dioxide in the atmosphere,
86

CARMS: cyclotron and resonance
masers, 7

CASCADE: ICF reactor design, 122

CEA: Commissariat à l'Energie
Atomique, 203

CEC: Commission of the European
Community, 35

CFFTP: Canadian Fusion Fuels
Technology Project, 203

Chernobyl, 87, 225

Ci: see curie

CIC: cable-in-conduct
(superconductor), 137

CIT: Compact Ignition Torus, 77,
314

Coal,
    as energy soruce, 83, 89
    price history, 92
COE: cost of electricity, 11, 67,
69, 164

Collective risk, 227

Commissariat à l'Energie Atomique
(CEA), 203

Commission of the European
Community (CEC), 35

Compact Ignition Torus (CIT), 77

Confinement time, 96, 97, 106

Cost of electricity (COE), 11, 67,
69, 164

Cost,
    coal, 92, 94
    comparative, fusion-fission
    reactors, 68, 71
    concepts, 323
    energy, 296
    fission energy, 11, 13
    fusion energy, 11, 13, 15, 325
    future energy, 39
    GENEROMAK reactor, 68
    ICF reactor, 123
    ITER, 284, 324
    learning curves, 40, 324
    NET, 281, 324
    oil, 92
    see also under unit cost
Creep rupture, 193

Curie: old unit for disintegration rate (1 Ci = 3.7 10^{10} Bq), 234

Current drive,
    figure-of-merit, 298
    inductive, 297
    non-inductive, 297

D-III-D: Doublet-III D-shaped tokamak, 96

D: deuterium, 238

DBE: design basis events, 42, 48

Decay heat, 191, 235

DEMO: demonstration reactor, 155

Department of Energy (DOE), 68

Desulphurisation, 86

Deuterium (D),
    helium-3 fuel cycle, 159
    in sea water, 38
Deuterium-helium 3 fuel,
    basic reaction, 159
    intrinsic features, 174
    reactor types, 162
Direct conversion, 7

Disruption, of plasma current, 97, 141

DISTRAIR: tritium-release prediction code, 210

DITE: tokamak at Culham, 100

Divertor, 138, 245

DKES: drift kinetic equation solver code, 112

DOE: Department of Energy (US), 68

Dose, radiological,
    at NET/ITER boundary, 328
    chronic, 313
    collective, 226, 313
    early and longterm, 313
    equivalent and effective, 312
    ingestion, 313
    limit, for most exposed individual, 226
    prompt, 313
    whole and total body, 312
Doublet-III D-shaped (D-III-D), 96

Driver, ICF,
    direct drive, 119
    electrons, 119
    heavy ions, 10, 120
    indirect drive, 117
    laser, 10, 11
    light ions, 10, 119
    reactor, 124
    various types, 118
DT: tritiated deuterium, 257

DTS: detritiation system, 150

EC: European Community, 35

ECE: electron cyclotron emission, 99

ECH: electron cyclotron heating, 113

Ecological considerations,
    fusion vs fossil energy, 320
    see also under environmental impact
Economic aspects,
    in the EC, 37
    see also under cost
ECRH: electron cyclotron resonance heating, 7

EEF: environmental and economic potential of fusion, 35, 37, 131, 231, 327

Electric Power Research Institute (EPRI), 68

Electricity,
    cost of generation, 325
    direct conversion, 329
    future price, 93
    share in final consumption, 83, 87
Electron cyclotron emission (ECE), 99

Electron cyclotron heating (ECH), 113

Electron cyclotron resonance heating (ECRH), 7

Energy confinement,
    global, 96, 106
    in JET, 105
    in reverse field pinches, 106, 110
    triple product parameter, 106

Energy consumption,
    OECD countries, 81
    savings, 85
    share of primary sources, 83, 89
    world up to 2050, 329
    world, 82, 87
Energy intensity, 85, 86

Energy inventory, in fusion
reactor, 234

Energy pay-off, 324

Energy sources,
    in fusion reactor, 56
    share in Western Europe, 83
    share in world, 89
Engineering safety feature (ESF),
53, 319

Environmental impact,
    comparative, fusion-fission
    reactors, 69
    fusion reactor, 296
    ICF reactor, 125
    in the year 2050, 320
EPRI: Electric Power Research
Institute, 68

ESECOM: Environmental, Safety and
Economic Aspects of Magnetic Fusion
Energy, Senior Committee, 12, 67,
285
ESF: engineering safety feature,
53, 319

ETF: Engineering Test Facility

European Community (EC), 35

European energy policy, 38

European Parliament, 36

Event categorization, in fusion,
47, 60

Exposure to radiation,
    individual, 223
    public, 234
    worker, 234

Failure rate (FR), 299

Fast breeder reactor (FBR), 24, 39

FBR: fast breeder reactor, 39

FBSA: function-based safety
analysis, 43, 48

FCTR: First Commercial-sized
Tokamak Reactor, 36

FED: Fusion Engineering Device

FER: Fusion Experimental Reactor,
42, 54

FFST: function of fuel storage, 45

FFTR: function of fuel transfer, 45

FHCI: function of material
circulation, 45

FHCO: function of heat conversion,
45

FHEG: function of electricity
generation, 45

Field-reversed configuration (FRC),
176

Fire accidents,
    beryllium, 249
    graphite, 249
    lithium, 194, 328
    tungsten, 249
First wall (FW),
    graphite armour, 138, 250
    heat fluxes, 169
    materials, 151
    neutron fluence, 70
    neutron flux, 70, 138
    neutron load, 194
    of ICF reactor, 8
    safety limits, 193
    thermal loads, 138
Fission reactor,
    fast breeder (FBR and LMFBR),
    24, 39
    high-temperature gas (HTGR and
    MHTGR), 68, 71, 75
    pressurized water (PWR), 39, 68,
    71, 75
    safety features, 319
    waste material, 24
FOB: free on board, 89

FORIG: radionuclide generation and
depletion code, 70

Fossil fuel,
    reserves, 329
    see also under coal, natural
    gas, oil
FPC: fusion power core, 12, 288

FPEX: function of particle exhaust,
45

FPFG: function of magnetic field generation, 45

FPHE: function of plasma heating, 45

FPSP: function of fuel separation and purification, 45

FPSU: function of fuel supply, 45

FPY: full-power years, 70

FR: failure rate, 299

Fracture mechanics, 300

FRC: field-reversed configuration, see also RFP, 176

Fuel cycles,
    advanced fusion, 18
    deuterium-helium 3, 159, 173
    reaction rates, 160
Fuelling of reactor, 167

Full-power years (FPY), 70

Fusion energy,
    cost break-down, 39
    policy, 316
Fusion Experimental Reactor (FER), 42

Fusion plant,
    non-electrical, 329
    overall cost, 325
    public safety limits, 192
    safety limit, 192
Fusion power core (FPC), 12, 288

Fusion power plant, see fusion reactor

Fusion reactor,
    activation products, 21, 36
    advanced-fuel, 26
    blanket designs, 183
    cost estimate, 15, 39
    design studies, 4, 12
    EEF study, 39
    high field compact, 7
    inertially confined, 8, 21, 122
    internal energy sources, 56
    key parameters, 162, 165
    magnetically confined, 4
    radioactive inventory, 24, 71
    radioisotope leaks, 44
    radioisotope sources, 44, 47
    radiological problems, 230
    reference case for ESECOM, 70

    safety objectives, 314
Fusion system,
    descriptive model, 45
    event categorization, 47, 60
    safety features, 48
FW: first wall, 151

FWPR: function of waste processing, 45

FWST: function of waste storage, 45

GA: General Atomic Technologies Inc., 77

GDM: general descriptive model, 44, 56

GDP: gross domestic product (see also GNP), 37

GEMSAFE: general methodology of safety analysis and evaluation for fusion energy systems (Japan), 41

GENEROMAC: generic magnetic fusion reactor, assessment cost, 12, 68, 285

GM: outage risk contribution, 299

GNP: gross national product (see also GDP), 85

Goldstone scaling, 97

Gray: absorbed radiation dose, i.e. 1 joule radiation energy per kilogram of matter (tissue), 223

Greenhouse effect, 38

Gross domestic product (GDP), 37

Gross national product (GNP), 85

Gy: see gray, 223

H-mode: high confinement mode, 106

H: hydrogen, 238

Hands-on-limit dose rates, 265

Hazard,
    in fission vs fusion, 22

in fusion reactors, 18, 19
   intruder potential, 24
   non-nuclear, 26
HBTX: reverse field pinch
experiment, 111

Heavy ion driver, 120

HELIOTRON: stellarator experiment,
110

Helium-3,
   cost, 164
   deuterium fuel cycle, 159
   resources, 166, 167
Helium,
   fire accident, 194, 328
   resources, 325
HFT: high field tokamak, 175

High level waste (HLW), 260

High-temperature gas cooled reactor
(HTGR), 68, 71, 75

HLW: high level waste, 260

HT: Tritiated hydrogen, 202

HTGR: high-temperature gas cooled
reactor, 68, 71, 75

HTO: tritiated water, 202, 255

HYB: hybrid fusion-fissile breeder
reactor, 71, 75

Hybrid fusion-fissile breeder
reactor (HYB), 71, 75

Hydroelectricity, 83, 89

IAEA: International Atomic Energy
Agency, 15

IBF: interface boundary between
blanket and fuel areas, 45

IBHP: integrated biological hazard
potential, 24

ICF: inertial confinement fusion,
4, 117

ICRP: International Commission on
Radiation Protection, 199, 208, 312

IFW: interface boundary between

fuel and waste areas, 45

IHP: intruder hazard potential, 24

Individual risk (Ri), 225

INEL: Idaho National Engineering
Laboratory, 72

Inertial confinement fusion (ICF),
4

Instabilities,
   ballooning, 97
   current disruption, 97
   ideal MHD, 97, 113
   kink-mode, 97
   trapped-particle, 111
International Atomic Energy Agency
(IAEA), 15

International Commission on
Radiation Protection (ICRP), 199,
208, 312

INTOR: international tokamak
reactor, 4

Intrinsic (or inherent) safety
feature (ISF), 53, 319

Intruder hazard potential (IHP), 24

Ionizing radiation,
   effects on individuals, 223
   individual risk, 225
IP: items to be protected, 44, 48

ISF: intrinsic (or inherent) safety
feature, 53, 319

ITER: International Thermonuclear
Experimental Reactor, 4, 69, 77,
111, 231, 314

IVF: interface boundary between
vacuum and fuel areas, 45

JAERI: Japan Atomic Energy Research
Institute, 42, 54

JET: Joint European Torus, 7, 96,
106

JRC: Joint Research Center, 219

JT60: Japanese Tokamak, 96

KFA: Kernforschungsanlage Jülich, 219

KfK: Kernforschungszentrum Karlsruhe, 219

L-mode: low confinement mode, 106

LA: low activation, 154

LAM: low activation materials, 148 153, 188

LANL: Los Alamos National Laboratory, 203, 219

Large Coil Task (LCT), 6

Laser Microfusion Facility (LMF), 120

Laser,
 krypton fluoride, 11, 120
 neodynium, 117
 NOVA facility, 118
 see also under driver
 solid state, 11
Lawson criterion, 96, 160

LBL: Lawrence Berkeley Laboratory, 118

LCT: Large Coil Task, 6

Lead,
 isotope, 151
 lithium compound, 150, 194, 235, 240
Level of safety assurance (LSA), 23, 73, 75

Licensing regulations, 19

Liquid Metal Fast Breeder Reactor (LMFBR), 24

Liquid metals,
 chemical reaction hazard, 150, 328
 in fusion reactors, 150
LITFIRE: lithium fire consequences code, 195

Lithium,
 blanket materials, 149
 fires, 194, 328
 lead compound, 150, 194, 235,

240
 reserves, 38
LLNL: Lawrence Livermore National Laboratory, 70, 77

LLW: low level waste, 154, 260

LMF: Laser Microfusion Facility, 120

LMFBR: Liquid Metal Fast Breeder Reactor, 24

LNG: liquid natural gas, 82

LOCA: loss of coolant accident, 23, 73, 184, 241, 243

LOFA: loss of coolant flow accident, 184, 241, 248

Loss of coolant accident (LOCA), 23, 73, 184, 241, 243

Loss of coolant flow accident (LOFA), 184, 241, 248

Loss of plasma confinement (LPC), 241, 251

Loss of vacuum accident (LOVA), 241, 249

LOVA: loss of vacuum accident, 241, 249

Low activation materials (LAM), 148, 153, 188, 328

Low dose effects, 224

LPC: loss of plasma confinement, 241, 251

LSA: level of safety assurance, 23, 73, 75

LSPB: large-scale prototype breeder, 68, 71, 75

Lunar helium-3 content, 167

Magnet system,
 see also under superconductor magnet
Magnetic confinement fusion (MCF), see MFE

Magnetic fusion energy (MFE) 4, 8, 67, 74

Magnetic system,
   failure, 252

MANET: martensitic steel type, 246

MARS: tandem mirror fusion reactor, 6, 14, 73

Mass power density (MPD), 6, 12

Materials,
   in blanket, 7, 149
   in shield, 148
   in superconducting magnets, 148
   low activation (LAM), 18, 153, 310
   recycling of, 310
   reduced activation steel (RAF), 68
   required development and testing program, 76
   structural, 7
Maximum exposed individual (MEI), 223

Maximum plausible dose (MPD), 21

MCF: magnetic confinement fusion, see MFE·

MDT: mean plant down time, 299

MEI: most exposed individual, 208, 233

Metten report, 36

MFE: magnetic fusion energy, 4, 8, 67, 74

MHD conversion, 169

MHD: magneto hydrodynamics, 3, 68

MHTGR: modular high-temperature gas cooled reactor, 68, 71, 75

MIT: Massachussetts Institute of Technology, 72

MLW: medium level waste, 154, 260

Moon, helium-3 resources, 167

Most exposed individual (MEI), 208, 233

MPD: mass power density, 6, 12

MPD: maximum plausible dose, 21

MPRF: maximum possible release fraction, 23

Natural gas (NG),
   as energy source, 83, 89
   long distance transport, 82
NBH: neutral beam heating, 7

NBI: neutral beam injection, 7, 110

NEA: Nuclear Energy Agency, 287, 297

NECDB: nuclear energy cost data base, 69

Neodynium glass laser, 117

NET,
   first wall, 152
   major parameters, 131
   radioactive waste, 261
   radioactivity inventories, 240
   shield, 149
   tritium inventories, 238
NET: Next European Torus, 111

Neutral beam heating (NBH), 7

Neutral beam injection (NBI), 7, 110

Neutron fluence, 70

Neutron flux, 70, 138

Neutron multipliers, 188

Next European Torus, see NET

NG: natural gas, 82

NI: non-inductive (current), 296

NIR: Niedersächsisches Institut für Radioökologie, 208

NOVA: laser facility at Livermore, 117

NRL: Naval Research Laboratory, 118

Nuclear Energy Agency (NEA), 287, 297

Nuclear fusion reactions, 159, 160

Nuclear quality control, 288

O&M: operation and maintenance, 11

OECD: Organisation for Economic
Cooperation and Development, 82

OH: ohmic heating, 110

Ohmic heating (OH), 110

Ohm's law, neo-classical, 95, 97

OHTDC: tritium-release prediction
code, 210

Oil,
    as energy source, 83, 89
    cost, 92
    price fluctuations, 85
    world production, 90
ONEDANT: neutronic code, 184

OPAS: overall plant accident
scenario, 241

Organisation for Economic
Cooperation and Development (OECD),
82

ORLIB: activation averaging code,
70

ORNL: Oak Ridge National
Laboratory, 69

Passive safety feature (PSF), 73,
319

PCSR: prototype commercial-sized
reactor, 286, 298

Pellet refuelling, 167

PIC: plant inner containment, 282

Plant load factor, 297

Plasma,
    confinement loss (LPC), 251
    disruptions, 141, 170
    pressure, 97, 104
    profile, 97
    refuelling, 167
    stability, instability, see
    under instabilities

Poloidal beta,
    from pressure balance, 95
    neo-classical value, 95
PPPL: Princeton Plasma Physics
Laboratory, 219

Pressurized water reactor (PWR),
39, 68, 71, 75

PRISM: power reactor inherent
safety module, 68, 71, 75

Probabilistic safety assessment
(PSA), 42

Proliferation issues, 174

PSA: probabilistic safety
assessment, 42

PSF: passive safety feature, 319

Public acceptance, 311

Public exposure, 234

Pump limiter, 7

PWR-BPE: pressurized water reactor,
best present experience, 68, 71, 75

PWR-ME: pressurized water reactor,
medium experience, 68, 71, 75

PWR: pressurized water reactor, 39

R-tokamak: reacting tokamak plasma
experiment, 54

Rad: old unit for absorbed
radiation dose (100 rad = 1 Gy)

Radiation collapse limit, 96

Radioactive materials,
    inventory of reactors, 71
    mobility categories, 71
    release fractions, 72
    see also under radioisotopes
Radioactive waste, 260

Radioactivity inventory (RAI), 71,
239

Radioisotope (RI),
    leaks in fusion reactor, 44

mobility classification, 48
see also under radioactive
materials
sources in fusion reactor, 44,
47, 57

RAF: reduced activation ferritic
steel, 68

RAI: radioactivity inventory, 71,
239

RCG: recommended concentration
guidelines, 20, 24

REAC: activation code, 184

Reaction rates, 160

Reliability,
    fusion reactor, 299
    mechanical, 300
    reactor subsystems, 302
Rem: old unit for radiation
exposure, 1 rem = 0.01 sievert

Reverse field pinch (RFP),
    experiments, 106, 110, 111
    reactor, 6, 12, 68, 71, 75,
    176, 184

RFP: reverse field pinch, 6, 12,
68, 106, 176

RFX: Reverse Field Pinch
Experiment, 110

RI: radioisotope, 44

Risk,
    acceptance, 315
    collective, 225
    exposure to ionizing radiation,
    220
    management, 222, 225
    perception, 315

S&E: safety and environment, 15,
231

S: severity (of interruptive
phenomena), 50

Safety assessment,
    GEMSAFE, 41
    probabilistic (PSA), 42
Safety assurance, level of (LSA),
23, 73, 75

Safety features,
    active, 73

advanced fuels, 177
engineering (ESF), 53, 319
in LWR, 319
intrinsic (ISF), 53, 319
passive (PSF), 73, 314, 319, 320

Safety,
    assurance levels (LSA), 72
    fusion reactors, 15, 18
    fusion-fission reactors, 69
    ICF reactor, 125
    limits, 192
    objectives, 314
SC: superconductor, 149

SCAN: system for cost analysis of
NET, 281

SCIROC: tritum-release prediction
code, 210

SCM: superconduting material, 148

SEAE: safety, environmental aspects
and economy, 330

Second stability regime (SSR), 6,
175

SF: safety features, 50, 53, 73,
314, 319, 320

Shallow land burial (SLB), 154,
260, 265, 270, 310

Sievert: absorbed dose equivalent,
where the dose is weighted for its
effects on humans, 225, 312

SLB: shallow land burial, 154, 260,
265, 270, 310

Solar power, 329

SOLASE: ICF reactor design, 122

SRL: Savannah River Laboratory, 219

SSR: second stability regime, 175

STARFIRE: commercial tokamak fusion
power study, 6, 12, 69, 285

Stellarator experiments, 110

STOA: Scientific and Technological
Options Assessment (European
Parliament), 36, 67

Strain ranges, cyclic operation,
300

SUPERCOIL: code for the computational design of tokamaks, 281

Superconducting toroidal coil management (TESPE), 242

Superconductor magnets,
   conceivable failures, 242
   cryosystem, 234
   fabrication procedures, 135
   high field, 169
   materials, 135, 148
Superconduting material (SCM), 148

Sv: see sievert, 225

Sweet report, 36

Synchrotron radiation, 176

T: tritium, 238

Tandem mirror fusion reactor (MARS), 6, 14, 73

Targets, in ICF,
   direct drive, 8
   fractional burnup, 125
   in reactor, 123
   indirect drive, 117
   multilayer, 119
TART: Monte Carlo code for neutron and gamma spectra, 70

TBR: tritium breeding ratio, 150

Tensile stress, ultimate (UTS), 192

TESPE: superconducting toroidal coil management, 242

TFTR: Tokamak Fusion Test Reactor, 7, 96, 106

Thermal diffusivity,
   in JET, 107
   see also under energy confinement
THIOD: thermal analysis code, 184

Three-mile-island (reactor) (TMI), 192

TITAN: reverse field pinch reactor, 6, 12

TMI: Three-mile-island (reactor), 192

TOK: tokamak, 71, 75

Tokamak Fusion Test Reactor (TFTR), 7, 96, 106

Tokamak,
   experiments, 7, 71, 75, 96, 106
   physics, 95
   reactor, see fusion reactor
TRILOCOMO: tritium-release prediction code, 210

Tritium (T),
   accidental release, 20
   breeding ratio, 7
   cancer probability, 199
   concentration guidelines, 20
   consumption in fusion reactor, 199
   fusion reactors inventories, 19, 71, 238
   half-life, 19
   inventory, 236
   release experiments, 202, 259
   release from fission reactors, 199
   release from fusion reactors, 237, 254
   system failures, 242
   test assembly (TSTA), 20
   tritiated vapor vs hydrogen, 201, 212, 257
TSTA: Tritium System Test Assembly, 20

TZM: molybdenum alloy, 153, 248

UCRL: University of California Radiation Laboratory, 77

Unit cost,
   current drive, 69
   electricity (COE), 69, 325
   reactor material, 69
UTS: ultimate tensile stress, 193

UWMAK: University of Wisconsin Tokamak Reactor, 12

Vanadium alloys, 150

W.VII AS: Wendelstein VII AS, 110

Waste, radioactive,
   fusion vs fission, 320

fusion, 260
  geological disposal, 310
  shallow land disposal, 154, 265,
    270, 310
WEC: World Energy Conferences, 82,
88

Wendelstein VII AS (W.VII AS), 110

Worker exposure, 234

World Energy Conferences (WEC), 82,
88

ZT-H: reverse field pinch
experiment, 110